Of the first edition

Volume Four

five hundred copies

have been signed by the author

and numbered.

This is book number 295.

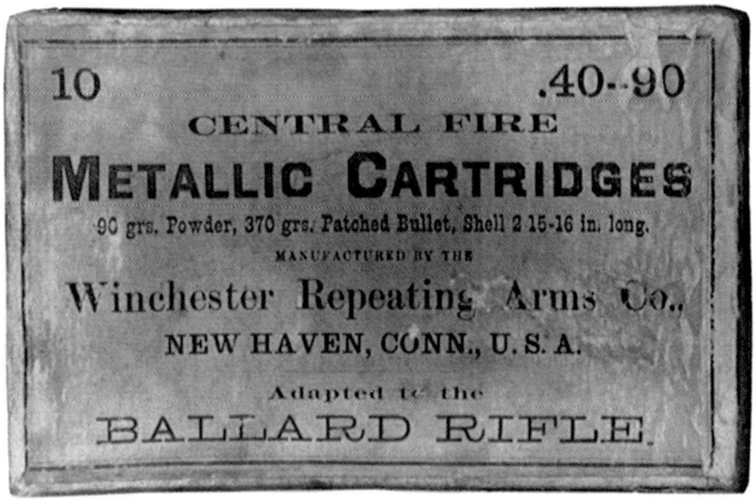

The HISTORY and DEVELOPMENT of SMALL ARMS AMMUNITION

Volume Four: Percussion and Centerfire American Sporting Rifle, and Primitive

WINCHESTER
REPEATING ARMS CO.
METALLIC AMMUNITION,
PAPER AND BRASS SHELLS,
FELT GUN WADS,
PRIMERS &c.
NEW HAVEN, CONN., U.S.A.
SAN FRANCISCO
Winchester Repeating
Rifles AND SHOT GUNS
AND SINGLE SHOT RIFLES

THE HISTORY AND DEVELOPMENT

OF

SMALL ARMS AMMUNITION

GEORGE A. HOYEM

Volume Four

Nineteenth Century American Percussion and Centerfire Rifle And Primitive Cartridges of Europe and North America

1999
ARMORY PUBLICATIONS

This book is dedicated

to the memory of

David R. Hughes, Frank C. Hicks

and Paul Parsons with fond

memories of the comraderie

and collecting discoveries we

shared for so many years.

17171 Bothell Way NE, # 276
Seattle, Washington 98155 USA

Library of Congress Catalog Card Number: 80-67532

ISBN-0-939683-16-4

ACKNOWLEDGMENTS

This book owes its existence to the generosity of people who have contributed information to it with the sole motive of sharing what they have found with their fellow collectors and historians. Some of you contributed many years ago. **Alva (Mac) McCoy** let me acquire two Sharps cartridges at a gun show back in 1962, and that set me on the trail of the rest of them. **Bill Woodin** x-rayed the needle fire specimens and **Beth Woodin** let me photograph her Dreyse shotgun. The cartridge auction catalogues issued by **Robert T. Buttweiler (R. T. B. Ltd.)** and **Pete de Coux** have been invaluable. I have found myself returning to them time and again. And I thank **John Witzel** not only for sharing rare cartridges but for his labors in producing his "**Index To The Buttweiler Auction Catalogues**," which has made the information therein accessible.

Guenther Wohl shared photographs and information about his needle fire collection. Finally, it was **Frank Barnes** who broke ground on the subject of American centerfire rifle cartridges in his "**Cartridges Of The World**," and made collectors out of so many people.

The following people have shared with me knowledge I did not have, opened their collections and libraries to me for photography and study, and contributed in other ways too numerous to list. Some of you may wonder what you did for this book. If I have omitted anyone, it was due to a lapse in memory.

Gilbert Oekermann
James McCarron
Frank J. Hicks
Judge Robert M. Holter
Otto C. Witt
Robert T. Borcherdt
Ronald Anderson
John R. Salzer
Robert C. Solomon
George H. Murphy
Leonard Messinger
Frank Pellett
Randy Long
George G. Kass
William E. Asher
Terry White
Richard D. King
Richard A. King
Edward A. Hull
Charles Osborne & Co.
Howard M. Madaus
Kenneth R. Knox

Lester P. Bassinger
Charles R. Suydam
Lynn Hughes
John Belton
Matthew (Mick) McLaughlin
Paul Molans
John R. Nelson
Michael Cobine
Paul Klatt
R. Peyton Hall
John Bartlett
Gaye Tillinghast
Peter and Milada Moucka
Daniel Shuey, Jr.
Roy M. Marcot
Manfred Beutter
Douglas Beneck
Robert Brotherson
John H. Hintlian
Dennis C. Brooks
Charles Adkins
Tom Rowe

PREFACE

Like its predecessors, this is a case type book. Its primary purpose is to organize American percussion and centerfire sporting rifle cartridges in families by their originators. A few metallic shotshells have been included because it would be unreasonable to leave the Maynard, Stevens and Wesson families of case types incomplete. Headstamps number in the tens of thousands, so I have left this subject for others to explore.

Centerfire rifle cartridges are covered up to the beginning of the smokeless powder era, but there is no clear end of the black powder era and the beginning of smokeless. It will take another volume to carry the study from approximately 1900 to WWII.

Some of our military case types are included under "Single Specimens" because they were used as sporting loads in surplus American rifles such as the .58 Musket and Carbine, the .50-70 Springfield, even the .43 Spanish. The headstamps, primers and load variations in these form an almost endless collecting specialty.

This book was an opportunity add primitive cartridges that were not available for study when Volumes One and Two were published back in the early 1980s. These you will find in Appendix A.

Some of the 19th century American patents for centerfire cartridge primers and case construction are found in Appendix B for the collector and researcher who wants to study the subject in depth.

Several of our colleagues are involved in other studies of American ammunition manufacturers' products, and headstamps. I trust that these projects will expand on my own efforts and join our very specialized body of historical literature.

George Hoyem

CONTENTS

INTRODUCTION

The development of centerfire metallic cartridges for the military rifle began in the United States shortly after the Civil War ended in 1865. The first centerfire cartridge cases were inside primed with the rim formed by folding the head. They were weak at this point. The folded head Martin case with integral primer pocket was used briefly for military ammunition, and inserted cups were tried to strengthen the case head. Folded-head cases were used for some years but ultimately the problem of strength was solved with the solid drawn head.

It was not until 1873 that the first practical centerfire ammunition for sporting rifles, and rifles for it, appeared. First came the Winchester Model 1873 magazine rifle, and the early single shots—Maynard, Remington, Wesson, Ballard and Peabody.

Winchester first adopted the Milbank primer but found it unreliable. Edward Maynard, with Massachusetts Arms Company, settled on the Berdan primer and a two-piece case with iron anvil screwed into the heavy head.

Josiah Meigs worked for years on primers and his efforts resulted in creation of the United States Cartridge Company (Vol. 2, pp. 22-27). The Meigs ideas were not the answer and United States Cartridge Company adopted the Farrington primer. The other primary suppliers of sporting ammunition--Union Metallic Cartridge Company and Winchester Repeating Arms Company--used the Berdan primer at first and ultimately replaced it with the Boxer-style primer with its self-contained anvil. Other systems, such as the Hobbs and Orcutt were tried out but soon dropped.

Black powder did not create high pressures so most early cases could be reloaded repeatedly, and a multitude of loading tools was available from the 1870s on. However, Maynard, Ballard (later Marlin), Wesson and Stevens offered thick-walled "Everlasting" cartridge cases that could be reloaded almost indefinitely. Some were nickel plated, and the Maynard Model 1882 cases were counter-bored, or reamed out, at the mouth to the exact dimensions of the proper bullet. Accuracy was the goal, and these old single shot rifles will still make astonishing targets with original equipment and black powder.

The favorite bullet for target shooting and long-range buffalo hunting with the big, heavy Sharps, Remington and Ballard rifles was paper patched, so the lead did not touch the barrel. Multi-grooved, lubricated lead bullets were found to be just as effective and less work to produce, providing you washed and scrubbed the bore free of powder residue and lead deposits.

The multitude of case types that appeared from 1873 to ca 1895 was the result of efforts to sell rifles, and from the demands of customers for something special. You could have anything you wanted in a Maynard rifle; thus the over 40 case types they made for one pattern of single shot rifle--the most of any maker ever.

The 19th century was only the beginning of American centerfire sporting rifle cartridge development. The search goes on today for new case types, practical or not, that will interest buyers and sell rifles. And so we will never run out of specimens to collect.

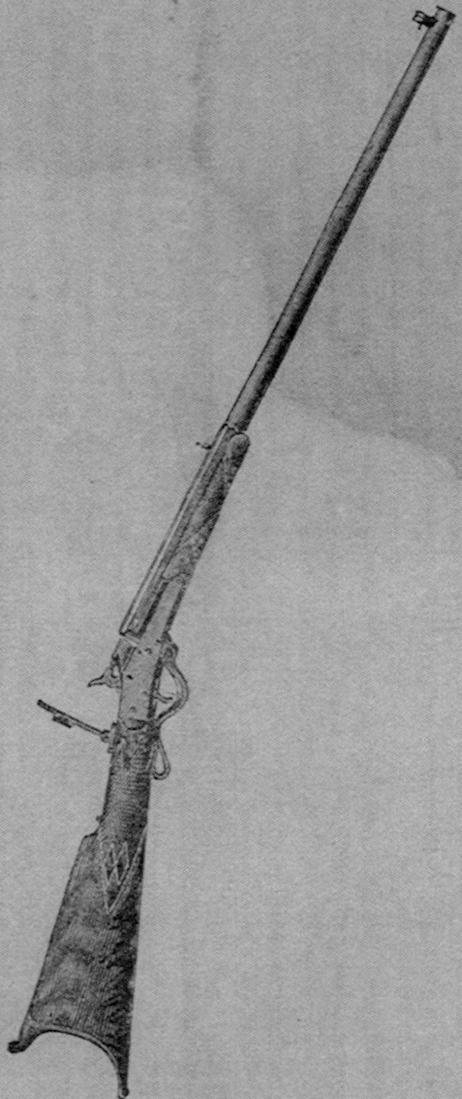

FULL SIZE

CENTRE FIRE CARTRIDGE.

.32 Cal., 10 grs. Powder.

THE MAYNARD.

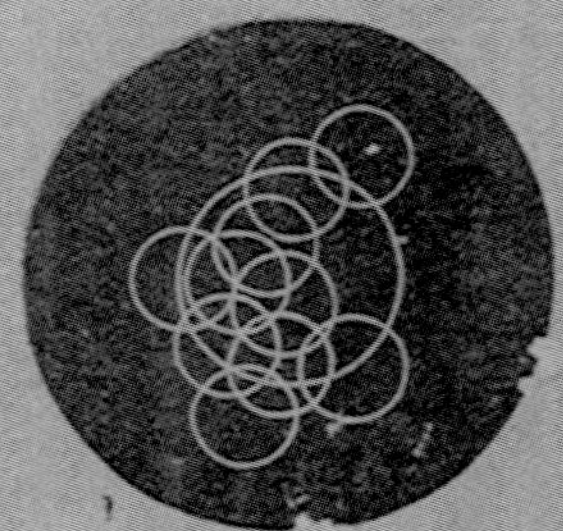

TEN CONSECU[illegible]E SHOTS.

100 yds., with a Mayna[illegible] ifle, 28 in., .32 calibre, 35 grs. o[illegible] er, with a patched bulle[illegible] grains.

DR. EDWARD MAYNARD'S FIREARMS AND THEIR CARTRIDGES

Dr. Edward Maynard, dentist at the City of Washington, District of Columbia, experimented with firearms and ammunition for over 30 years. He was granted his first patent for the tape primer and mechanism to use it on September 22, 1845, and his last—claiming a cartridge index for magazine guns—was issued June 18, 1886.

More cartridge case types were made for the Maynard, hinged, break-open, single shot action produced by Massachusetts Arms Company than for any other firearm in American history—over 40 for the three models, 1859/65, 1873 and 1882.

Edward Maynard was born April 26, 1918 in Madison, New York. On July 1, 1831 he entered the United States Military Academy but resigned before the end of his plebe year because of ill health.

He then studied dentistry and moved to Washington, D. C. where he practiced. He was an innovator in his chosen profession, introduced anesthesia to dental surgery, and received national and foreign honors and awards for that work. At the same time he was busy with his firearms and ammunition developments. He died in Washington, D. C. on May 4, 1891 and is buried at the Congressional Cemetery.

Dr. Maynard's tape primer was adopted by the United States Ordnance Department and several thousand muskets were altered and equipped with the device. He received $15,000 for this work on February 3, 1854. Secretary of War Jefferson Davis arranged to buy out the entire interest in the Maynard tape priming device for $50,000. Many other muzzle-loading and percussion breech-loading firearms were either altered for the tape primer or manufactured with it.

There were two distinct generations of Maynard breech-loading firearms. While similar in operation, each generation had components and dimensions which differed from the other. The first generation universally utilized the Maynard priming device. A delicate, small hammer was offset from the center of the bore by a pleasing cant to the right. The Maynard tape primers automatically centered upon a nipple similar in appearance to a regular percussion cap nipple. However, the nipple was manufactured so that it was part of a tube held into the receiver by a separate screw. The Maynard priming device fitted beneath a cover inletted into the receiver on the right side thereof. The buttstock was thin and slablike, to which was fitted a unique buttplate and cap box.

The first-generation Maynards were manufactured under various patents held by Dr. Maynard, the earliest of which was May 27, 1851. A few were sold to the military. It is known that Jefferson Davis sought the shipment of some of these early model Maynards to the South just before and at the time of secession. In all, it is thought that about 4,800 first-generation rifles were produced.

In the early days of the Civil War, the Massachusetts Arms Company burned. It was not until late 1863 or early 1864 that production was resumed. This renewed production resulted in the second generation of Maynard arms. Gone was the tape priming device with its unique cover. The earlier slab stock with its buttplate and cap box gave way to a heavier, straight stock with a conventional buttplate. No cap box was used.

This second generation of production was manufactured with a beef-up receiver equipped with a standard, screw-in percussion nipple. The hammer was still canted to the right, but was considerably heavier than the first model. Massachusetts Arms Company also did away with the takedown feature in the first-generation model and the profile of the buttplate was changed. They manufactured 22,002 of the second-generation .50 cal. carbines for the Union army.

After the Civil War the heavier, second-generation receivers became the basis for production of Maynard sporting models. The takedown feature again became the standard. The immediate postwar production was percussion cap fired; subsequent production became the basis for the multitude of self-contained cartridges that have delighted cartridge collectors for generations.

Designated variously as Models 1865, 1873 and 1882, Maynard rifles and shotguns continued to be marketed until the 1890s. Massachusetts Arms Company converted many of the second-generation percussion arms, including the military carbines, to fire self-contained ammunition.

In the 1890s the Massachusetts Arms Company was purchased by J. Stevens Arms and Tool Company. Stevens manufactured an inferior .22 rim fire designated "The Maynard Junior" until World War I which utilized a break-open action entirely dissimilar to the Maynards.

References: **The Cartridges of Edward Maynard**, 1982 by Richard D. King and Illustrator Richard A. King.
Dr. Edward Maynard—The Man, His Patents, His Firearms, His Cartridges, George H. Murphy.
The Rifle In America, Philip B. Sharp.

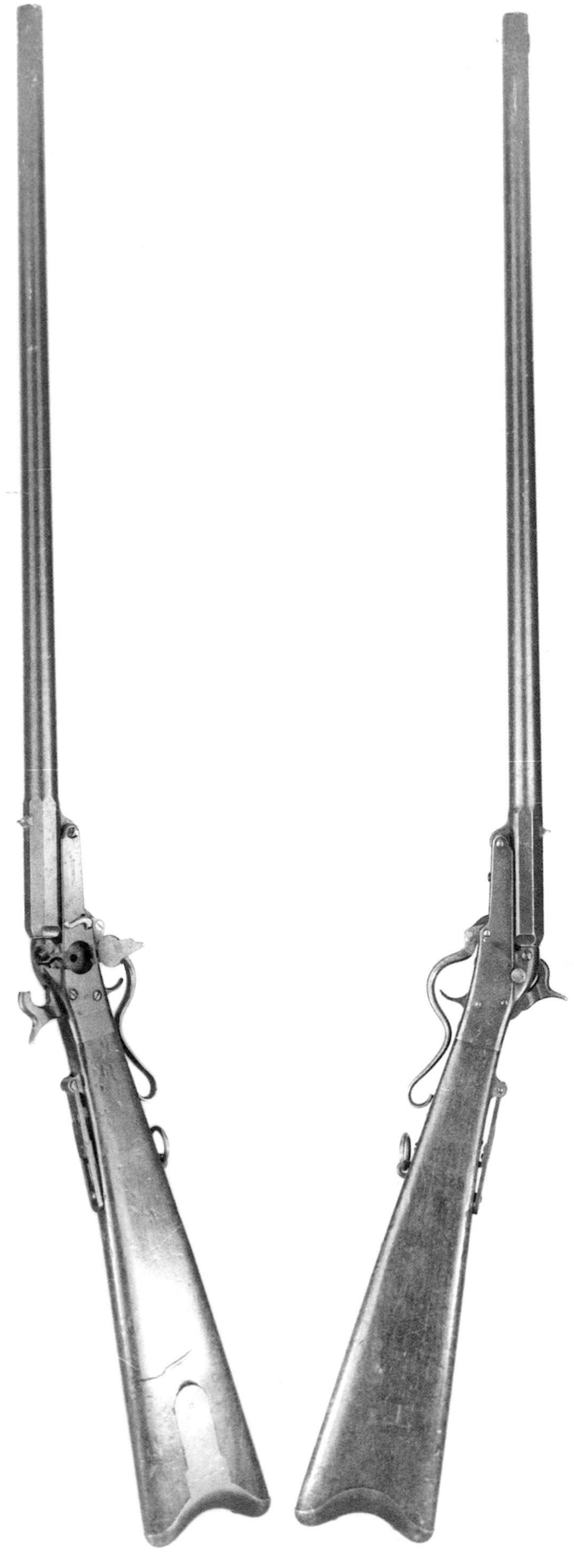

At left are two views of a first generation Maynard rifle in .50 caliber. Note the open cover of the tape primer magazine, the delicate hammer and unique buttstock, buttplate and patch box. This rifle was manufactured before the Civil War. It came from southwestern Virginia where it had been preserved by one family for generations. Its sling swivel tells us it was a Confederate weapon. The only unusual markings are the carved initials "W. F. H" on the left side of the buttstock. At right is one of the 22,002 second generation Maynard carbines made for the Union army from 1863 to 1865.

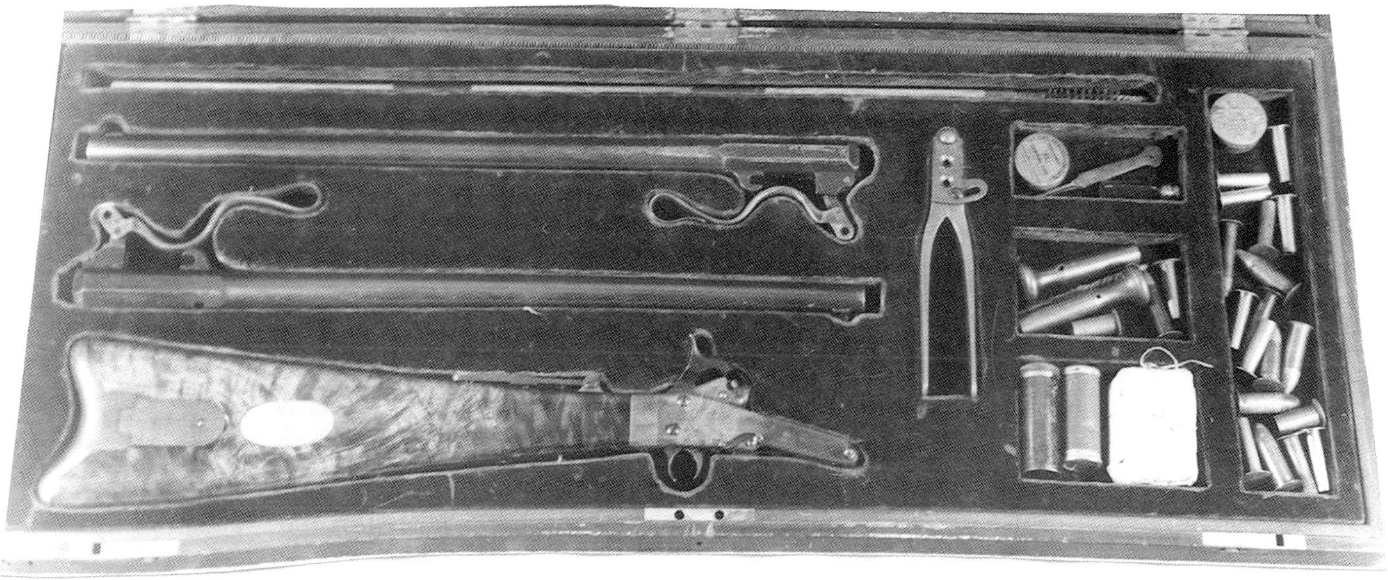

Above is a mahogany-cased two-barrel, first-generation Maynard set--.35 and .50 calibers and 20" barrels. While this is a first model, it was made late in the production of first models, probably just before the Civil War. There is an engraved presentation plate in the buttstock. The case contains the original cleaning rod, .35 and .50 cal. cartridge cases, a ten-round packet of .50 cal. cartridges, an original tin of tape primers, a two-cavity bullet mould, lubricant and assorted loading tools. The patch box behind the presentation plate contains a spare nipple in its own fitted slot, with space for rolls of the tape primers. Note the rare double-set triggers. This particular model of case was known as "French fitted." Below is an assortment of Maynard loading tools. From left: 1873 Berdan primer de-capper, 1882 de-capper, 1873 re-capper, re- and de-capper for 1882 cases, and a bullet seating tool.

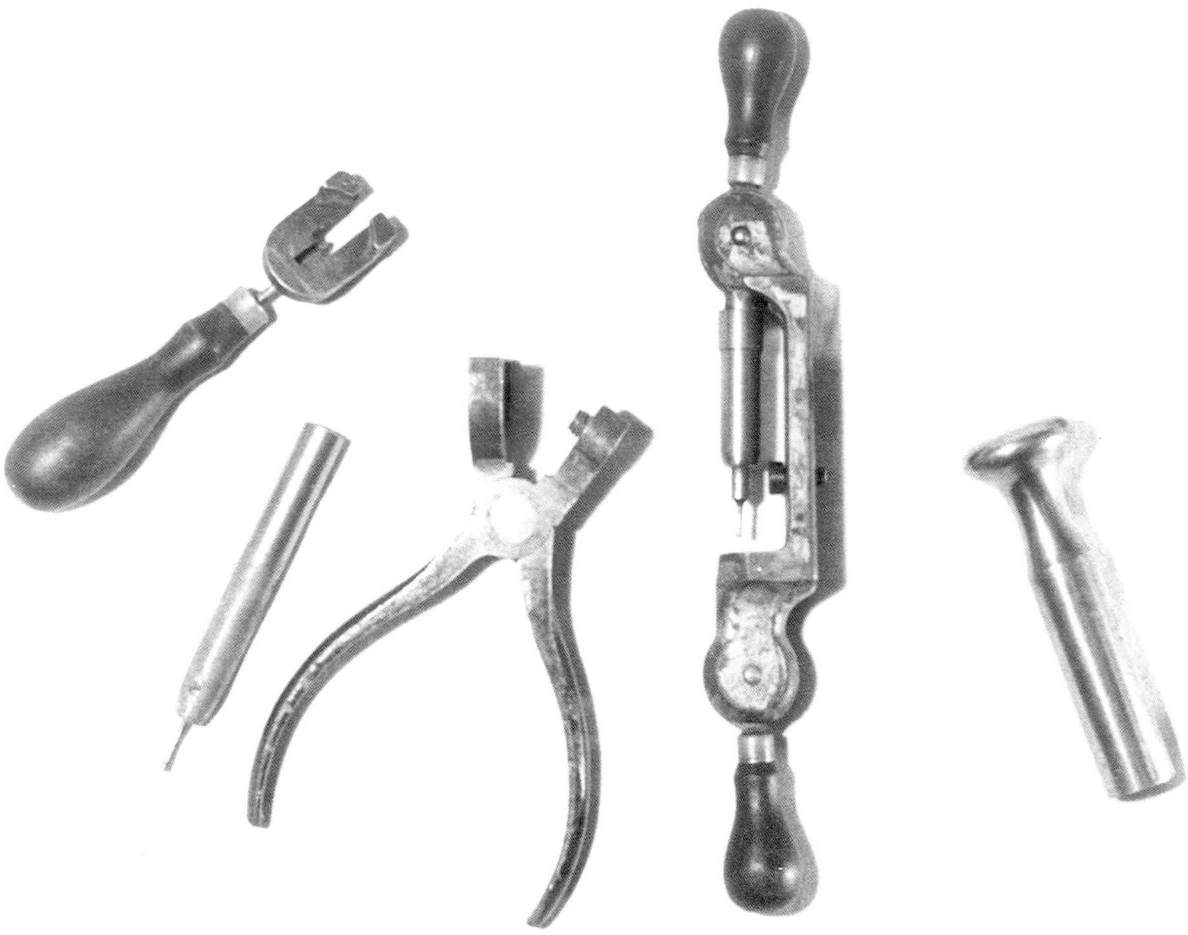

.46 Rimless, 1856. The early experimental for which Dr. Maynard provided a protective cap that fit over the head to prevent moisture from getting to the powder via the central hole in the base.

.69 Thong or Cord Extractor, 1858 trials, for muzzle-loading rifled musket converted to breech loading. This specimen has the percussion cap attached to the lanyard for convenience in priming the rifle. A central hole in the base admitted the flame from the exploding percussion cap to fire the cartridge.

.69 Flop Ear, 1858, for muzzle-loading rifled musket converted to breech loading. The arm served as the manual extracting device. This has the same central perforation in the base to admit the ignition flame to the powder charge.

.69 Capsule Shape, 1858 trials, with perforation in the rounded base for converting the muzzle-loading rifle/musket to breech loading.

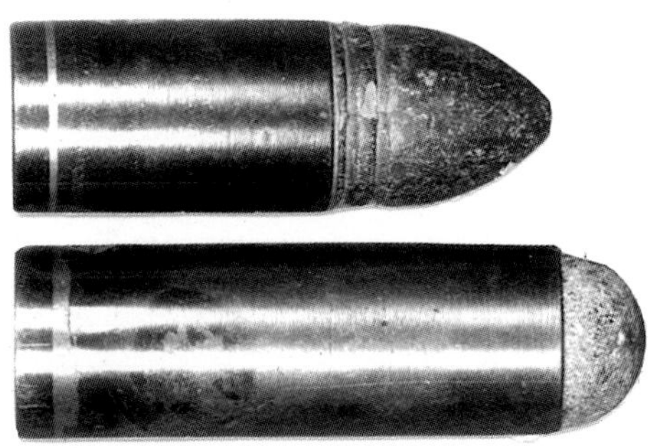

.69 Ball and Buck and Ball, 1858 trials.

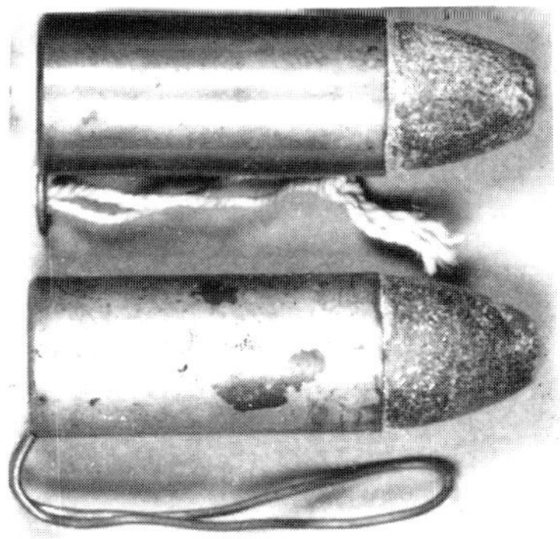

.50 cal. cord and wire extractors, 1863.

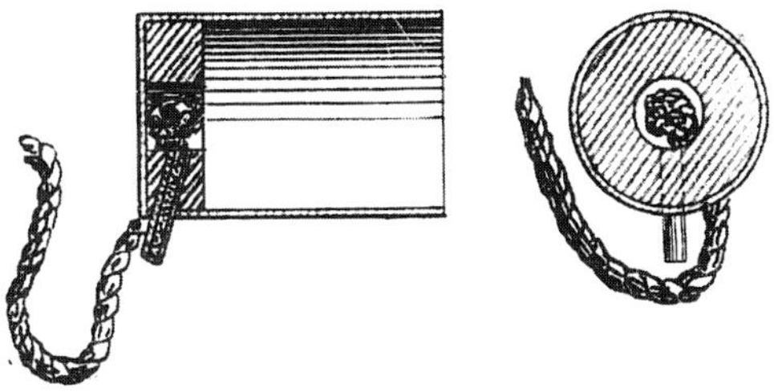

.69 Tube Ignition, 1864, U. S. Patent 45,420. For this experimental Dr. Maynard used a tube filled with percussion powder which was inserted into the case at the head and held in place by a papier mache wad. The hammer crushed the tube, igniting the priming and exploding the black powder charge within. The cartridge is also fitted with a cord extractor. No actual specimen is known to exist.

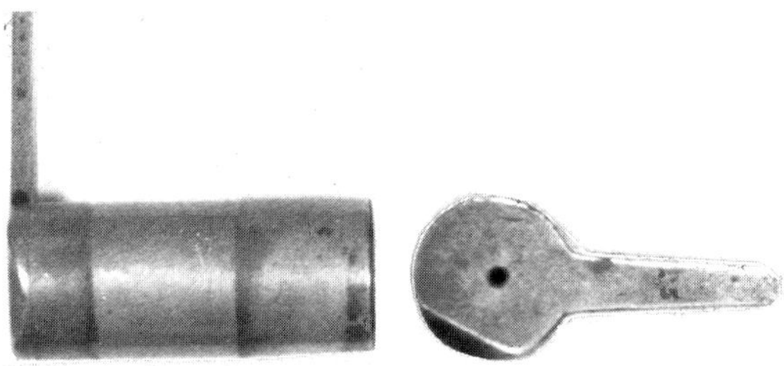

A specimens of .58 "Mule Ear" with long extractor arm and rim beveled on one side. The arm comes off the head in the center, or opposite the central ignition orifice. This form of extractor arm is illustrated in Maynard patent no. 59,044 granted October 23, 1866.

(See Appendix B for U. S. Maynard patents and the British patent for the 1859 rifle action).

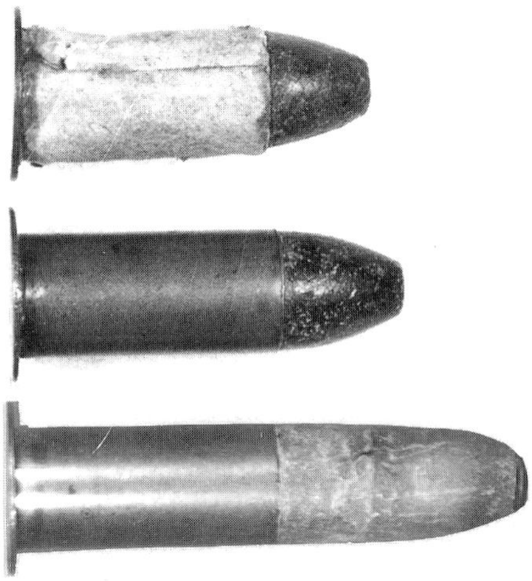

.50-50 Poultney & Trimble paper-covered brass foil case and the same cartridge with drawn brass case body. The rim is soldered to the case. Some are reinforced by a brass rivet with the central perforation to admit the flame from the percussion cap. Below is a specimen with Meigs explosive bullet. The .50-50 is the chambering for the carbines used by the Union during the Civil War and for the pre-war first generation rifles.

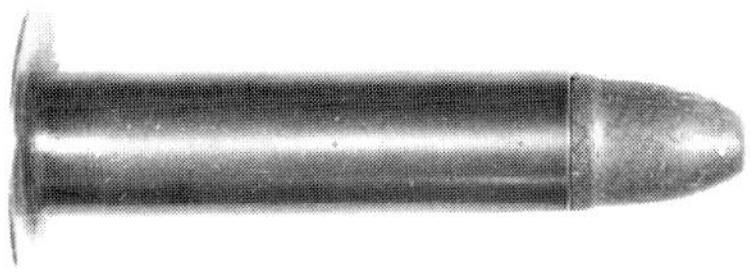

.35-30. Smallest of the 1865, this round was also made with the brass rivet, and with a nickeled or tinned case.

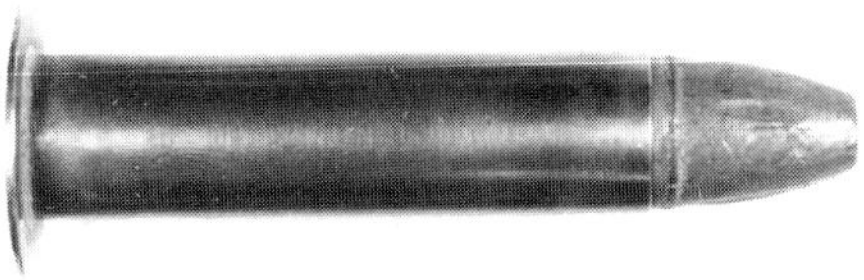

.35-40. Same construction as above. Found only with the plain base and brass case and rim.

.40-40. Same as above. This case is found only with the plain base.

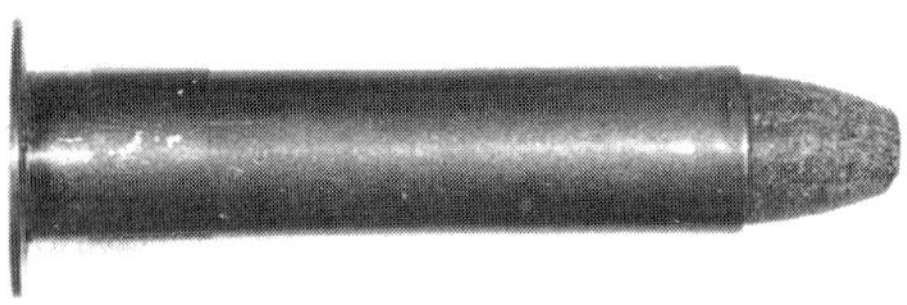

40-60. Same as above, but also found with the brass rivet.

.40-70. Same as previous but found only with the plain brass base.

.47, 1inch case. Found with iron or brass rim. This was an experimental only, never produced in quantity.

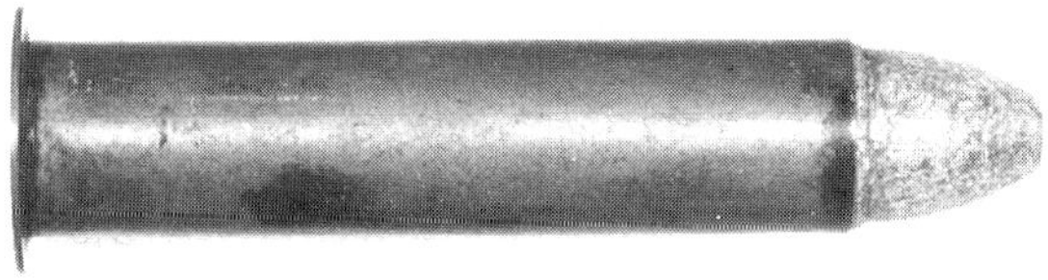

.50-100 with 2.35" to 2.5" case, for rifle, plain brass head only.

.55 Shot with2.27" case, no rivet. Found also in tinned or nickeled case.

.64 Shot, 2.4" case, found with copper rivet and with nickeled or tinned case.

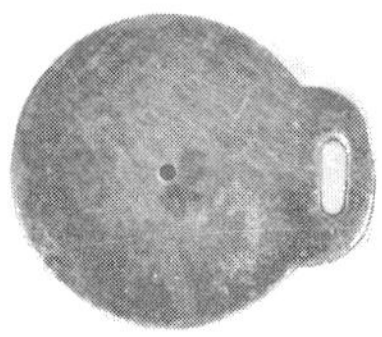

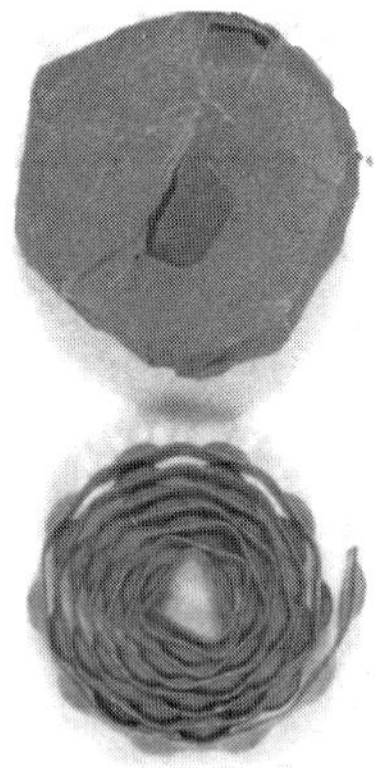

The Maynard percussion carbine, rifle and shotgun were provided with "stopper" cases so that the shooter could, in an emergency, load from the muzzle with loose powder and ball. The lanyard was tied to the extended loop on the rim, and with a hook on the other end, it could be attached to the rifle or shotgun. These were made in .35, .40, .50, .55 and .64—a bit shorter than the chamber. Usually the lanyard and hook is missing from these rarities because the lanyard has rotted off. Below are all five provided with lanyards and hooks. At left are the Maynard percussion caps for use in the inventor's patented tape priming device, still used in today's cap pistols.

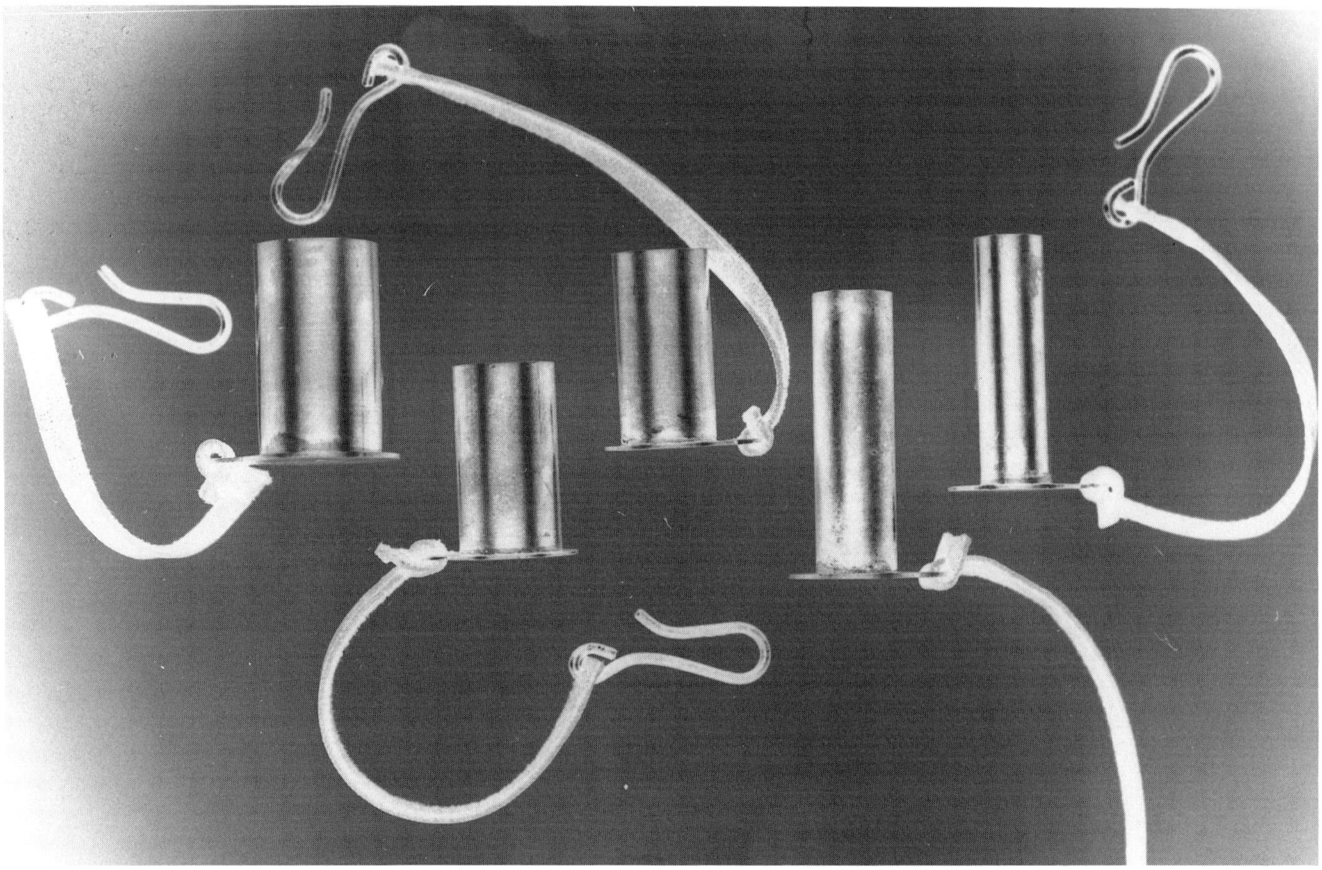

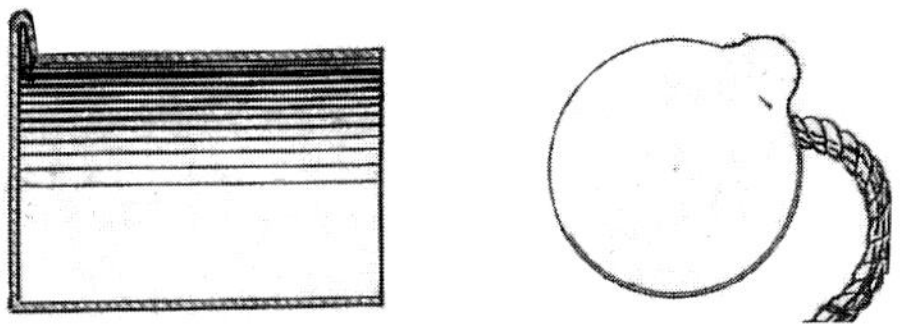

.58 Lip Fire. A small projection on the base of the case holds priming composition and would be struck by the hammer to crush and ignite it. This experiment appears on Dr. Maynard's patent no. 45823 for the rifle-musket conversion. It does not appear on the cartridge patent. Ethan Allen had already protected this idea in his U. S. patent 30109 granted September 15, 1860.

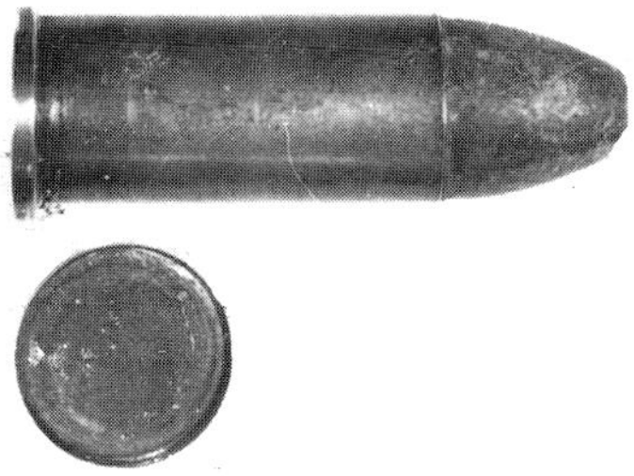

.50 Model 1866, one-piece drawn brass, rimmed with protective cap to fit over the head.

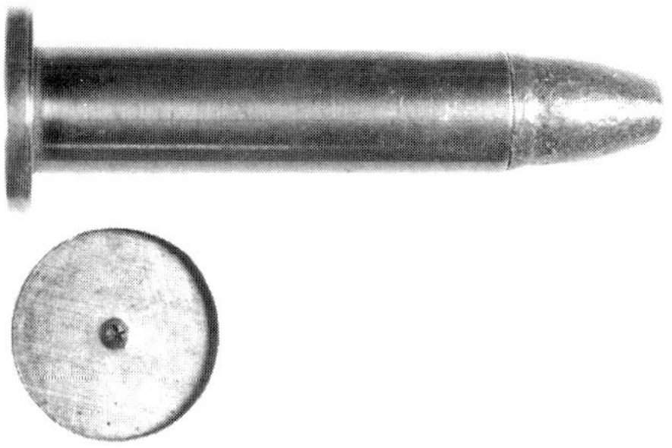

.40-40 rifle, Model 1867, made under patent no. 61223, January 15, 1867 which describes the thin, shallow-cup copper disc with priming in the inner center cavity . This fits snugly over the head to create a self-contained centerfire cartridge.

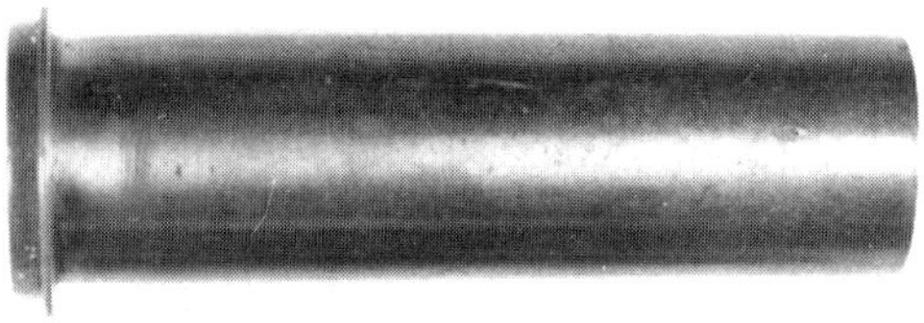

.64 Shot, Model 1867, used the same type of priming cap but the head was made with a thin rim projecting out from the raised portion of the base.

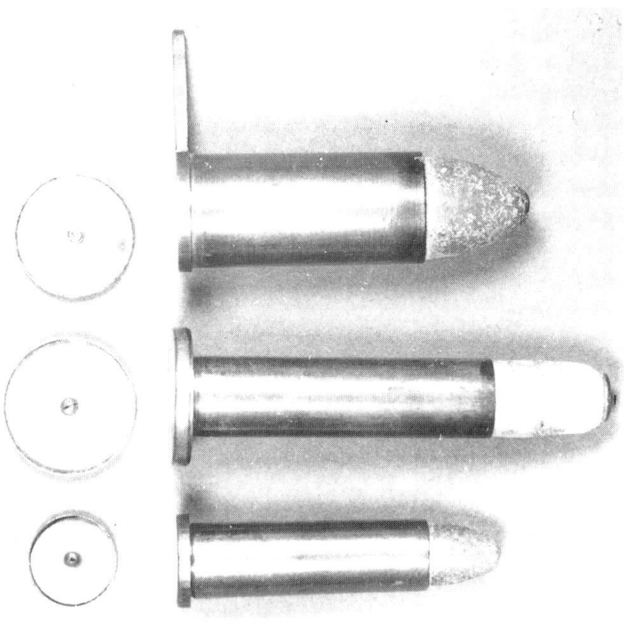

Three other experimental specimens made to use the 1867 primed cup. The upper is the .58 flop ear. The cap should have a notch to accept the extraction arm. The next is the .40-40 rifle shown above but with explosive bullet. The third is .35 caliber, never produced.

35-30. 1.635" brass two-piece case with head soldered to the case tube ¼" from the base, Berdan primer, iron anvil, with the Logan & Hart-patented decapping groove on either side of the primer pocket. No headstamp.

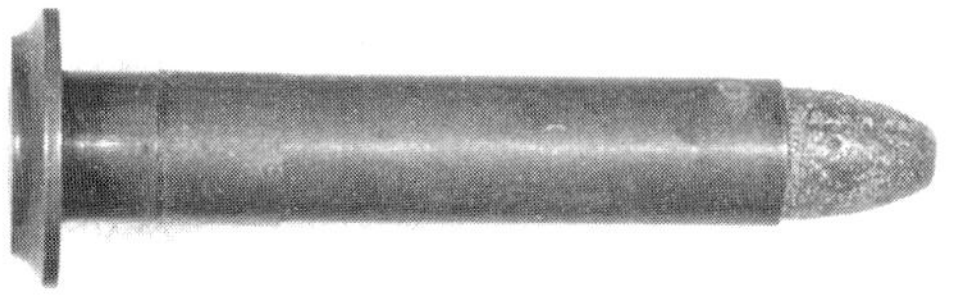

35-40. 2.125" two-piece brass case, same as above.

.40-40. 1.86" two-piece brass case, same but also reported with a nickel-plated case. May be found with conical or the heavy cylindrical bullet shown here.

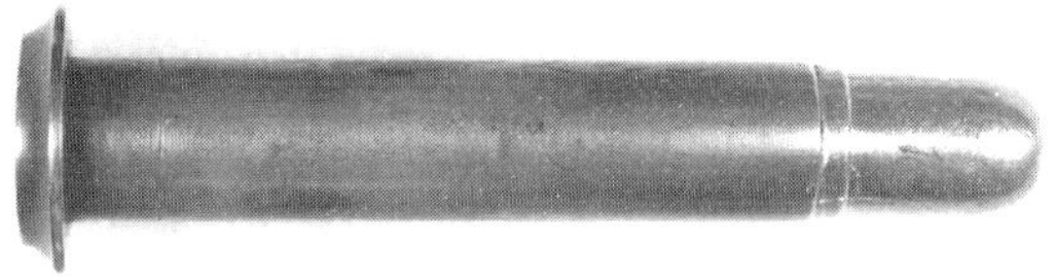

.40-60. 1.86" brass case, also reported with nickel-plated case. These cartridges were intended for repeated loading so bullets vary as do case lengths as they were worn and trimmed down after repeated use.

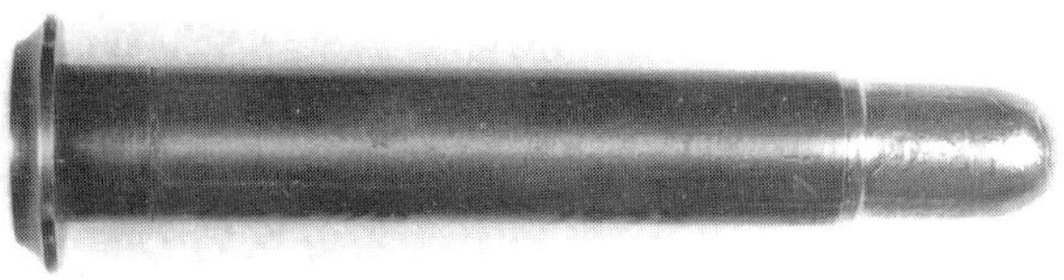

.40-70. 2.75" brass case only. Specimens of the .40 Model 1873 are found in odd case lengths, obviously cut down from the long 40-70. Since rifles for just these five case lengths were made, different case lengths must be viewed as spurious.

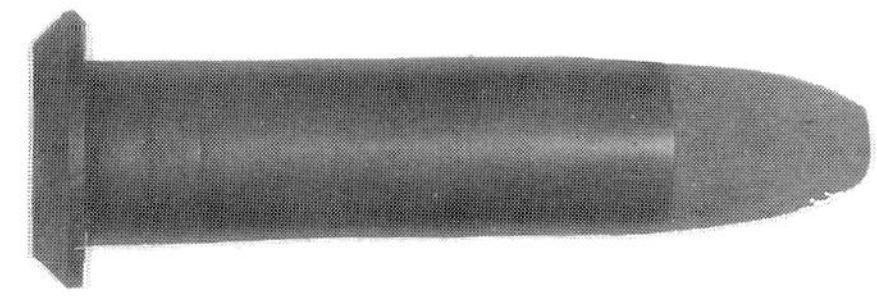

.44-60, 1.75" case, brass only.

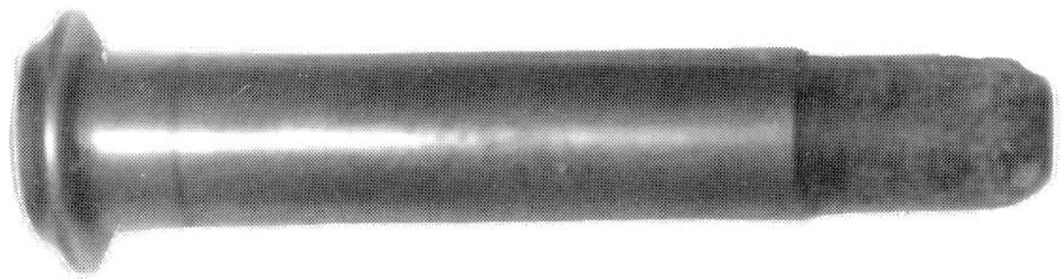

.44-70, 2.19" case, brass only.

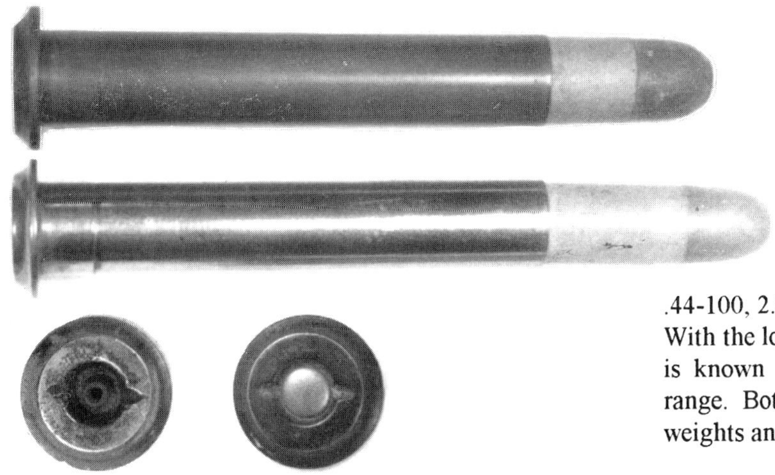

.44-100, 2.89" brass case, also nickel plated. With the long paper-patched bullets, this round is known as the Creedmoor after the target range. Both are hand loads with varying bullet weights and powder charges.

.50-50, 1.565" to 1.47" brass case.

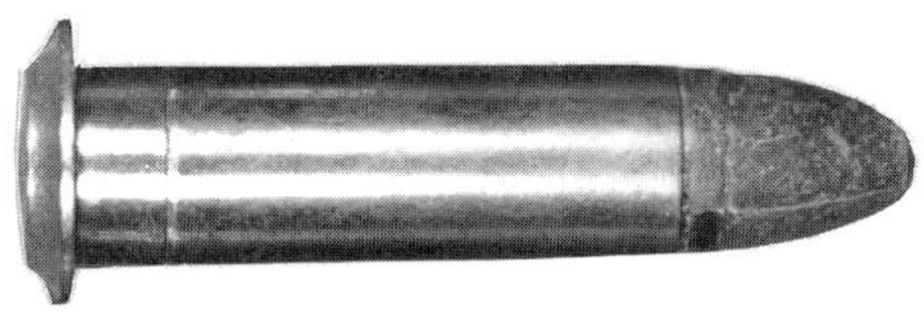

.50-70, 1.84" brass case.

50-100. 2.39" brass case. May be found with the conical or heavy cylindrical bullet.

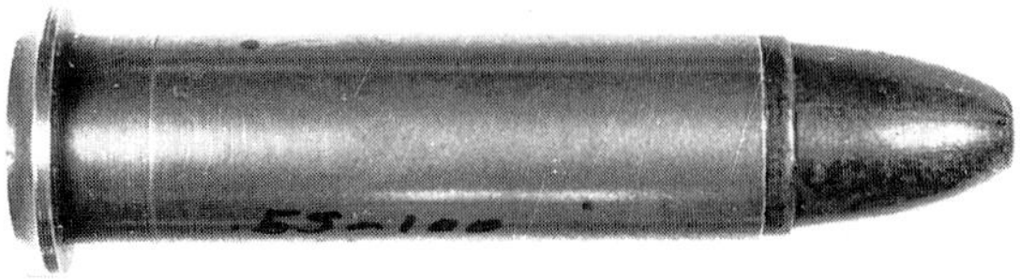

.55-100. 2.15" brass case. Also found with either conical or cylindrical bullet.

.55 Shot. 2.38" brass case.

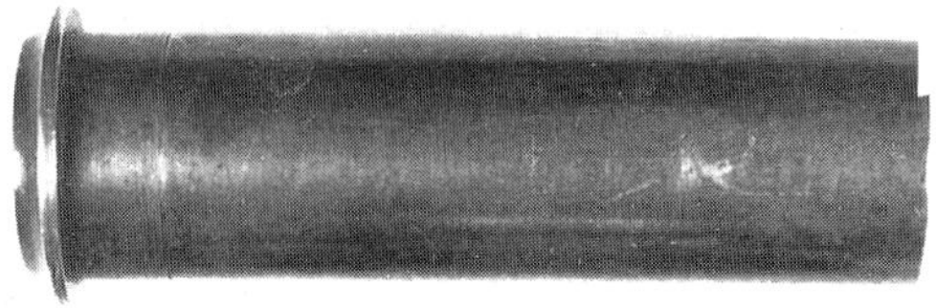

.64 Shot. 2.53" brass case.

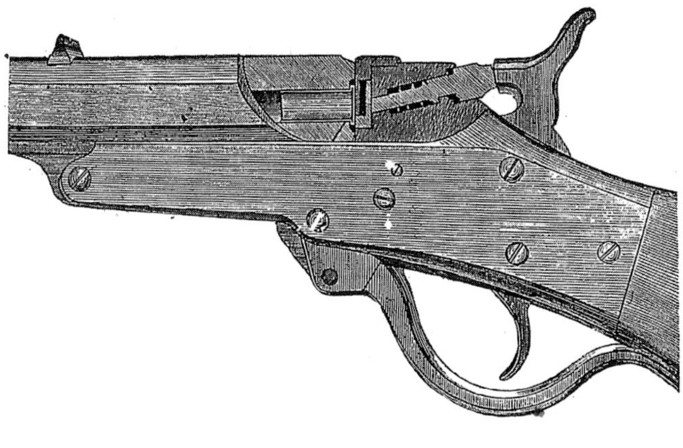

Hadley's Patent Device for using Rim Fire Cartridges in the Maynard Rifle.

APPLIED TO "MODEL 1873" OR TO "MODEL 1865" WHEN CHANGED TO "MODEL 1873."

By referring to the cut, it will readily be seen how simple this device is. It consists of a cap attached to the breech-piece by two small screws, in which is a disk with a firing pin projecting through the cap at such a point from the center as to strike the rim of any cartridge from 22-100 to 38-100 calibre. This disk plays freely in the cap, and is driven forward by the firing-pin in breech-piece. To use our central-fire cartridges "Model 1873," you have only to remove the cap, change firing pins, and you are ready to insert the regular Maynard center-fire cartridge, "Model 1873."

Barrels have to be especially chambered for using the standard fixed ammunition.

The attachment *cannot* be applied to "Model 1882."

For applying attachment to "Model 1873,"..$3 00
For changing "Model 1865" to "Model 1873" and applying attachment,.........10 00

At left is a Model 1873 rifle, cal. .44-100, 32" barrel with rack and pinion rear sight and target front. It fits the specifications of the No. 14 Creedmoor except for the buttstock which is the ordinary sporting design. The other rifle is a very rare 1873 Model 15 in .40-60 caliber. Fancy wood, checkering and the gentle "roll" at the rear portion of the pistol grip distinguish this as a rifle with a lot of extras. Model 15s came with both hunting and schuetzen buttplates, which interchanged. This one has the sporting buttplate in place.

Massachusetts Arms Company used many of the 22,002 Civil War-generation percussion-fired carbines, which they evidently re-acquired, to produce the early Model 1873s. The sporting percussion rifles were also altered. The percussion nipple was replaced by a firing pin in a manner that can easily be seen on these early centerfire Maynards. Later some of these Model 1873s were reworked to chamber Model 1882 ammunition.

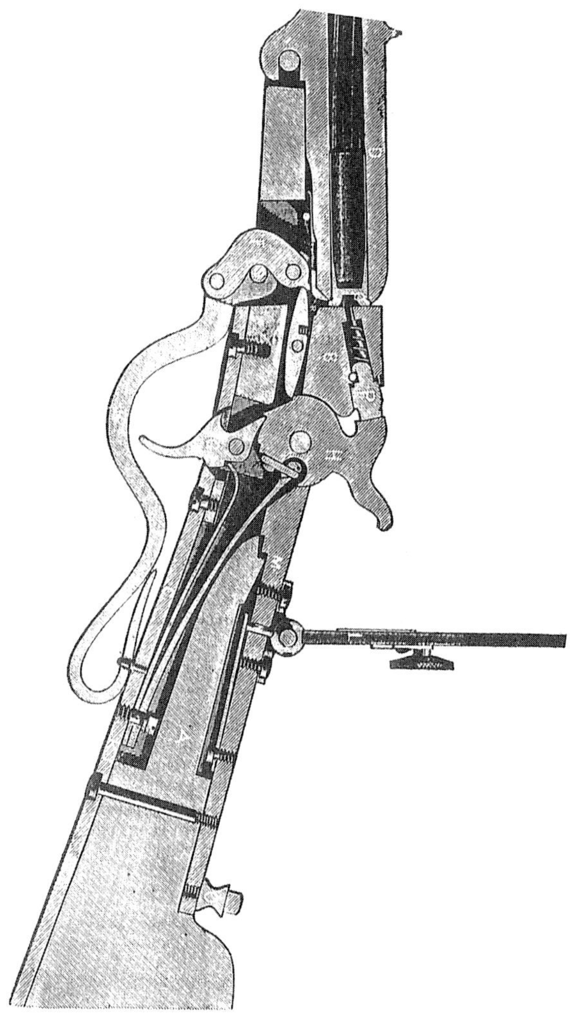

Full Size Shells, "Model 1882."

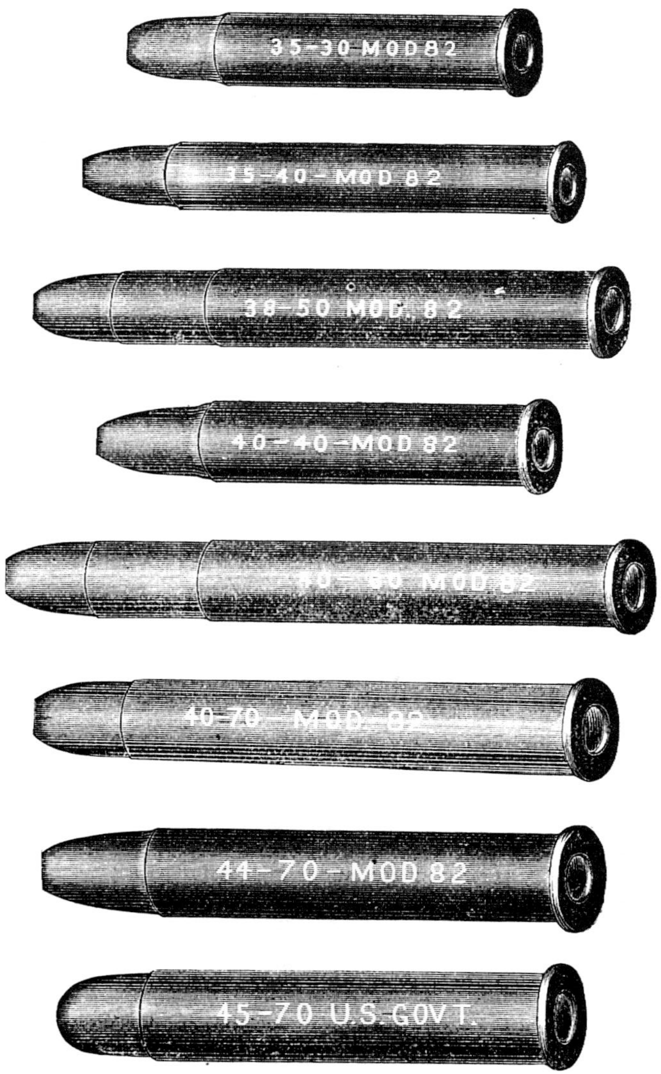

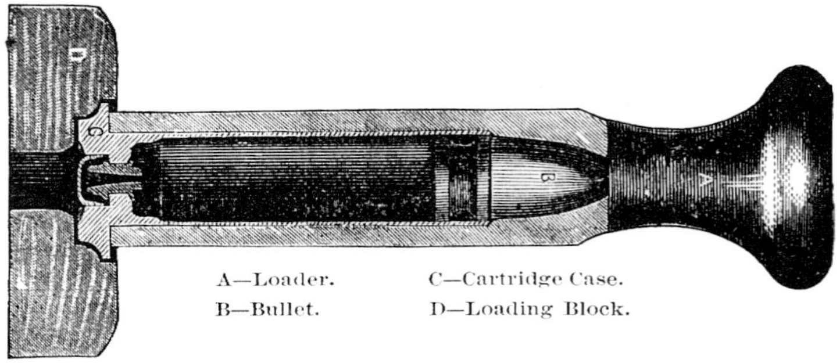

A—Loader. C—Cartridge Case.
B—Bullet. D—Loading Block.

An inspection of the foregoing cut will exhibit to the practical sportsman some of the peculiar advantages of the Maynard arms.

The cartridges are loaded with an implement like a starter which insures placing the bullet with mathematical precision, where it remains in the same position through the barrel until delivered. The position of the primer and bullet renders the cartridges perfectly water-proof. The quality of powder is subject to the sportsman's own selection, and the expense for ammunition is covered by the bare cost of powder, lead and primers, as the cartridge cases can be reloaded for an indefinite number of times. (See directions for loading on next page.)

Calibre of Cartridges for Rifles and Shot Guns.

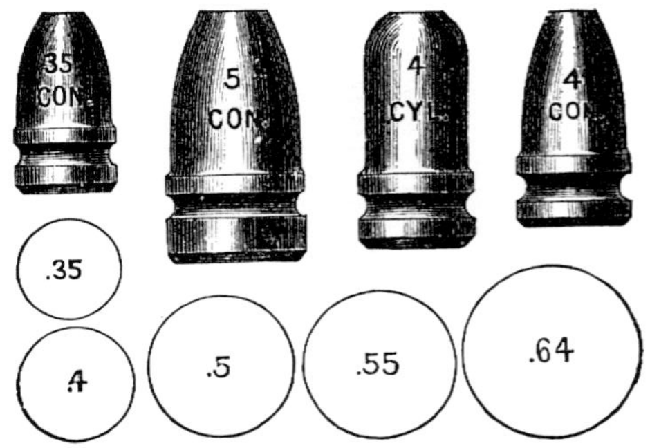

In addition to the above we now furnish .22, .32, .38 and .44 calibres.

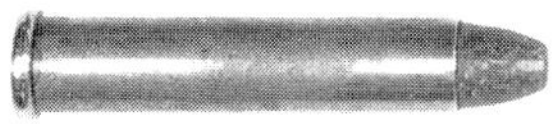

.22-10-45. 1.168" case. Plain head. Ten grains of powder, 45-grain four-groove lead or paper-patched bullet.

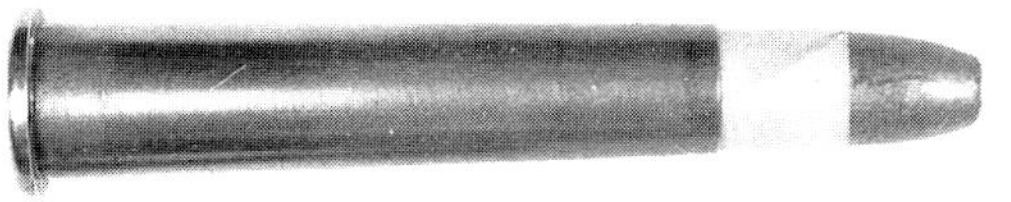

.32-35, 1.86" case, slight taper toward the mouth, approximately .060. The rim is slightly rounded at the base with a sharp edge at the front. Rim diameter is between .052" and .063". This and other Maynard Model 1882 cases may be easily confused with similar Ballard, Wesson and Stevens cartridges. No headstamp occurs on any of the Model 1882 cartridges. All were produce by Massachusetts Arms Company and most were the heavy-walled everlasting type, and were bored out, or rebated, at the mouth down into the case to accept the exact diameter bullet required.

35-30. 1.56" case. Same as above but .025 taper from head to mouth. The early Model 1882 rounds have the large Berdan primer, those made later were loaded with smaller Winchester boxer primer.

.35-40. 2.058" case. Same as above, but average .007 taper.

.38-50. 2.1" case. The Massachusetts Arms Co. 1885 catalogue lists barrels for this cartridge and either 255-gr. paper-patched or 245 gr. grooved bullets. The existence of this cartridge has been doubted but it is illustrated in the catalogue, full size, the barrels are offered, and the two weights of bullets seem to confirm that it was, in fact, made. The .35-40 case may be slightly shorter but repeated reloadings resulted in trimming so this is not a good determining factor. Bullet size is the only way to tell the 35-40 and .38-50 apart.

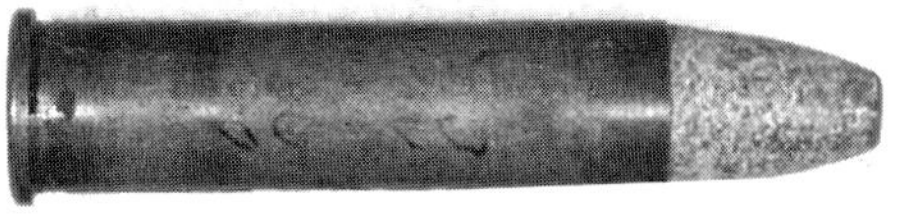

.40-40. 1.75" to 1.78" case. Same as above. May be loaded with the heavy 340-gr. cylindrical bullet, the 330-gr paper-patched, or the 270-gr. grooved.

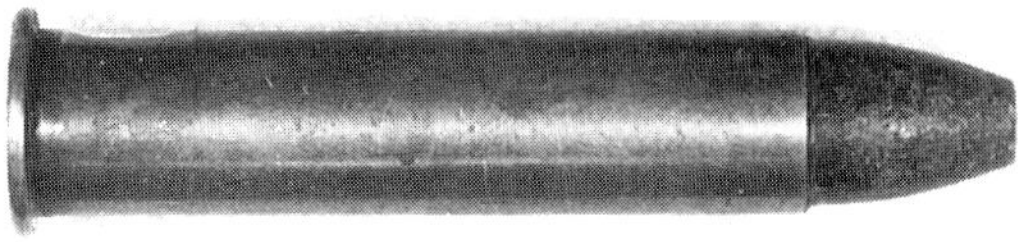

.40-60. 2.19" average case length. Same bullets as above.

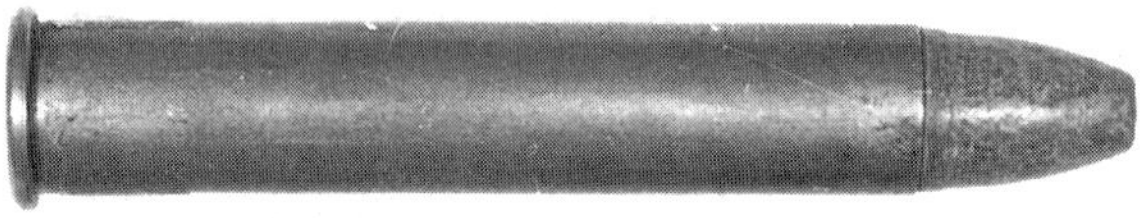

.40-70. 2.4" case. Same bullets as above.

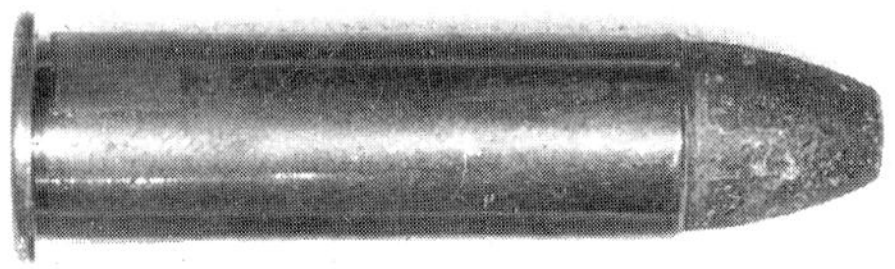

.44-60. 1.92" average case length. The 1885 catalogue lists a 430-gr. cylindrical bullet but not the conical in this specimen.

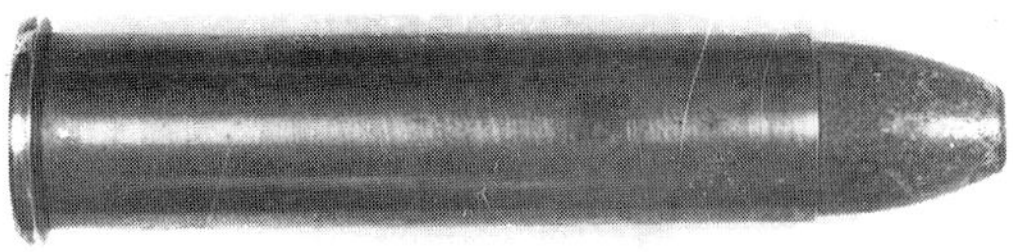

.44-70. 2.195" case.

.45-70. 2.1" case. The catalogue lists the Maynard Improved Hunters' Rifle in .45-70 U. S. Government but offers no bullets or bullet moulds for this caliber. The bullet in the specimen shown here is the Maynard flat-nosed conical type but has a shoulder at the case mouth, so interior dimensions are .45 and exterior at the mouth is .44.

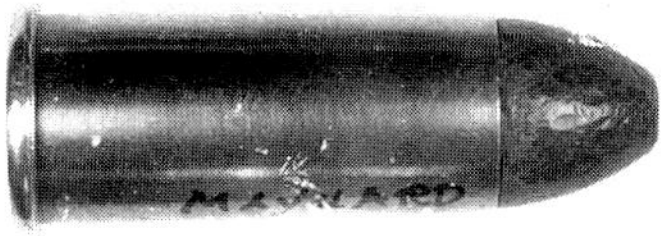

.50-50. 1.39" case. Massachusetts Arms Co. listed a .50 caliber No. 1 bullet, but no weight for it, a 470-gr. conical and 513 gr. cylindrical. The bullet in this short case could be the No. 1.

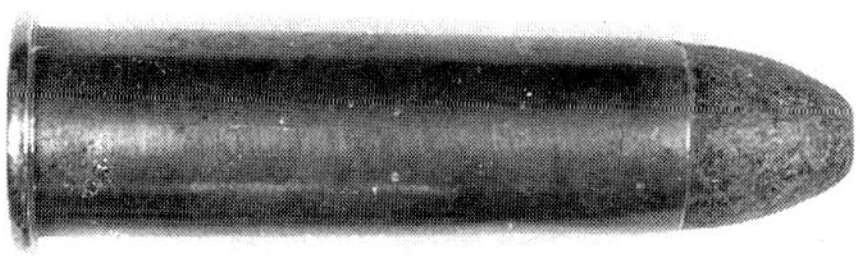

.50-70. 1.77" case. Same bullets used in this case as above. The catalogue also lists a .50 caliber explosive, No. 2 bullet, no specimen to show here.

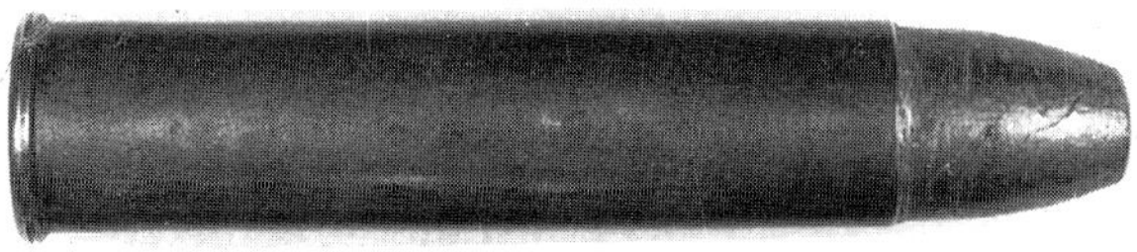

.50-100. 2.31" case. This specimen is loaded with the 470-gr. flat-nosed conical bullet. The maker described these rifles as for "large and dangerous game."

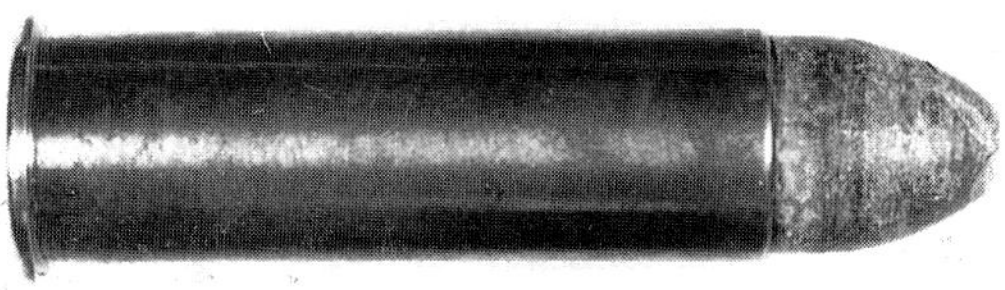

.55-100. 1.93" case. The catalogue does not list a .55 caliber bullet but Maynard Improved Hunters' Rifles were chambered for this case. It is shorter than the .55 Shot case.

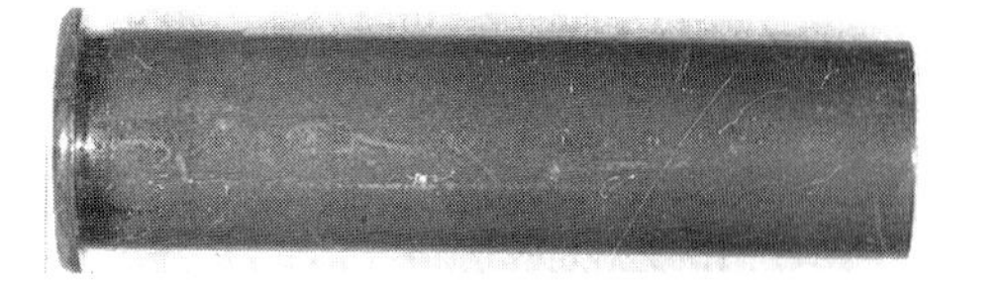

.55 Shot. 2.31" case. Maynard offered .55" shot barrels in 20" and 26" lengths which could be interchanged on any action and stock by opening the action, turning and removing the hinge pin, lifting out the barrel, and replacing it with a shot barrel or any other caliber of rifle barrel.

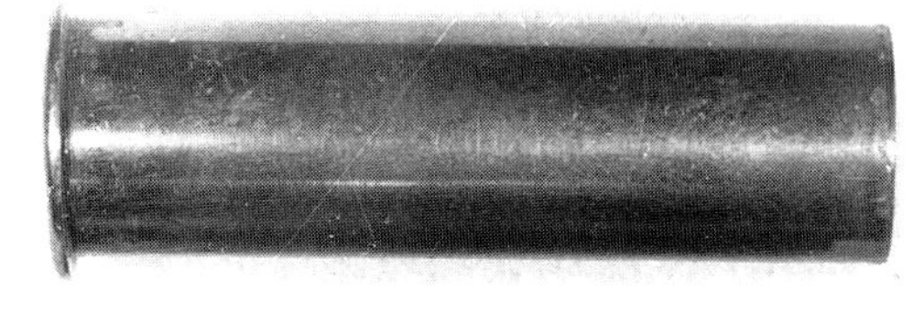

.64 Shot. 2.46" case. Maynard offered .64 shot barrels in 26", 28" and 32" lengths for use on any Model 1882 action and stock. The cost in 1885 was $9.50 to $12.00 depending on length of barrel.

List of Prices.

RIFLE BARRELS.

20 inch, 4 inch octagon, .22 calibre, for rim-fire fixed ammunition,.............$9 00
24 inch, 4 inch octagon, .22 calibre, for rim-fire fixed ammunition,.............10 00
24 inch, 4 inch octagon, .32 calibre, for rim or central-fire fixed ammunition,..10 00
26 inch, 4 inch octagon, .32 calibre, for rim or central-fire fixed ammunition,...10 00
26 inch, 4 inch octagon, .38 calibre, for rim or central-fire fixed ammunition,.. 9 50
28, 30 or 32 inch, 9 inch octagon, .38 calibre, for rim or central-fire fixed ammunition, ...12 00
20 inch, 4 inch octagon, .35, .40, .44 or .50 calibre, for special ammunition,...... 8 00
26 inch, 4 inch octagon, .35, .40, .44 or .50 calibre, for special ammunition,...... 9 50
26, 28, 30 or 32 inch, 9 inch octagon, .35, .38, .40, .44, .50 or .55 calibre, for special ammunition,..12 00
26, 28, 30 or 32 inch, full round, .40, .44, .50 or .55 calibre, for special ammunit'n,..12 00
26, 28, 30 or 32 inch, full round, .45 calibre, for U. S. Government ammunit'n,....12 00
26, 28, 30 or 32 inch, 9 inch octagon, .40 calibre, 70 grains, central-fire fixed ammunition,..12 00

SPECIAL BARRELS FOR THE .22 AND .32 CENTRAL-FIRE SHELLS.

24 and 26 inch, 4 inch octagon, .22 calibre, 10 grains, central-fire,..............$10 00
28 inch, 9 inch octagon, .22 calibre, 10 grains, central-fire,...................... 12 00
26 inch, 4 inch octagon, .32 calibre, 35 grains,...................................... 10 00
28 inch, 9 inch octagon, .32 calibre, 35 grains,...................................... 12 00

Barrels for using the standard fixed ammunition may be chambered as follows:

.22 calibre, for short or long cartridges.
.32 calibre, for short, long or extra long cartridges.
.38 calibre, for long or extra long cartridges.

Barrels for using the Maynard special ammunition are chambered as follows:

.35 calibre, for 30 or 40 grains of powder.
.40 calibre, for 40, 60 or 70 grains of powder.
.38 calibre, for 50 grains of powder.
.44 calibre, 60, 70 or 100 grains of powder.
.50 calibre, 50, 70 or 100 grains of powder.
.55 calibre, for 100 grains of powder.

SHELLS,—"Model 1882":

.22 calibre, 10 grains, each,........	05
.32 calibre, 35 grains, each,........	07
.35 calibre, 30 grains, each,........	07
.35 calibre, 40 grains, each,........	08
.38 calibre, 50 grains, each,........	08
.40 calibre, 40 grains, each,........	07
.40 calibre, 60 and 70 grains, each,........	08
.44 calibre, 60 and 70 grains, each,........	08
.44 calibre, 100 grains, each,........	10
.50 calibre, 50 and 70 grains, each,........	08
.50 calibre, 100 grains, each,........	10
.55 calibre, 100 grains, each,........	10

This page from the 1885 Maynard catalogue lists 38 caliber barrels and .38 caliber cartridge cases taking 50 grains of powder.

This is a Model 1882, caliber .40-70, which preserved some of the attributes of the Model 1873 rifle from which it was altered. The receiver is dated 1873 on the left side and 1882 on the right. Note the high hammer which distinguishes the 1873 models. The buttstock is characteristic of the early 1873 Model 10.

The cartridge for which this rifle was chambered was the ever-lasting type with an interior rebate at the mouth about 5/16" into the case to facilitate loading of the correct diameter bullet.

Sights for Maynard Rifles.

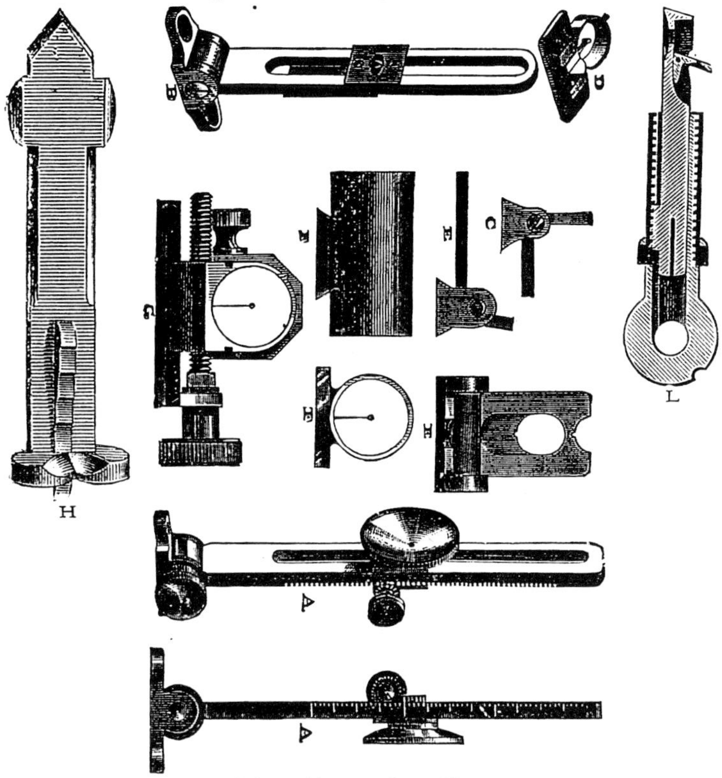

Directions for Use.

TO CAST THE BULLETS.—Melt the lead in a large-sized ladle or other vessel, and when QUITE HOT, so that it will run freely, dip it with a small ladle, and fill the molds, which should be previously heated almost hot enough to blue them. After filling the molds, empty the lead remaining in the small ladle into the larger, keeping it at as even a degree of heat as possible, and repeat the operation. From 1½ to 2 per cent of tin mixed with pure soft lead will improve the bullets. Accurate shooting can only be done with perfect bullets.

TO LUBRICATE THE BULLETS.—Melt a small quantity of the lubricator in a shallow vessel, and dip the bullet in it so as to fill the grooves, or prepare as a substitute a mixture of one part of beeswax to three parts of tallow, and apply as above. By means of the Lubricating Pump, more satisfactory results are obtained by using the lubricator cold.

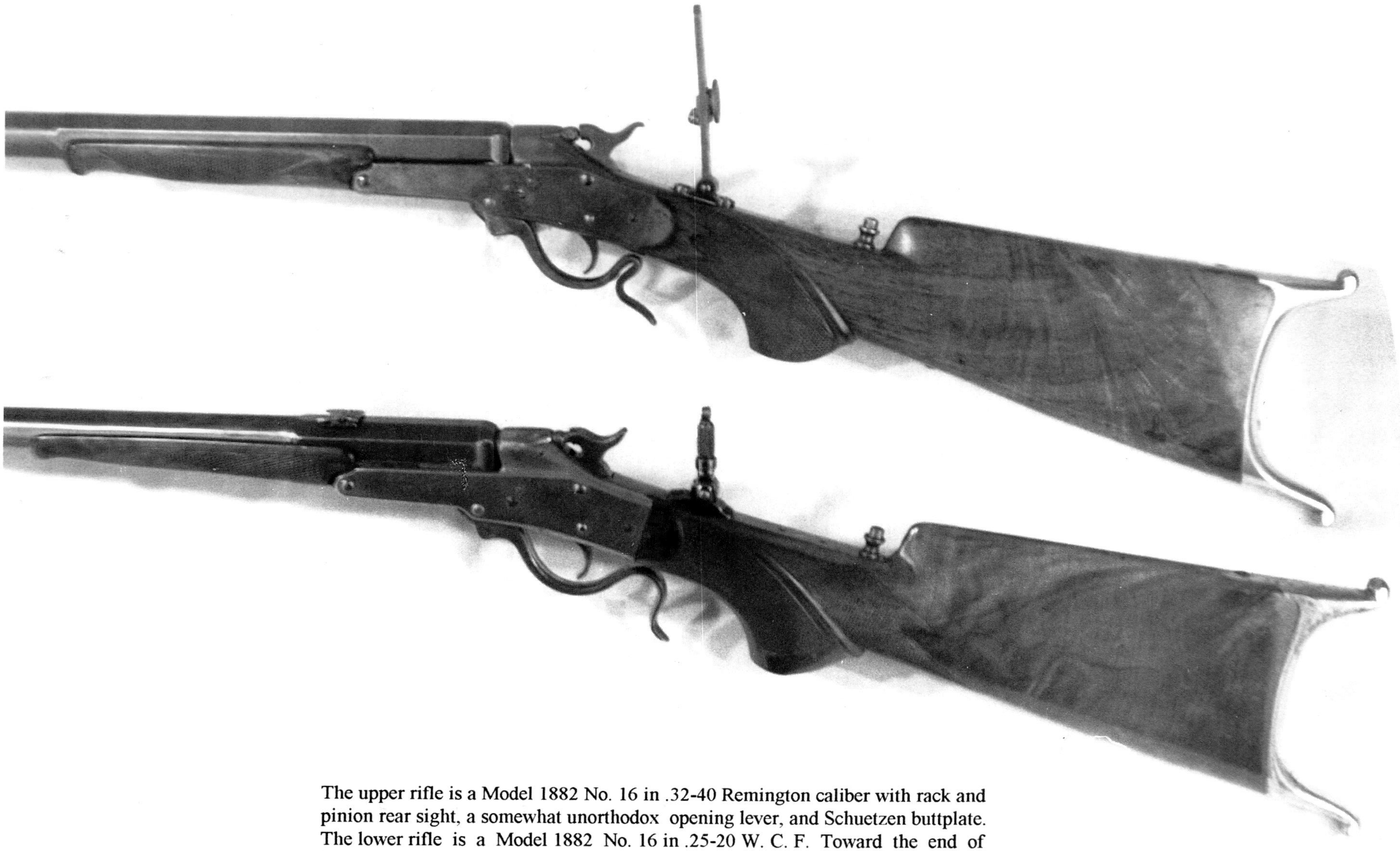

The upper rifle is a Model 1882 No. 16 in .32-40 Remington caliber with rack and pinion rear sight, a somewhat unorthodox opening lever, and Schuetzen buttplate. The lower rifle is a Model 1882 No. 16 in .25-20 W. C. F. Toward the end of Maynard rifle production, Massachusetts Arms Company chambered barrels for Winchester and other cartridges such as .22-15-60 Stevens. Evidently the buyer could have anything he wanted. This specimen was also fitted with Lyman sights by the factory.

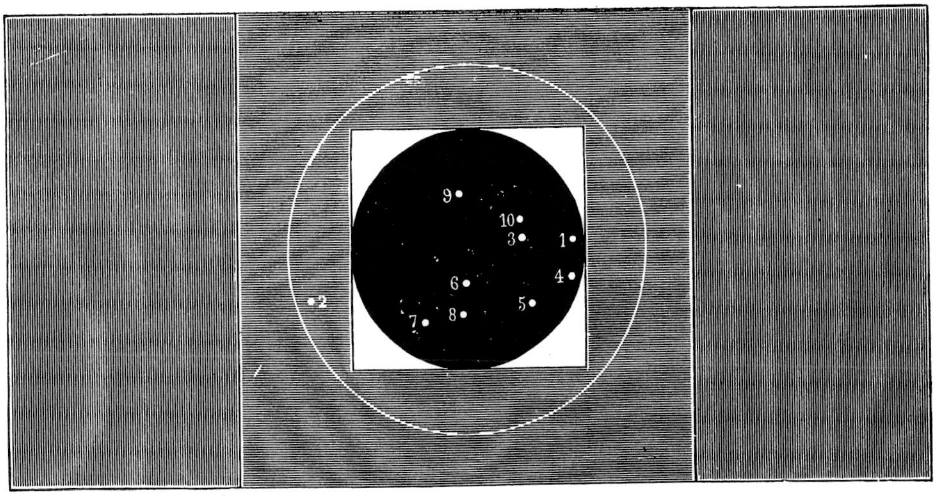

5 4 5 5 5 5 5 5 5 5—49 out of a possible 50.

SCORE OF TEN CONSECUTIVE SHOTS MADE WITH A SHARPS CREEDMOOR RIFLE,

December 26th, 1874,

At his FIRST trial of the arm, by SIR HENRY ST. JOHN HALFORD, BART., at Wistow, Leicester, England, without rest and without wiping. Distance 800 yards; condition of weather, unfavorable. New Wimbledon Target; Bull's Eye 3 feet in diameter.

Twenty-Five Years of Use, and Constant Progress!

THE BEST IN THE WORLD!!

SHARPS RIFLE CO.

MANUFACTURERS OF

Sharps Patent Breech-Loading Metallic Cartridge Military, Sporting, and Creedmoor Rifles.

For SAFETY, ACCURACY, SIMPLICITY and PENETRATION, combined with UNEQUALLED WORKMANSHIP,

THEY STAND UNRIVALLED!

Every part is carefully and thoroughly inspected, each barrel is severely tested for accuracy and strength, and no work that is not as ABSOLUTELY PERFECT as human skill and experience can produce, is permitted to leave the Armory.

In using this weapon the hunter or marksman cannot be injured by accidental discharge while loading, the Sharps in this respect possessing a virtue shared by few other breech loaders. The exploded shell is always infallibly removed by the lever.

At the Creedmoor Range, near New York city, in the Summer of 1873, the Sharps Rifle made the following record:

Winner of the Amateur Club Gold Medal for 500 yards, three times, with scores of 26, 27, 28, against 25, 27, 25, 28, by other rifles. Winner's average with Sharps 27 out of a possible 28. Winner's average with other rifles, 26, 25.

Winner of four out of six prizes in the match for military rifles at 500 yards, June 6th. Winner's average with Sharps Military Rifles, 25, 25. Winner's average with other rifles, 24, 50.

Match for the Amateur Club, Saturday, June 20th. *Distances* 800, 900, *and* 1,000 *yards.*

Nine Sharps and ten *other* rifles entered. Sharps won the gold medal. Average, 25.88, and average for the *others*, 23.50. Odds in favor of Sharps Rifles *over two and one-third points per man.*

Match of same club, same distance, June 28, eight Sharps and ten *other* rifles entered. Average for Sharps 28.37. Average for the *other* 23.77. Odds gained by using Sharps 4.60.

Match for Turf, Field, and Farm Badge, same day, 200 yards off hand. Sharps average 13.80. The others 12.35.

WON DIAMOND BADGE Saturday, July 11, at 500, 800, and 1000 yards, making score of 71 out of a possible 84.

In the International Match at Creedmoor, September, 1874, between the Irish and American teams, (won by the Americans), the Sharps Long Range Rifle made the highest score of any breech loader used as such, at 800, 900, and 1,000 yards, and on the same range, one week later, The Sharps Rifle Company were the only American manufacturers who dared accept the challenge of Mr. J. Rigby, of the Irish team, for a test of breech loaders against muzzle loaders, twenty-five rounds each, at 1,000 yards, *without wiping*. Here, also, the Sharps proved its great superiority.

At the Fall Meeting, beginning September 29th, 1874, of the National Rifle Association, at Creedmoor, the Sharps Rifle won the following prizes:

ARMY AND NAVY JOURNAL MATCH.

500 yards. Open to teams from all regularly organized Military organizations in the United States—180 entries—only 12 *Sharps Rifles used.*

First Prize,—Silver plate, value $750.00. Second Prize,—Silver cup, value $50.00.

The Sharps Rifle also won in this match the First and four other of eight prizes given for the best individual scores.

Note.—The Seventh New York Regiment, whose team won the First Prize, $750.00, in this match, are armed with Remington Rifles, but a majority of the team laid aside their own rifles, and used the Sharps.

Of the team of the Twelfth New York Regiment (also armed with Remington Rifles), that won the second prize, (Cup, value $50.00,) five men out of the twelve used the Sharps Rifle in preference to their own weapons.

ALL COMERS' MATCH.

500 and 600 yards. 12 prizes in all—200 Entries. The Sharps Rifle won the 1st, 2d, 3d, 4th, and three others (in all 7 out of 12).

PRESS MATCH.

500 yards. The Sharps Rifle won first prize, Gold medal, value $50.00; also, 2d, 7th, and 9th prizes.

BENNETT MATCH.

800, 900, and 1,000 yards. The Sharps Rifle won 13 prizes.

Weight of Rifles, 8 to 12 lbs. Charge of Powder, 50 to 95 grains. Length of Ball, $\frac{7}{8}$, $1\frac{1}{8}$, $1\frac{1}{4}$, $1\frac{3}{8}$ inches. Calibre, $\frac{40}{100}$—$\frac{44}{100}$—$\frac{45}{100}$—$\frac{50}{100}$.

THE BEST ALWAYS THE CHEAPEST.

SHARPS SPORTING RIFLES.

Weight, 8 to 12 lbs. $\frac{40}{100}$—$\frac{44}{100}$—$\frac{45}{100}$ and $\frac{50}{100}$ calibre, polished stock, open sights, single trigger.

Inch.		Inch.	
26 Octagon Barrel,	$35.00	26 Half Oct. Barrel,	$33.00
28 " "	36.50	28 " "	34.50
30 " "	38.00	30 " "	36.00

SHARPS HUNTERS RIFLES.

Weight 8 to 10 lbs. $\frac{40}{100}$—$\frac{44}{100}$—$\frac{45}{100}$ and $\frac{50}{100}$ calibre, oiled stock, plain finish, open sights, single trigger.

26 inch, Round Barrel, - - - - -	$30.00
28 " " " - - - - -	31.50
30 " " " - - - - -	33.00

SHARPS CREEDMOOR RIFLE No. 2.

$\frac{44}{100}$ calibre, 32 inch barrel, long, straight, hand-made, polished stock, weight just under 10 lbs., trigger pull 3 lbs., peep, rear sight, with Vernier scale, by means of which a register may be kept of the elevation required for a given distance under varying circumstances; wind gauge, with interchangeable globe and split-bar front sights, - - - - - - - - - - - $90.00

SHARPS CREEDMOOR RIFLE No. 1.

Designed exclusively for very long ranges. $\frac{44}{100}$ calibre, 32 inch barrel, long, straight, hand-made *pistol grip*, polished stock, checkered grip and fore-end, regulation weight and trigger pull; peep, rear sight, with Vernier scale, allowing the necessary elevation for 1,300 yards; wind gauge, with interchangeable globe and split bar front sight, with morocco sight case. (This is the very best rifle manufactured, and will be found to give splendid results at the longest ranges,) - - - $125.00

Spirit Level attached to Front Sight, extra.

Special arms of any weight, length, or calibre made or engraved to order at the Armory.

Reloading implements, per set, - -	$6.00	Double Triggers, extra, - - -	$4.00
Globe and Peep Sights, extra, - -	5.00	Telescope Sights " - - - -	40.00

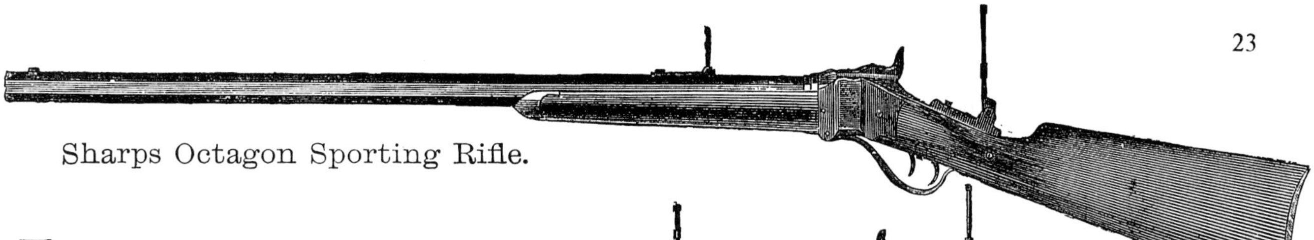

Sharps Octagon Sporting Rifle.

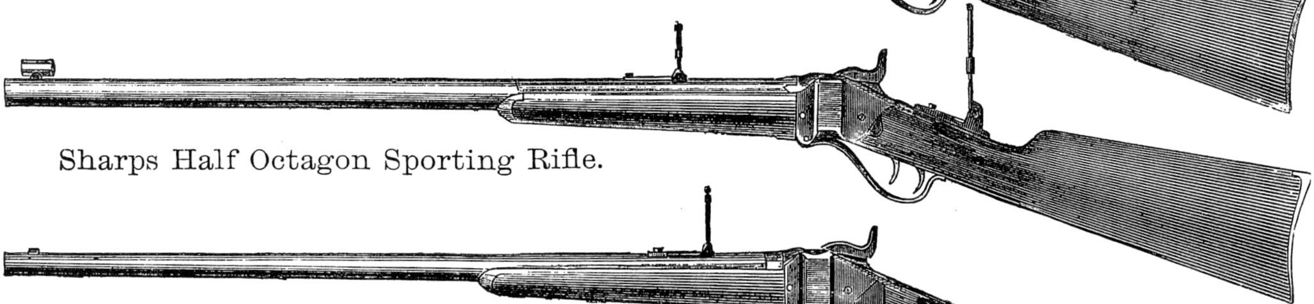

Sharps Half Octagon Sporting Rifle.

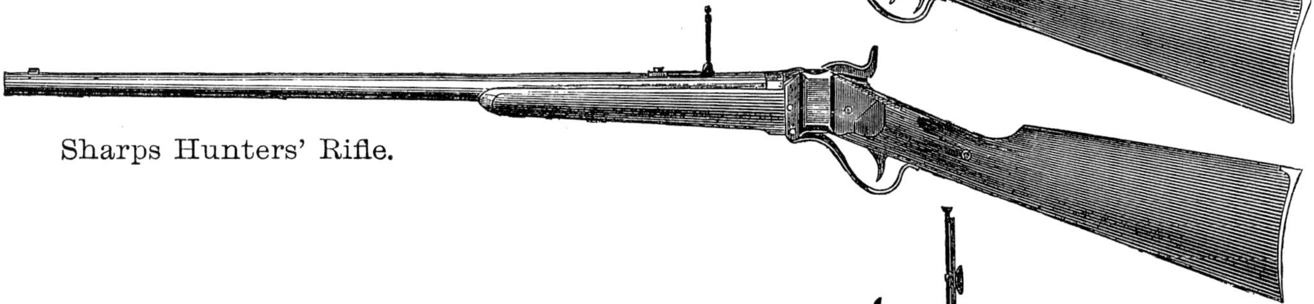

Sharps Hunters' Rifle.

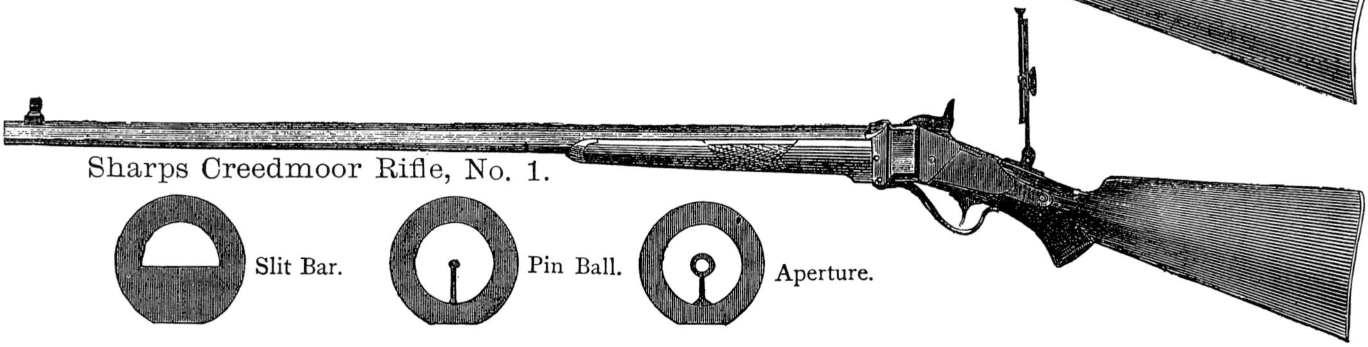

Sharps Creedmoor Rifle, No. 1.

Slit Bar. Pin Ball. Aperture.

ACTUAL SIZE OF CARTRIDGES.

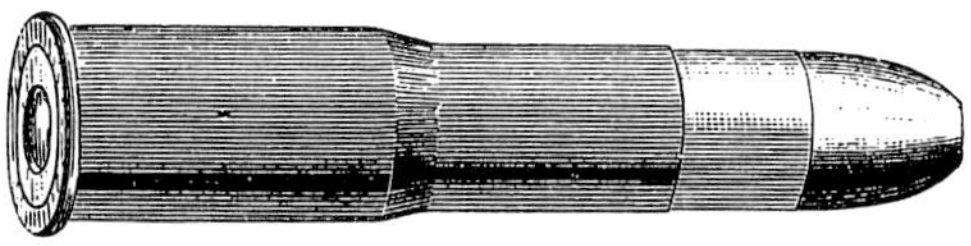

$\frac{40}{100}$ calibre, 50 grains powder.

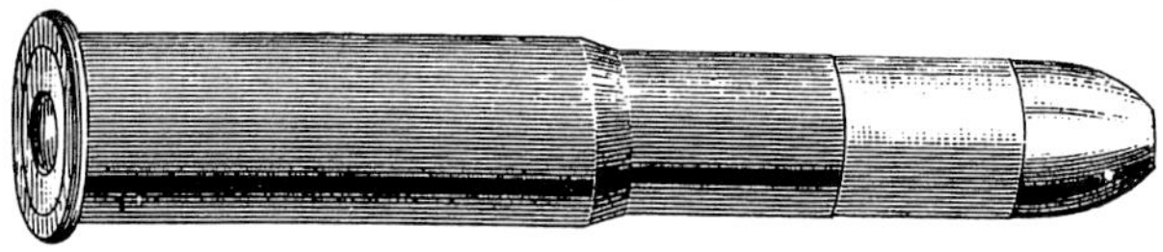

$\frac{44}{100}$ calibre, 70 grains powder.

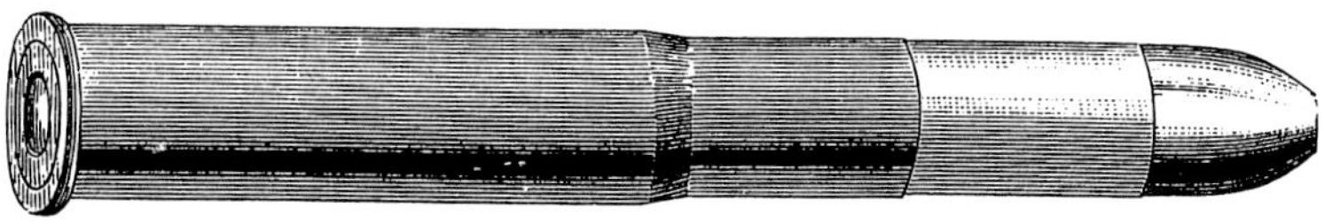

$\frac{44}{100}$ calibre, 95 grains powder.

4, 4, 4, 4, 4, 4, 4, 4, 4, 4.

40 OUT OF A POSSIBLE 40.

Scale, ¼ Inch to Foot.

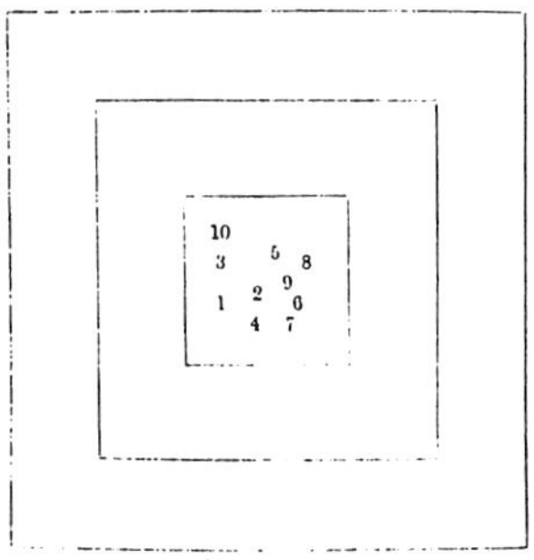

Score of Ten Consecutive Shots at 500 yards, made with a SHARPS MILITARY RIFLE, six lbs. trigger pull, without rest, by Col. H. A. GILDERSLEEVE, Attorney and Counsellor at Law, New York City, in a Match at Creedmoor, Dec. 26, 1874.

3, 4, 4, 4, 4, 4, 4.

27 OUT OF A POSSIBLE 28.

Scale, ¼ Inch to Foot.

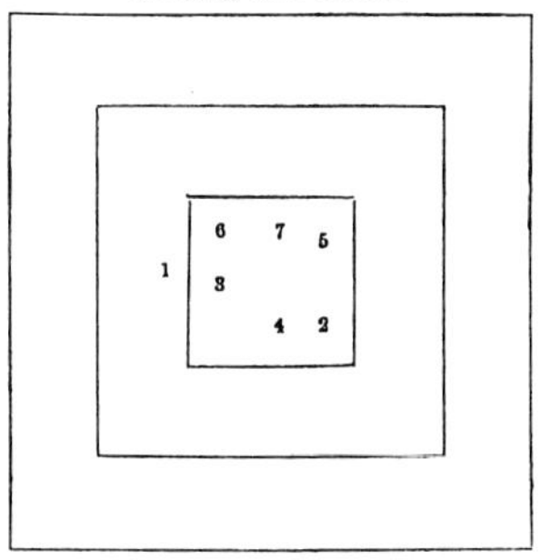

Score of Seven Consecutive Shots at 500 yards, made at Creedmoor, June 6, 1874, in Military Match, open to "Any Military Rifle," by Col. H. A. GILDERSLEEVE, Attorney and Counsellor at Law, New York City, with a SHARPS MILITARY RIFLE, six lbs. trigger pull, without rest, taking the FIRST PRIZE. *Fifty-eight Competitors.*

4, 4, 4, 4, 4, 4, 4, 4, 3, 4, 3, 4, 4, 4, 4.

58 OUT OF A POSSIBLE 60.

Scale, ¼ Inch to Foot.

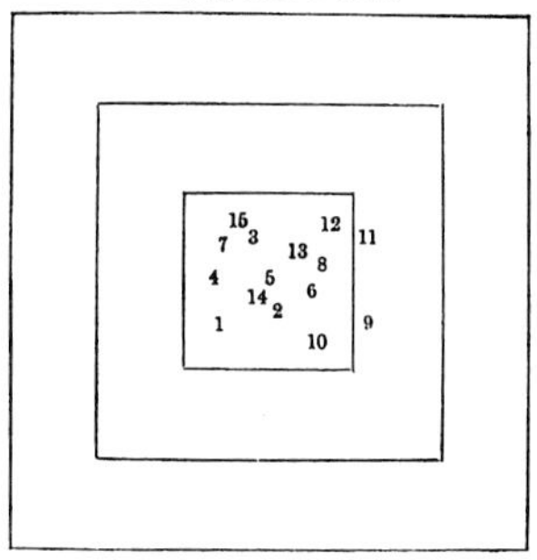

Score of Fifteen Consecutive Shots at 500 yards, made with a SHARPS MILITARY RIFLE, six lbs. trigger pull, without rest, by Col. H. A. GILDERSLEEVE, Attorney and Counsellor at Law, New York City, July 31, 1874.

CARTRIDGES.

$\frac{40}{100}$ calibre, centre fire, reloading shells, 50 grains powder, naked ball (2 cannelures), weight 246 grains, per 1,000, - - - - -	$40.00
Same, with smooth, patched ball, weight 314 grains, per 1,000, - - - - -	43.00
$\frac{44}{100}$ calibre, centre fire, reloading shells, 70 grains powder, naked ball (2 canneleurs), weight 312 grains, per 1,000, - - - - -	42.00
Same with patched ball, 380 grains, per 1,000, - - - - -	45.00
" " " 422 " " - - - - -	45.00
$\frac{50}{100}$ calibre, centre fire, reloading shells, 70 grains powder, naked bail (3 canneleurs), weight 457 grains, per 1,000, - - - - -	42.00
Same, with smooth, patched ball, weight, 500 grains, per 1,000, - - - - -	45.00
$\frac{44}{100}$ calibre, (Creedmoor,) centre fire, reloading shells ($2\frac{5}{8}$ inch), 95 grains of powder, smooth patched ball, weight 545 grains, per 1,000, - - - - -	52.00

Ball Swedges, all sizes and calibres, to order.

BRASS, CENTRE FIRE RE-LOADING SHELLS.

$\frac{40}{100}$ calibre,	$1\frac{11}{16}$ inches long, per 1,000, -	$22.00	$\frac{45}{100}$ calibre,	$2\frac{1}{4}$ inches long, per 1,000, -	$26.50
$\frac{40}{100}$ "	$2\frac{1}{4}$ " " - -	24.00	$\frac{45}{100}$ "	$2\frac{5}{8}$ " " - -	32.00
$\frac{40}{100}$ "	$2\frac{5}{8}$ " " - -	30.00	$\frac{50}{100}$ "	$1\frac{3}{4}$ " " - -	22.00
$\frac{44}{100}$ "	$2\frac{1}{4}$ " " - -	26.50	$\frac{50}{100}$ "	2 " " - -	24.00
$\frac{44}{100}$ "	$2\frac{5}{8}$ " " - -	32.00	$\frac{50}{100}$ "	$2\frac{1}{2}$ " " - -	30.00

With proper care in re-loading, these shells may be used for an almost indefinite number of firings. In testing them at the armory, single shells have been re-loaded and fired as high as *five hundred times* without rendering them useless.

Sharps Powder, - - -	per lb.,	$1.00	Ball Moulds, Creedmoor, - - - - -	$4.00
Lubricating Disks, - - -	per 1,000,	1.25	Wad Cutters, - - - - - - -	.75
" Material, - -	per lb.,	1.00	Cap Awls, - - - - - - -	.25
Ball Moulds, Sporting, - - - -		3.00	Lubricating Moulds, - - - - - -	.50

A LIBERAL DISCOUNT TO THE TRADE.

Goods will be forwarded from the armory to any part of the country, by express, with bill to collect on delivery. Persons not known to the company are required to remit by Post-office Order, draft on New York, or otherwise, at least one-fourth of the probable cost of the goods ordered, as an evidence of their good faith.

Arrangements have been made with the different Express Companies for special minimum rates on all matter forwarded from The Sharps Rifle Company.

For further particulars address the Company.

ARMORY AND OFFICE:
HARTFORD, CONN.

E. G. WESTCOTT,
PRESIDENT.

The following are a few extracts, selected almost at random, from a multitude of letters of commendation received by the Company from persons who are using the improved Sporting and Long Range Rifles. Of the thousands who have recently purchased, they have yet to hear of one who is dissatisfied:

From R. W. SNYDER, *Buffalo, Kansas, November* 20, 1871.

"The gun I must say, is a success,—and the Pet of the Plains,—in fact, has no equal, to my knowledge. I killed twelve Buffaloes with it in thirteen shots, the third day after it arrived, which is much better than I have ever done with any other gun."

From R. W. SNYDER, Esq., *Buffalo, Kansas, December* 18, 1872.

"The man that I sold my old 44 to, killed 119 Buffaloes, in one day, with it. That beats me with my big 50,—as 93 is the most that I have ever killed in one day."

From H. P. WEAVER, *Findley's Lake, N. Y., May* 22, 1872.

"I have had my gun long enough now to know how I like it, and I now write to let you know that it is the best Rifle that I ever drew to my face, and that is saying a good deal, for I have shot some splendid Rifles. It is the longest range and the most accurate shooting Rifle I ever saw. Every one that has seen it shot, says that its equal has not been made."

From JOHN BODINE, Esq., *Highlands, N. Y., June* 3, 1872.

"It is the best long-range Rifle which has yet come under my hands, and I think is capable of very superior work at one thousand yards. I think I could batter down a brick house with your gun at a thousand yards. It breaks stones in a wall like a sledge-hammer."

From M. L. CAMPBELL, *Wallace, Kansas, October* 8, 1872.

"I must brag a little on the gun you sent me. It will shoot through all the Buffaloes I can get in range. I killed a bull last Tuesday, at fully six hundred yards,—and that beats any gun on the Plains. I put two balls in him. I thought it pretty good."

From SUMNER BEACH, *Ellsworth, Kansas, Sept.* 12, 1874.

"I think that the 'Sharps' that we use for killing buffalo need no improvement, as they are a perfect gun. I saw Zack Light (who is one of the best buffalo hunters, and acknowledged to be the best shot on these Western prairies), last winter, while on a buffalo hunt, one day, squat down in a buffalo wallow (after crawling and getting a stand on a herd of buffalo), and killed 74 head of buffalo, just where he sat in the wallow. He killed that winter twenty-three hundred head of buffalo. The rest of us killed about twelve hundred head. He used a 'Sharps' 50 calibre."

From COL. H. A. GILDERSLEEVE, *A. A. G., 1st Div. N. G. S. N. Y., New York, Oct.* 10, 1874.

"I have used your rifle since June last, with great pleasure and success. I regard it as second to no rifle in the world for simplicity of mechanism, excellence of finish, and accuracy of execution."

From WM. M. TREADWAY, *Port Henry, New York, Oct.* 26, 1874.

"I have severely tested the Sharps Rifle recently purchased from you, and find it the most perfect shooting gun I ever saw. Send me two more at once, for friends who are delighted with the excellence of my rifle. I will probably order still another in a day or two."

From GRANVILLE STEWART, Esq., *Deer Lodge, Mont. Ter., Dec.* 29, 1874.

"The two Rifles ordered by me arrived on the 25th, and are beautiful specimens of the perfection reached by you in the manufacture of target guns. They are far superior to anything ever seen here before, and give satisfaction in every particular."

From J. T. ROTHROCK, M. D., *Wilkesbarre, Pa., Feb.* 12, 1875.

"That Sharps' Rifle is the perfection of a hunting gun. I have never seen anything in the rifle line, that so perfectly satisfies me."

SHARPS RIFLE COMPANY

The rifle shown on the dust jacket, and here full length, is the Model 1874 Sharps Long Range Target No. 3. This specimen is chambered for the .45-2.4" cartridge. The half-round barrel is 34" long, the rifle weighs ten pounds. The folding tang target sight can be moved for back position competition to the sight base at the heel. The buffalo horn pistol grip is an extra that Sharps offered. The distinction between this rifle and the Creedmoor No. 1 shown in the brochure following is the straight grip of the No. 3.

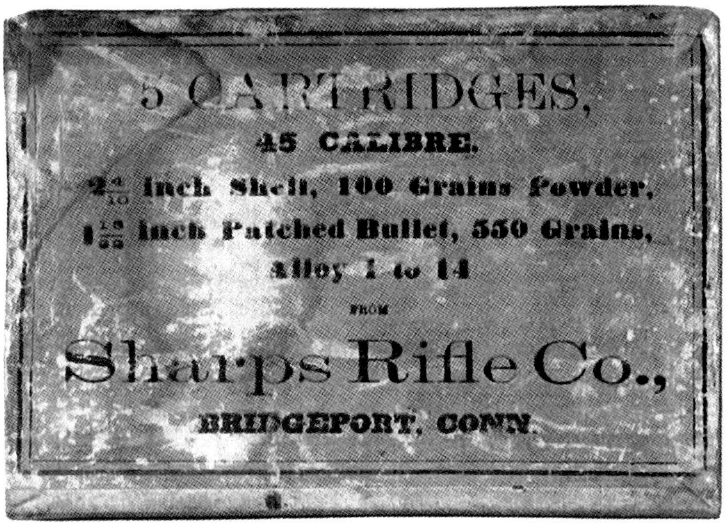

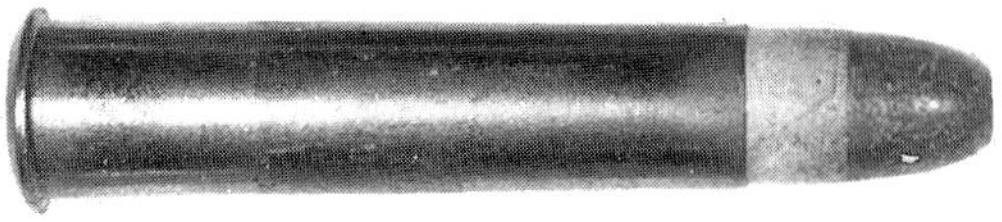

.40-50-265 Sharps Straight, 1.88" case.

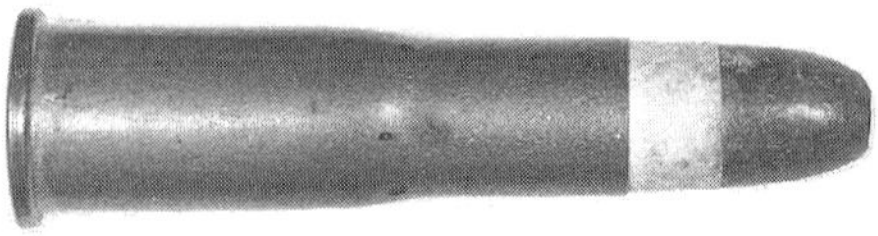

.40-50-265 Sharps necked,1.72" (often headstamped "Regular" by Winchester).

.40-70-330 Sharps Straight, 2 ½" case.

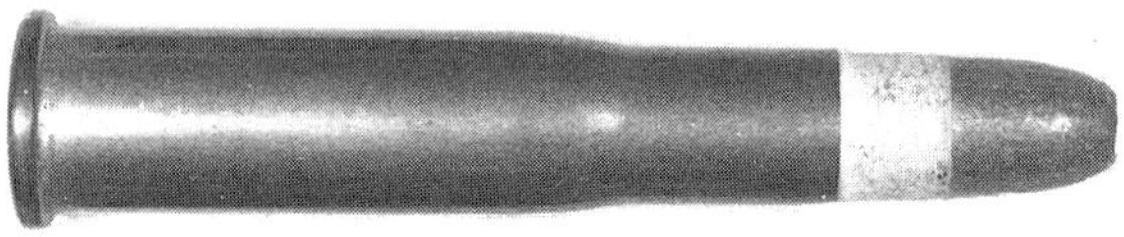

.40-70-330 Sharps necked, 2 ¼".

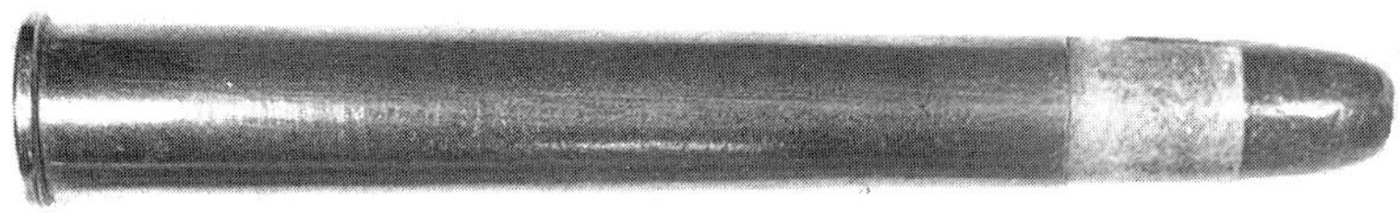

.40-90-370- 3 ¼" S S. (Sharps Straight).

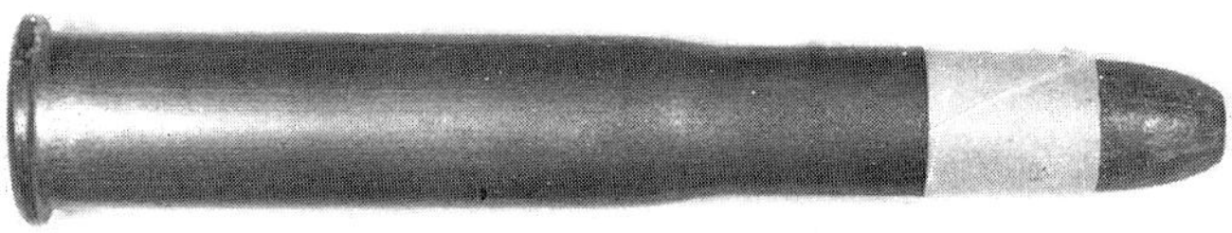

40-90-370 Sharps necked, 2.63".

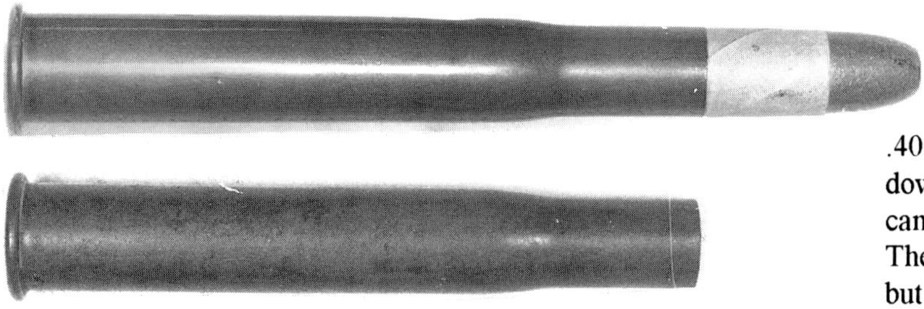

.40-90-3.277". This case is necked down from the Sharps 45-3 1.4" and came out a bit longer than its parent. There is no known rifle reported yet, but several specimens have been examined, all with the early unheadstamped Winchester head and Boxer primer.

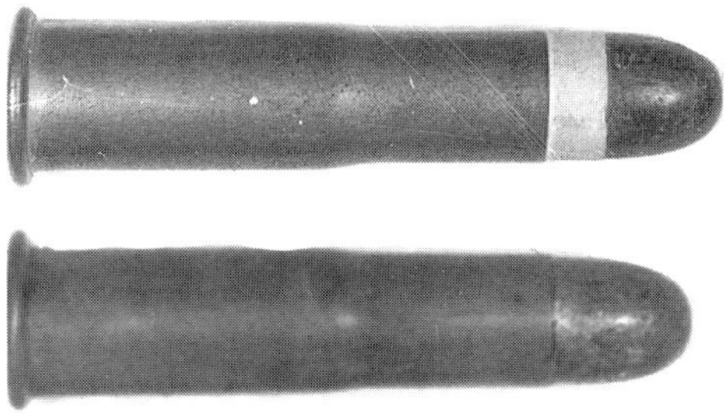

.44-60-396, 1.88”. Also found with the unpatched lead bullet.

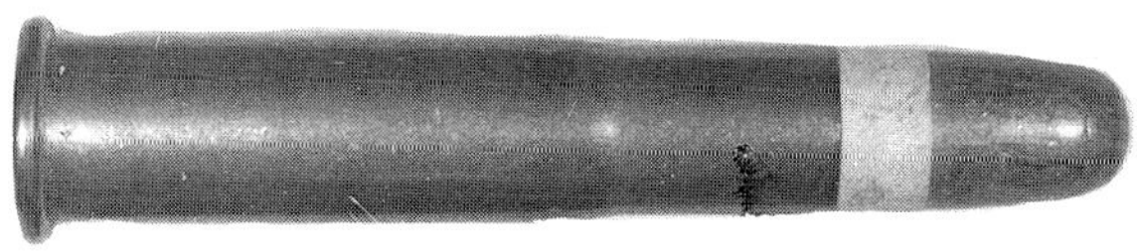

.44-77-300 to 470, 2 ¼”. The Remington rifles were also chambered for this case.

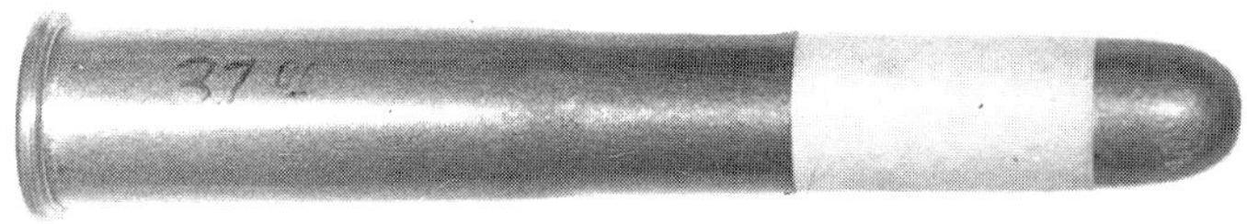

.44-90-500, 2 ¼”. This is the same case as above. The charge occupies a good part of neck so the 500 grain bullet in this one is seated way out. Other bullet weights will be found.

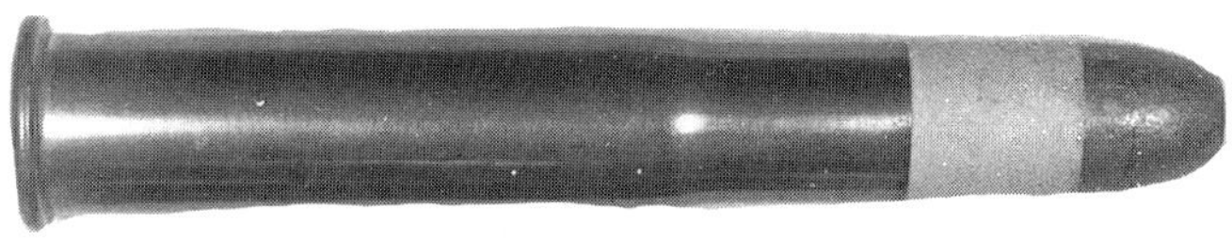

.44-90-520, 2 5/8”.

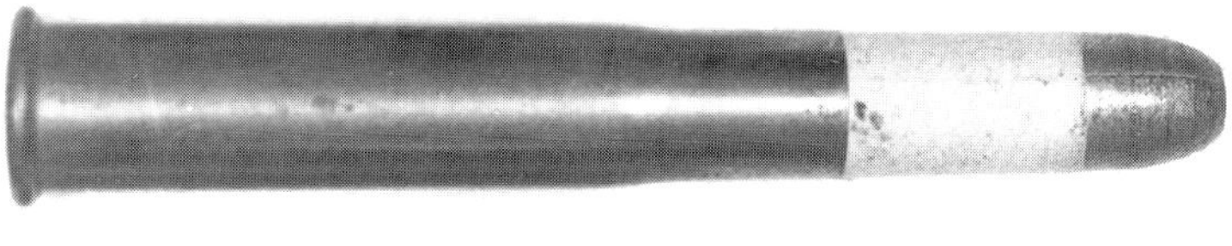

.44-100 or 105-2 5/8” The same case as that above. This one has the 500 grain bullet but other weights may be found.

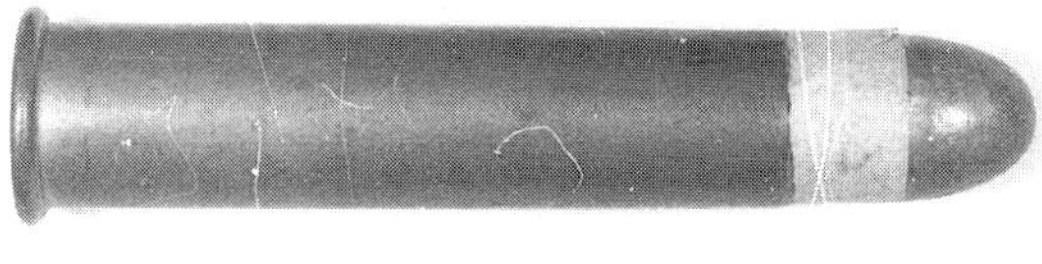

.45-75-420. 2.1”. This is the same case as the .45-70 Springfield, loaded for the Sharps rifle with paper-patched bullet.

.45-80-500, 2.1". This is also the .45-70 case. See the U. M. C. box label for this Sharps load.

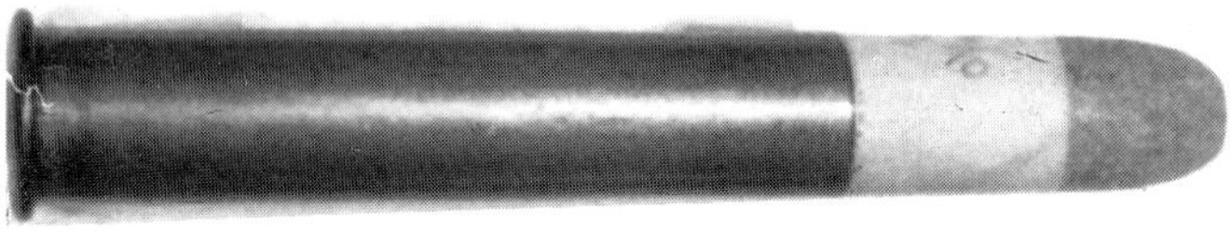

.45-2.4". The box labels establish that the .2.4" case was loaded with 90 and 100 grains of powder with a 550-grain bullet.

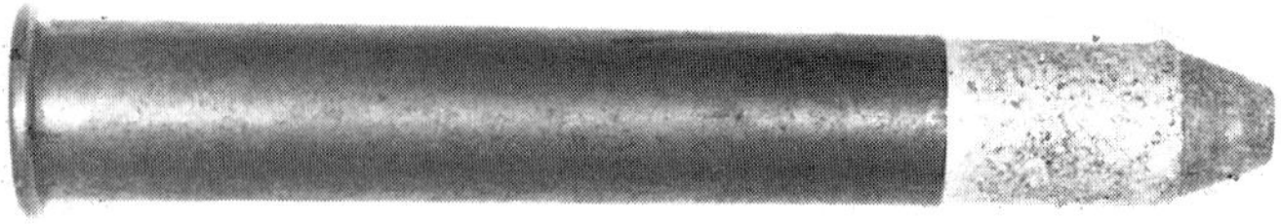

.45-2.6". The Sharps box labels preceding show that 90- and 100-grain loads were produced with 500-grain bullet. Other weights will be encountered. This specimen by U. M. C. has the patch all the way to the beginning of the bullet's cone.

.45-2 7/8". The box labels show 100 or 105 grain powder charges in this case with the 550-grain bullet, and a 120-grain charge with 500-grain bullet.

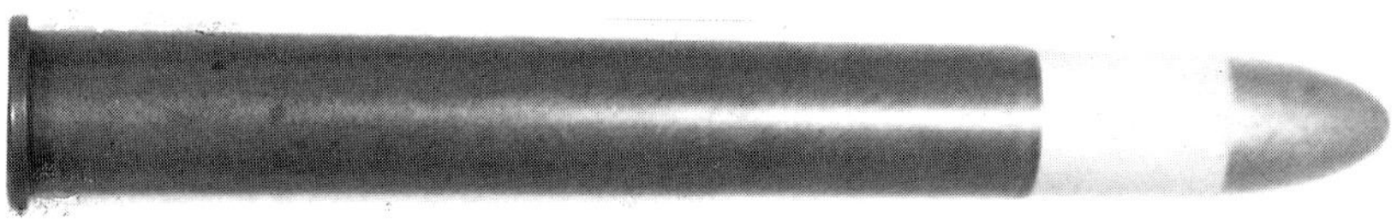

.45-3 ¼". This long case could be loaded with any powder charge from 110 to 130 grains, and with 500 or 550-grain bullets at the whim of the loader. The greater the powder charge the farther out the bullet would be seated.

A .45-3.6" Sharps primed case has been examined in the past but no specimen could be located to show with this study.

.50-70 or 80-450, 1 ¾". This is the .50-70 Springfield case loaded for Sharps with the paper-patched bullet.

.50-85 or 90-450, 2". This is the most elusive of the Sharps cases because it was used for only a few years. Only naked lead bullet factory loads have been noted.

.50-2 ½". This cartridge is commonly referred to as the .50-90 but it was also loaded with 100 and 110 grains of powder and 473- to 500-grain bullets.

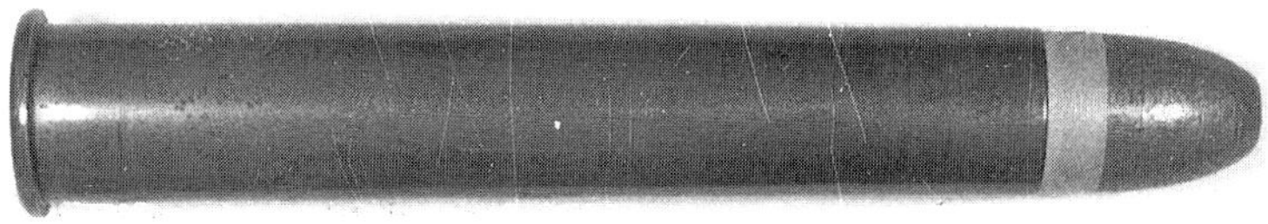

.50-140-700-3 ¼". This rare cartridge seems to have been an experiment which did not last long. Lighter bullet weights may be found.

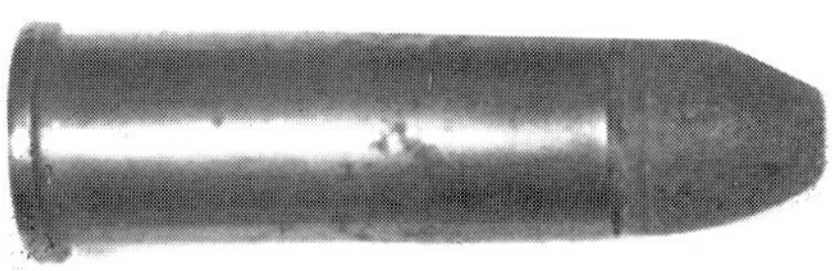

.52-70-1 ¾". This cartridge is believed to have been produced for some .52 Sharps percussion breech-loading military rifles converted to centerfire using their original barrels.

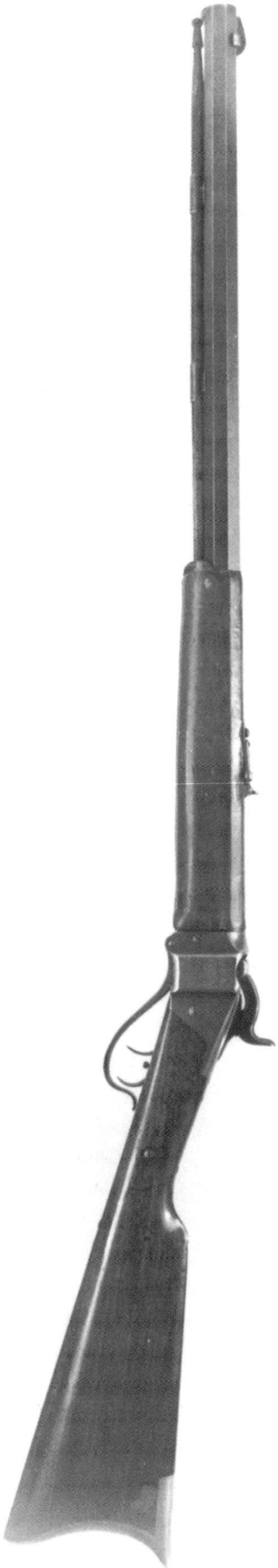

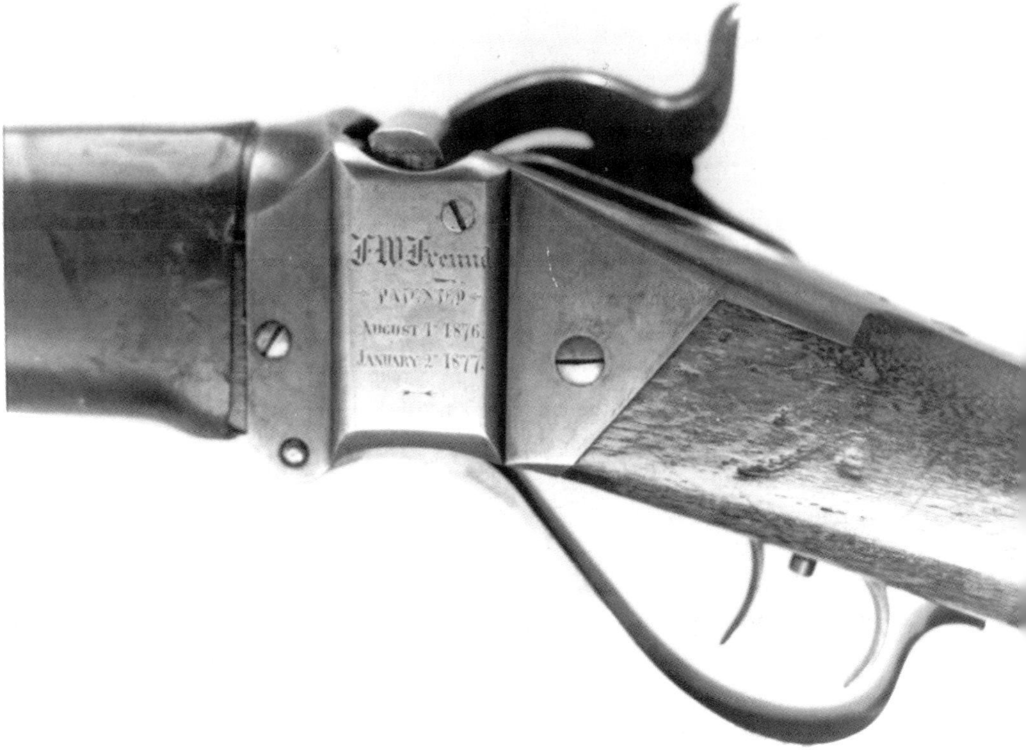

AN F. W. FREUND RIFLE

Gunsmith Frank W. Freund, Denver, Colorado, received patents for modification of the Sharps hammer rifle action on August 1, 1876 and another on January 2, 1877. This specimen is chambered for .45-2 7/8" Sharps. The rifle has a Freund breechblock, half cock device and front sight, and double-set triggers. The original wiping rod is Freund's design and the forend and barrel above it are covered entirely with leather to protect the hand from cold, apparently a concession to the original owner. It is stitched together with linen thread along the bottom. The buttstock and buttplate, curved to fit up over the shoulder, is also a special feature—designed by Freund or for his customer. The man who originally owned this rifle was a civil engineer and surveyor in the western states during the 1870s. A surveyor during those years might have to go from his transit to his rifle in a big hurry! Below is an ancestor—a double-set trigger percussion Berdan Sharps rifle. (See patents appendix for an 1875 Freund cartridge case design).

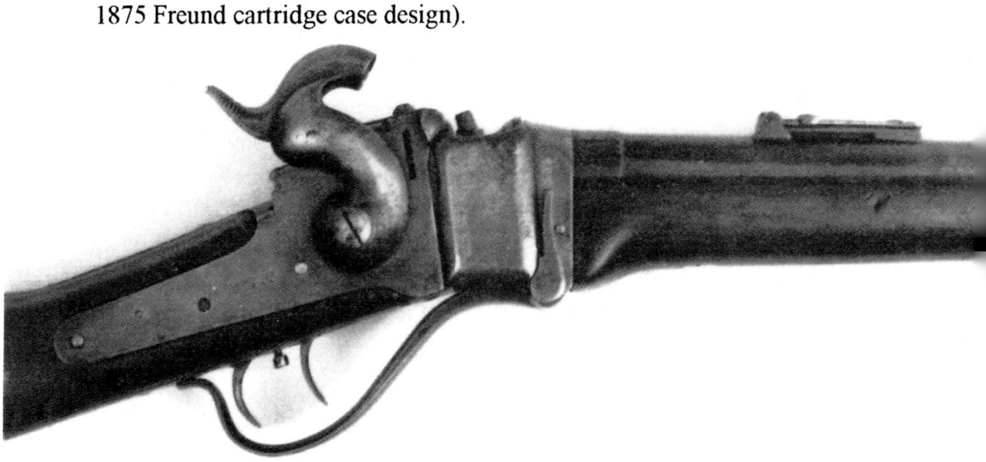

REMINGTON · ARMS · COMPANY

ILION, · NEW · YORK

MAY, 1896.

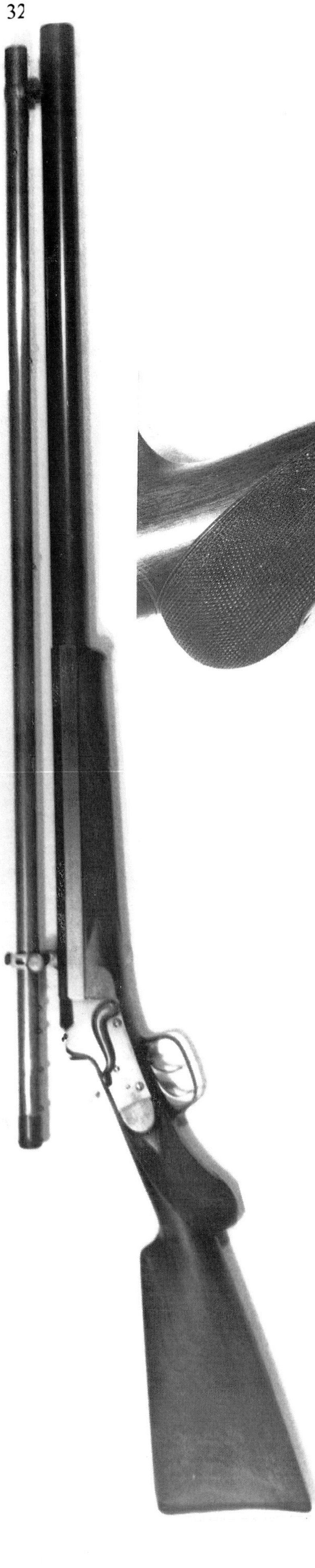

REMINGTON ARMS COMPANY

The rifle shown here has an unusually heavy 32" half-round barrel made on the Remington-Hepburn falling block action chambered for .40-70 2 ½" Straight—a Sharps cartridge. The 35" Malcolm telescope sight with windage and elevation adjustments in the rear mount is evidently original as the rifle was never provided with other sights. The whole outfit weighs 13 pounds. Pushing forward the front trigger sets the rear for letoff.

According to Roy M. Marcot, this action was invented by Lewis L. Hepburn, a gunsmith and Remington employee, for which he was granted a patent on October 7, 1879. The breechblock is dropped for extraction or loading by lowering the lever on the right side of the action.

This rifle was produced toward the end of Remington Hepburn production. The scope, unmarked, is Malcolm late production. Earlier models had a somewhat smaller tube. It was no doubt a target rifle special order since it does not conform to any of the Creedmoor or match rifles illustrated in Mr. Marcot's book **Remington**, the official history published by Remington Arms Company in 1998.

REMINGTON LIGHT "BABY" CARBINE.

44-CALIBRE CENTRAL-FIRE.

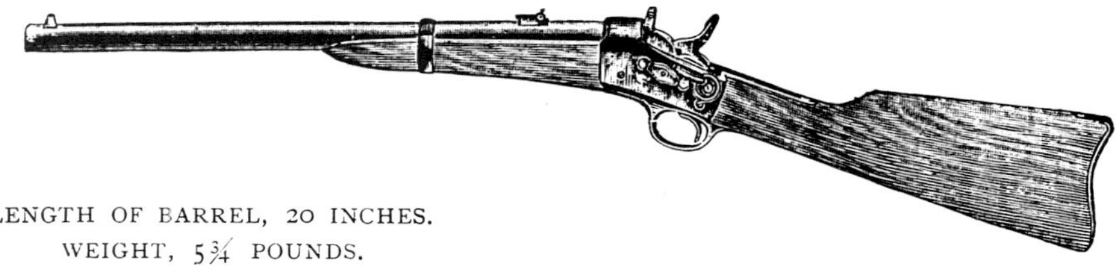

LENGTH OF BARREL, 20 INCHES.
WEIGHT, 5¾ POUNDS.

Attention is called to the Remington Light Carbine, weighing only 5¾ pounds, being only about double the weight of any army revolver. For sporting, police and saddle use, it is just the arm required. Its lightness and superior shooting qualities will recommend it to all.

Carbine as above, blue finish $13 25

Carbine as above, nickel-plated $14 00

.45-60 Remington Black Hills Rifle.

In his book, ***REMINGTON*** **America"s Oldest Gunmaker,** author Roy Marcot describes (p. 100) the Black Hills Rifle, made only in 1877 and 1878, based on the No. 1 military receiver with 28" barrel, standard sporting stock and forend, and chambered for the slightly necked .45-60 cartridge shown here, or military .45-70. This .45-60 is not related to any other cartridge and was used in no other rifle. The Headstamp is "E. REMINGTON & SONS (with two diamonds). Dimensions of this specimen are: head .518, rim, 619, case mouth .473, bullet at mouth .440, case length, 1.751, overall 2.282.

REMINGTON ••• FIREARMS

REMINGTON - SCHUETZEN

Match Rifle

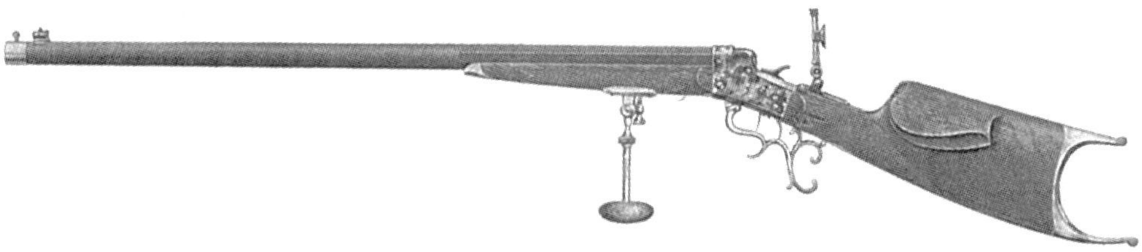

NEW SPECIAL MODEL

THIS Rifle is made to meet the demand for a match rifle of most improved expert design of high-class workmanship, material and finish, and of extreme accuracy. It is nearer popular expert requirements than any Schuetzen rifle made in this country. The action is a modification of our No. 3 with under lever and sliding breech block so arranged as to force home a shell partially inserted in the chamber.

New design finger lever guard, fancy walnut stock with cheek piece, finely checkered, new improved and carefully adjusted hand-made double set triggers, special rear wind gauge Vernier sight and hood front sight with interchangeable pin head and aperture discs, specially designed Schuetzen Swiss butt plate, barrel 30 or 32 inch half octagon, weight from 11 to 13 pounds			$60.00
Remington-Walker barrel, breech loading, extra	$12.00		
Palm rest, extra	5.00	$17.00	
Remington-Walker barrel, muzzle loading, with patent muzzle, starter, wooden rod, lubricator, sizer, bullet mould, etc., extra	$15.00		
Testing for guaranteed group within 2 1/4 inch circle at 200 yards	8.00	23.00	40.00
			$100.00

Price of parts furnished on application.

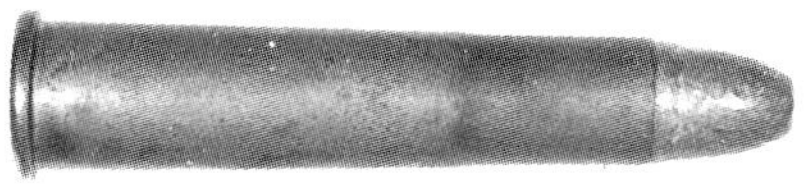

.32-30-125, 1.64" case. Chambered in the Remington-Hepburn rifle. Bullet diameter is actually .312".

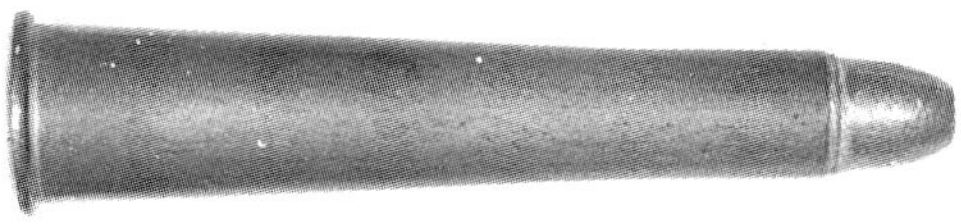

.32-40-150, 2 1/8" case. A distinctive case type with larger head and taper than the .32-40 Ballard-Marlin-Winchester. Chambered in the Remington No. 1 rolling block and No. 3 Hepburn single shot rifles.

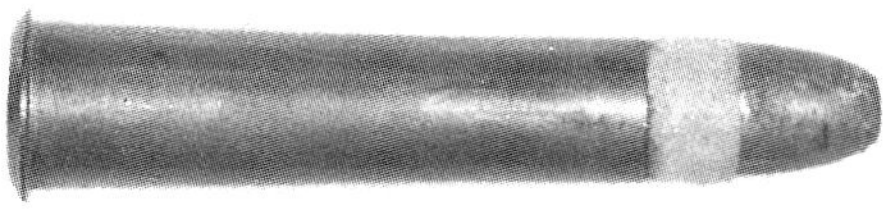

.38-40-245, 1.77". Remington No. 1 and Hepburn. It is found with paper-patched and plain lead bullet.

.38-50-255, 2.23" case. Remington Hepburn match rifles were chambered for this cartridge.

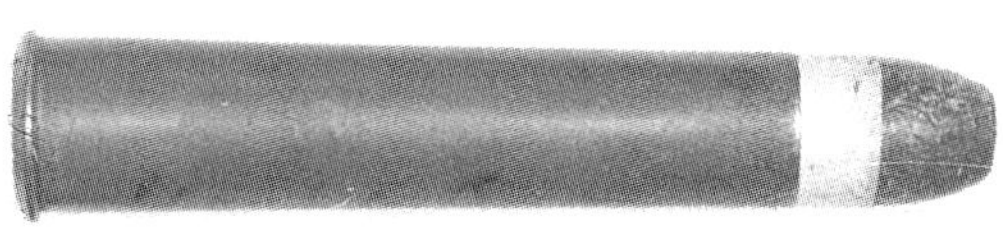

.40-45-2.1" case. Remington head with beveled rim and small copper primer. No headstamp. This is a little-known cartridge about which no other information has been found other than the identity supplied by the original owner many years ago, so the source is not available.

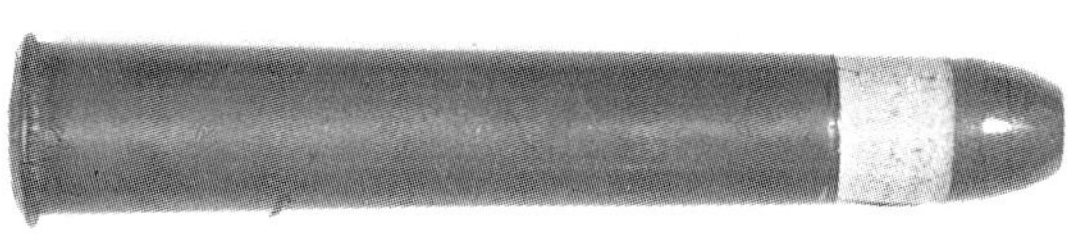

.40-60 Remington Straight, 2.25" case. Another uncommon Remington cartridge with the same head as the .40-45 above but with the slightly longer case. Presumably these two are short-lived match cartridges.

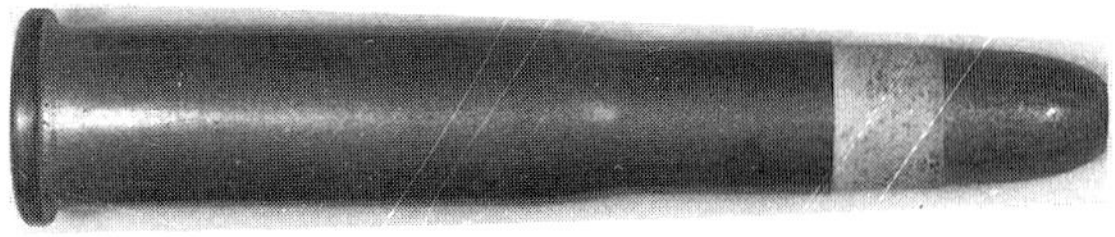

.40-70-330. Remington chambered its rolling block and Hepburn rifles for this Sharps case. The cartridges are identical. The early Remington-made cartridges have raised "E. Remington" headstamps, or are plain with rounded head or beveled rim.

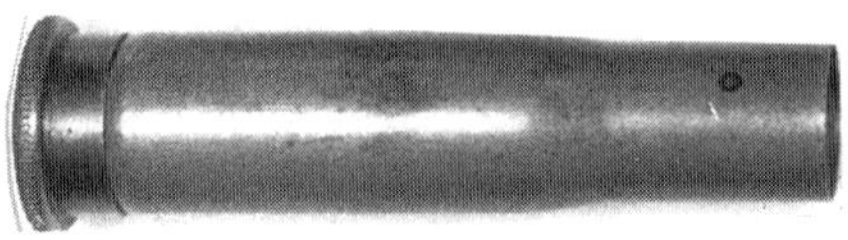

Next is the .40-70 Remington Reloadable. This is the .40-70 Sharps case made with screw-on steel head. The hollow screw which attaches the head to the case serves as the anvil for the Berdan primer. After firing the head was screwed off to remove the spent primer, a fresh one placed on the anvil, and the head screwed back on for the next shot. This device was also made for the .44-90 2 7/16" necked chamber.

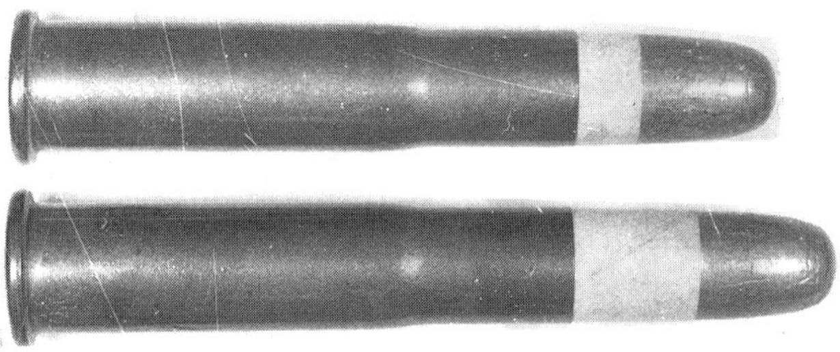

.44-77-365 and 44-90-520, both 2 ¼". This is a Sharps case for which Remington chambered their Hepburn and No. 3 rifles. The 44-90 has the bullet set out farther to take the larger powder charge.

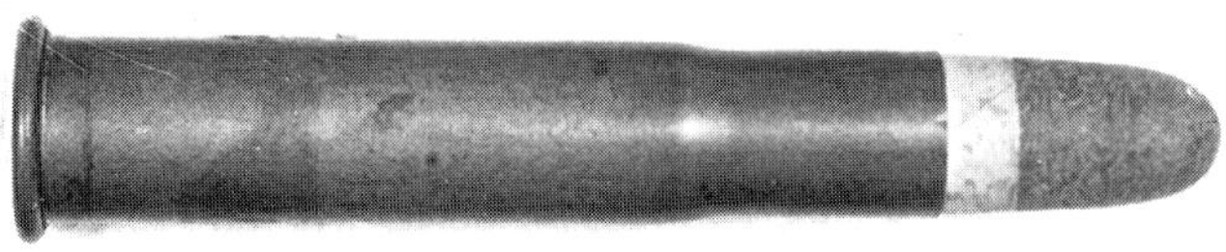

.44-90-550 2 5/8" case. This is the Sharps 44-90 case for which Remington chambered the rifles mentioned above. The bullet weight is Sharps.

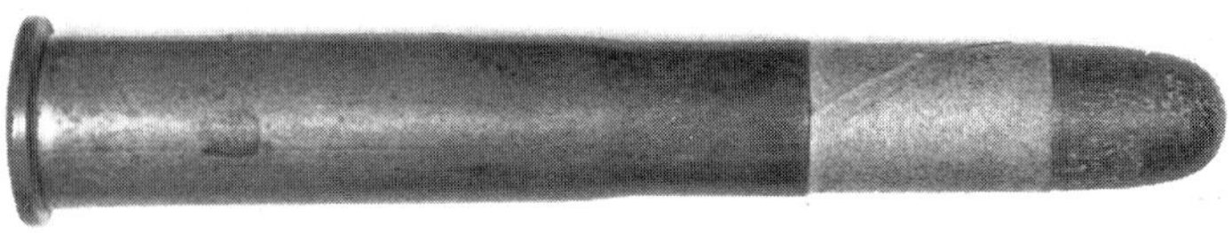

.44-90-550 Remington Special, 2 7/16". This case is not Sharps pattern. This one is slightly smaller in rim and head dimensions from those above.

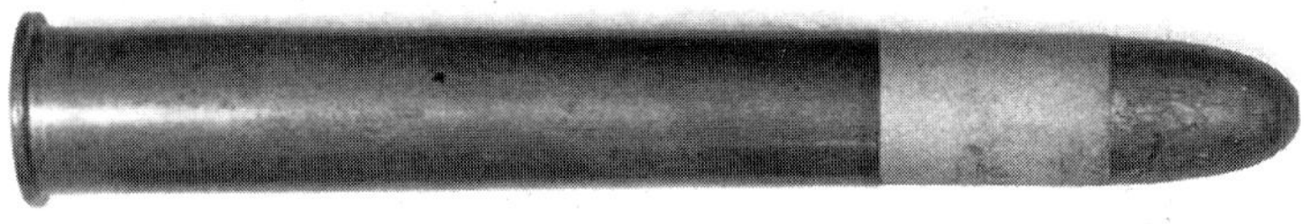

.44-90-550 2.6" Remington Straight. Also known as the 44-100. The cartridge was used in Remington Hepburn No. 3 long range match rifles.

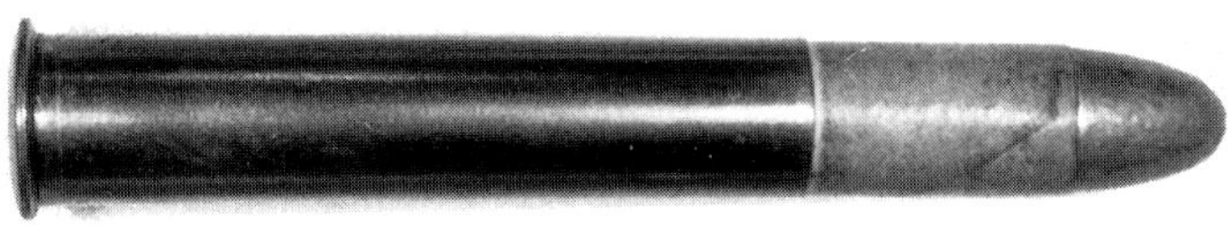

.45-90-2.4" Remington Creedmoor. Here is another Sharps case with rounded plain head for Remington target rifles (see Remington box label with "6-10" struck out and "4" penned in).

FRANK WESSON

The Frank Wesson single shot rifles made at Worcester, Massachusetts beginning in the early 1870s were the falling block and break open types. The two-page broadside printed in 1876 shows one design for the Long Range Rifle at top, which appears to be the action patented by Alexander Henry, Edinburgh, Scotland January 12, 1869, which Wesson may have used under license, or purchased from Henry directly. Pushing forward on the opening lever brought the breech block straight down to expose the chamber.

The No 2 Long and Mid-Range and Hunting Rifle, and the No. 1 below it were based on the U. S. patent 193,060 issued July 10, 1877 to Franklin Wesson and Charles N. Cutter, drawing shown below. The complete patent is in Appendix B under 1877. This is also a falling block system, the trigger guard serving as the opening lever. The No. 2 Mid-Range and the New Model rifles are double-trigger style in which pulling the front trigger released the locking lugs to let the barrel drop and the breech lift for loading or extracting the empty case. The long and mid-range rifles are all fitted with a rear sight attachment at the heel of the buttstock for target shooting in the back position with barrel steadied between the feet.

Frank Wesson designed his own cartridge cases. They are not interchangeable with those of any other single shot rifle maker. They were produced for him by United States Cartridge Company, Lowell, Massachusetts, Union Metallic Cartridge Company, Bridgeport, Connecticutt, and Winchester Repeating Arms Company, New Haven, Connecticutt.

The rifle below is a .32 Long fitted with the rimfire or centerfire adjustable hammer patented by Frank Wesson on April 9, 1872. This patent was also used by the Ballard Rifle Company. The front trigger in this rifle action disengages the locking lug and lets the hinged barrel drop to expose the chamber.

Information and documents provided by Frank Pellett.

In addition to the cartridges illustrated and described on the following pages, the Wesson rifles will be found chambered for:

Rimfire:
.22 Short and Long
.32 Long
.38 Long
.44 Long
.56-46 Spencer

Centerfire:
.22 Extra Long Maynard
.38-55 Ballard
.40-63 Ballard
.42 Russian
.44-90 Remington 2 7/16"
.44-90 Sharps, 2 5/8"

F. Wesson's Long Range Rifles, B Loading.

THE CHAMPION RIFLE OF THE WORLD.

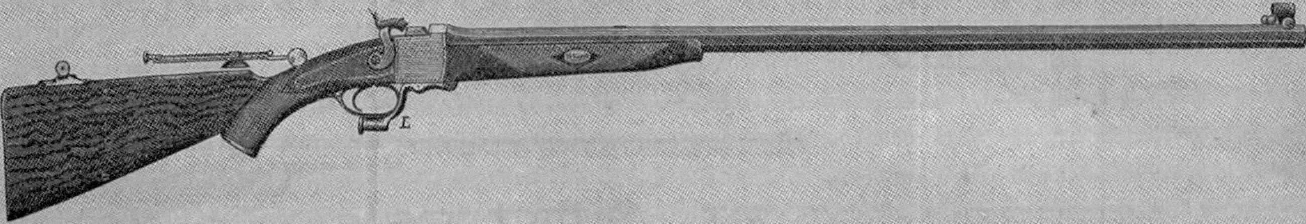

No. 1, Long Range Rifle, National Regulation.

Calibre, 44- and 45-100. Weight, 10 lbs. and under.

Hammer outside of Lock Plate. Rebounding Lock, full Vernier Peep Sight, with Spirit Level on Wind-gauge Muzzle Sight. Rifle complete, with full outfit, $125.00.

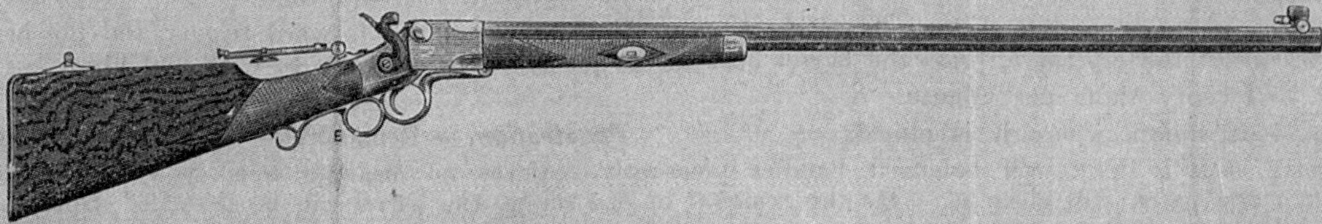

No. 2, Long and Mid-Range and Hunting Rifle.

Full Vernier Peep Sight, Wind-gauge Muzzle Sight with Spirit Level; with 100 Solid head Shells, and full set of Reloading Tools. $85.00.

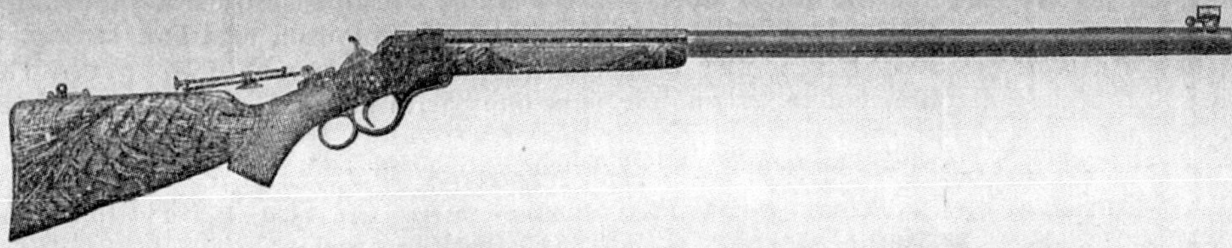

No. 1, Mid Range and Sporting Rifle.

Hammer in line of Barrel. Calibre, 38-, 40- and 45-100.

Full Pistol Grip, Checkered Grip and Forestock, Vernier Peep and Muzzle Sight with Spirit Level. Full set of Reloading Tools, with 100 Solid head Shells, Central fire. $65.00.

No. 2, Mid-Range Sporting and Gallery Rifle,

With Tilting Barrel, half Pistol Grip, Calibre 22-, 32-, 38- and 40-100. Barrel 28 to 30 inch.

This Arm combines all the good qualities of first class Rifles, with adjustable Hammer to fire both centre and rim cartridges. Full Globe and Open sighted. Price, $55.00.

THE ABOVE ARMS RECEIVED

THE HIGHEST CERTIFICATE OF AWARD

OF THE

United States Centennial Commission

At the International Exhibition held in Philadelphia in 1876.

SHOT GUNS BORED CHOKING, TO DRIVE SHOT CLOSE, AND REPAIRING DONE TO ORDER.

A LIBERAL DISCOUNT MADE TO THE TRADE. TERMS CASH.

ADDRESS FRANK WESSON, 20 MANCHESTER STREET, WORCESTER, MASS.

SEE OTHER SIDE.

WESSON'S
CELEBRATED
BREECH-LOADING RIFLE.

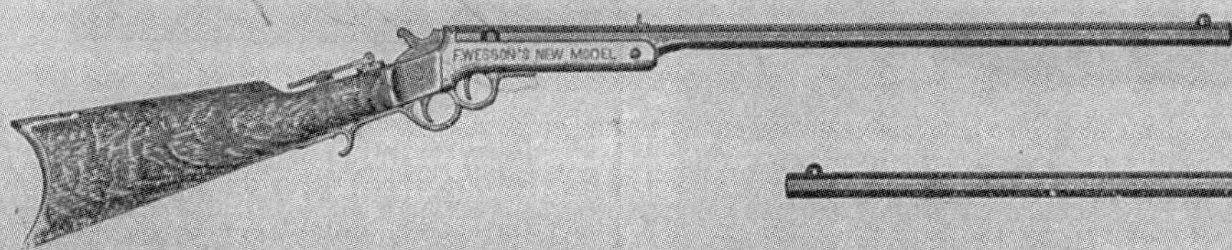

THE NEW MODEL RIFLES

Are all made to fire both Rim and Central Fire Cartridges, (*free of extra charge,*) and both pin fire by the use of my Patent Adjustable Hammer; Patented April 9, 1872.

Their great superiority consists in their

SIMPLICITY, RAPIDITY, ACCURACY, PENETRATION, CONVENIENCE, DURABILITY.

Simplicity. — It is impossible to load this Rifle wrong. Half cock, pull the forward trigger, and the barrel will rise at the breech; insert the cartridge, press the barrel back to its place, and it is ready to cock and fire.

Rapidity. — Twenty shots per minute

Accuracy. — It shoots where it is aimed. ***Penetration.*** — Remarkable — all that is desired.

Convenience. — It is light, well balanced, handles pleasantly, requires no cleaning until you have done your day's work, and then one patch will clean it. By the removal of one screw, the barrel can be detached from the stock, and the Rifle packed in a small space.

Recent Improvement. — Patented April 9, 1872, which consists of an adjustable hammer or cock, so arranged as to fire both rim and central fire cartridges. This improvement has long been sought for, and by the use of the central fire cartridge, which is reloadable many hundred times, the very best of Rifle shooting is obtained.

Durability. — The barrels are made of the finest steel, and the whole workmanship is without fault. There is no complicated machinery to get out of order. While the Rifle is light, everything is compact, firm and strong. In a word, it is justly pronounced "THE BEST BREECH-LOADING RIFLE YET CONSTRUCTED FOR HUNTING OR SPORTING PURPOSES." I also make a Rifle Barrel to insert in double guns, thus converting at pleasure one barrel into a Rifle and the other for Shot, both Breech-Loading. Price, $20.00.

FOUR CALIBRES. — CARTRIDGES WATER-PROOF.

These Cuts Represent the Exact Size of the Cartridge.

22–100 32–100 38–100 44–100

Length of barrel is 24, 26, 28 or 30 inches.

PRICE. — 24 inch barrel, open sight, $25.00. Each additional inch in length of barrel, 50 cts. Globe sight, $5.00 extra. Swivels and straps, $2.50. Full set Reloading tools complete, with 50 Shells, central fire, and box of primers, $5.00.

F. WESSON'S BREECH-LOADING POCKET RIFLE.

NEW MODEL, PATENTED MAY 31, 1870.

"Sportsman's Jewel."

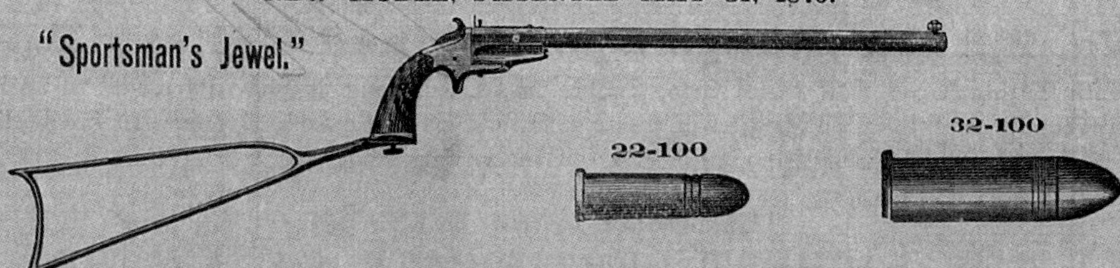

22-100 32-100

Embraces many advantages over the old style of tip-up barrel, which is liable to become loose with wear, shoot astray, and even cause accident; objections which are entirely overcome in the New Model. The barrel is firmly attached to the frame, thus insuring the greatest possible accuracy in firing; and, being simple in construction, it is less liable to get out of order. It operates easily. By pressing a stop under the frame, the barrel is readily turned aside to receive the cartridge and eject the empty shell. The Pistol is symmetrical in appearance, and can do better shooting than any on the tip-up principle, besides being safer and more serviceable. Gallery and Dueling Pistols furnished to order, without Extension Breech.

All are now made to explode the cartridges by a Firing Pin, either central or rim firing cartridges, as desired.

PRICES.— 10 inch octagon barrel, 22 or 32 calibre, Rosewood Stock, Plated Rest and Frame, - - $12.00
12 " " " " " " " " " " - - - 13.00
15 " " " " " " " " " " - - 14.50
18 " " " " " " " " " " - - - 16.00
20 " " " " " " " " " " - - 17.00

SEE OTHER SIDE.

.30 Extra Long, 1.598" case. Two specimens, the upper with belted, pointed bullet, the other with blunt bullet, both with small brass primer.

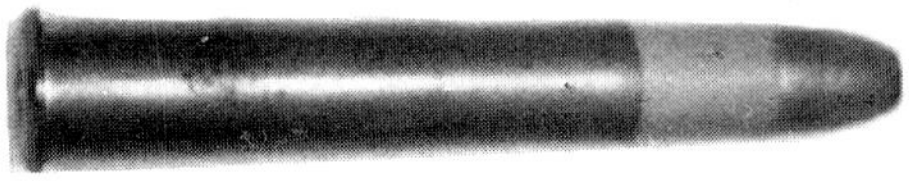

.30-30, 1.65" straight-tapered case. This is a target load chambered in the falling block rifles Frank Wesson designed and patented.

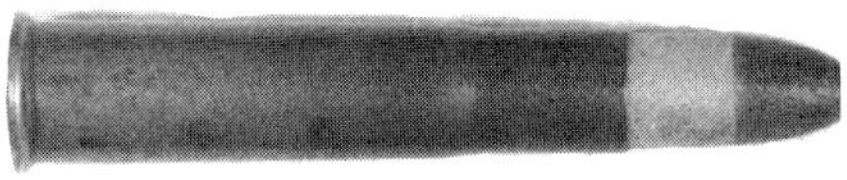

.30-40, 1.641" necked case. Small brass primer. Same use above.

.32 Extra Long, 1.25", copper primer, the headstamp reads "W. R. A..32 Ex.L."

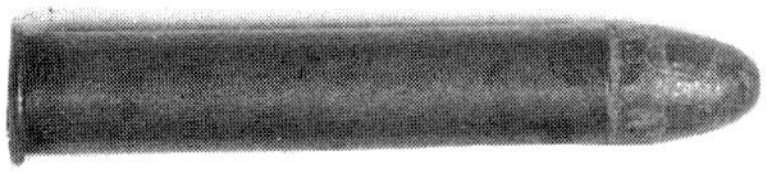

.32 Extra Long, 1.626" case with 1/16" belt at case mouth around the round-nosed lead bullet. Small copper primer, no marks.

.35-2.505" case. Boxer size primer pocket, no headstamp.

.38 Long, 1.62". Long belted bullet, copper primer.

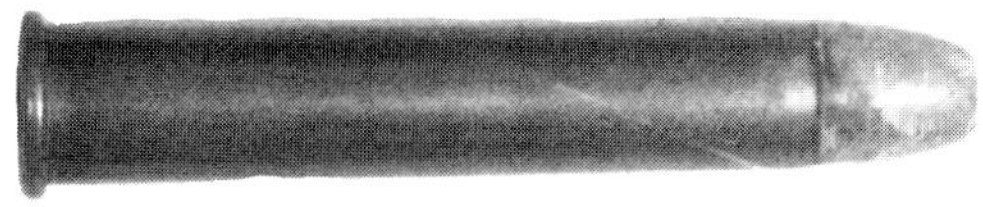

.38 Extra Long, 2.13" case. Copper primer, belted bullet.

.38 Extra Long, 1.891" case. Same copper primer, belted bullet. This is a U. S. Cartridge Company cartridge display board dummy. The wire attachment has been removed from this photo.

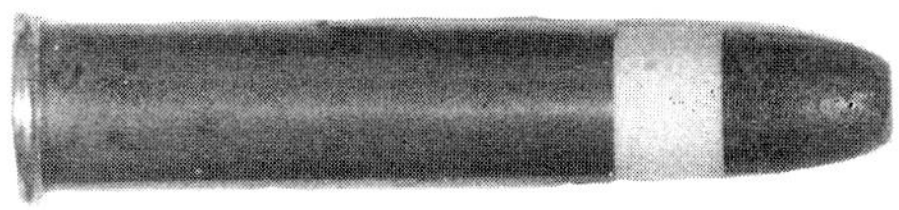

.40-40, 1.622" with copper primer, paper-patched flat-nosed bullet. This will also be found loaded with the typical belted Wesson lead bullet.

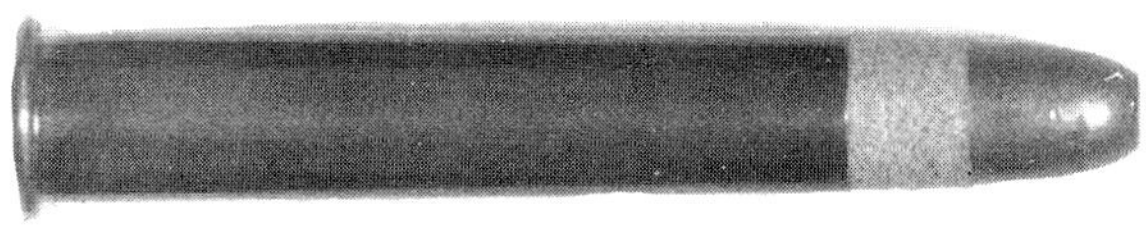

.40-60, 2.251".

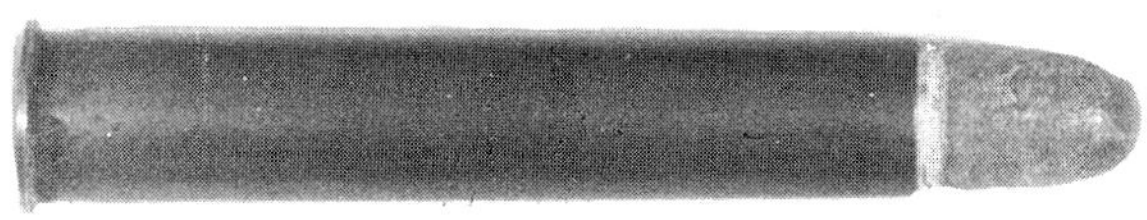

.40-70, 2.481" case, single-groove bullet.

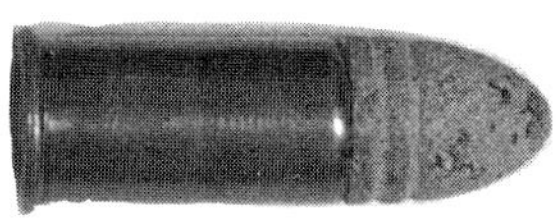

.44 Short, .886" case.

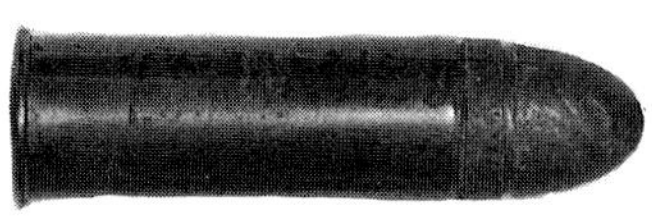

.44 Long, 1.21" case.

.44 Extra Long, 1.63" case, two-groove lead bullet. This case will also be found loaded with the paper-patched bullet.

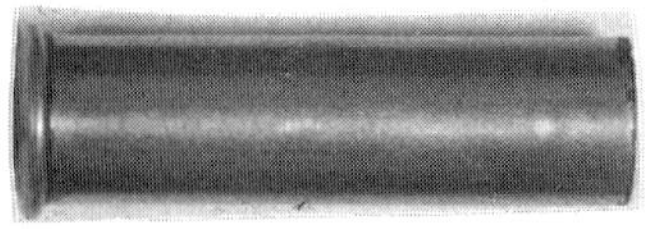

.44 Extra Long, 1.648". Appears to be the same as above but the case is nickeled and dimensions are a bit larger. (See Dimensions).

.44-85 (also known as .42-85), 2.510" case with the belted bullet. Also found with flat-nosed, paper-patched bullet.

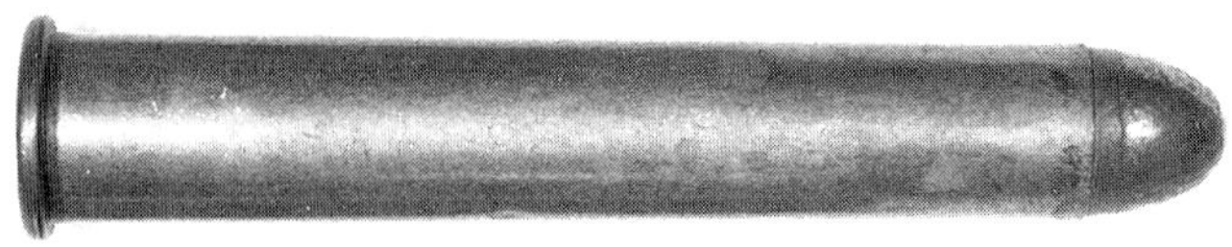

.44-100 (also known as .42-100), this specimen with 3.05" case, Wesson belted bullet. Specimens are reported with a 2.91" case, which could be the result of trimming due to reloading.

.44-100, 2.921" Ryder patent, two-piece nickel plated case with the steel screw-off base and percussion nipple for easy replacement of the primer.

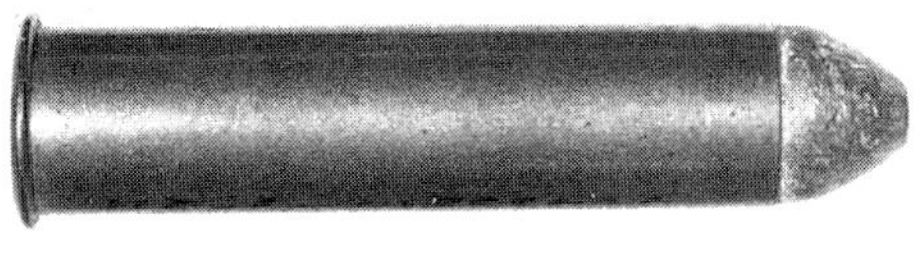

.45-60, 1.975" case. Factory loaded by U. S. Cartridge Company with the short, flat-nosed bullet.

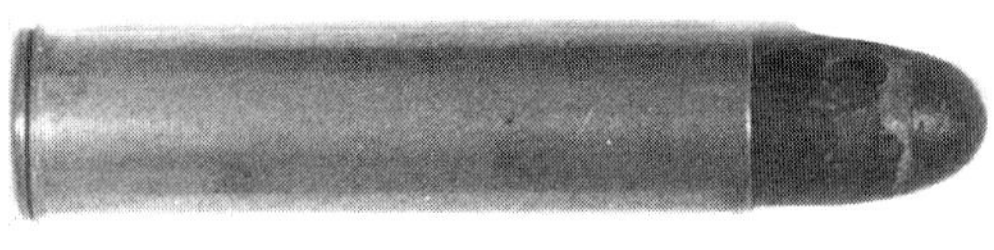

.45 Cooks Patent, 1.955". This is a nickel-plated tube fitted with a .44 short rimfire blank as a primer, this one loaded with a round-nosed paper-patched bullet. This was thought to be an easier way to reload. (See the 1875 patent in Appendix A).

.55 Wesson Shot, 2.009 nickel-plated case, small primer pocket.

J. STEVENS ARMS & TOOL COMPANY

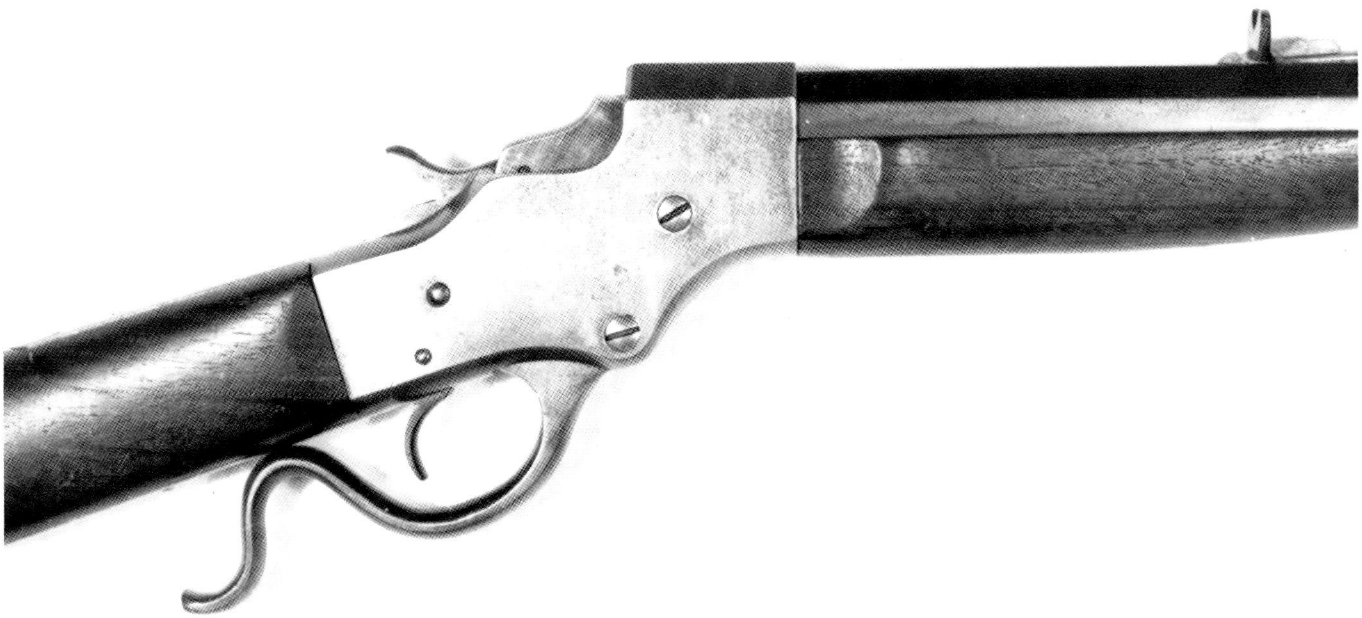

STEVENS RIFLES

STEVENS Single Shot Rifles
"IDEAL"

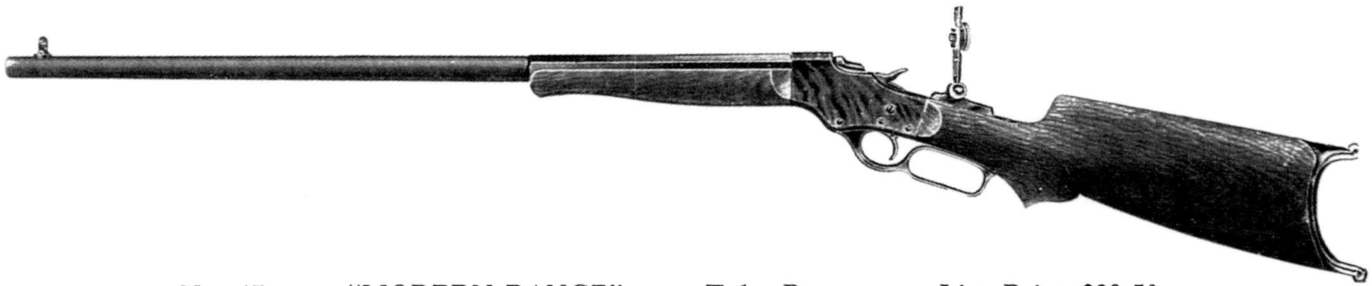

No. 47 "MODERN RANGE" Take-Down List Price $29.50

Barrel No. 3; half octagon; length 28 inches.

Frame Drop forged; casehardened.

Action No. 3 loop lever; AUTOMATIC EJECTOR IN REGULAR .22-CALIBER ONLY; plain extractor in other calibers; drop-down breech block.

Stock Walnut; pistol grip; No. 3 Swiss butt plate; length 13 inches, drop $3\frac{1}{4}$ inches; walnut fore-end.

Sights No. 205 Beach combination front and No. 102 Vernier peep (D sights).

Weight With No. 2, 28″ barrel about 8 lbs. 10 oz.

Ammunition Rim fire — .22 short, .22 long rifle, .22 W. R. F., .25 Stevens, .32 long.

Center fire — .22/15/60, .25/20 Stevens, .25/21, .25/25, .28/30/120, .32 long, .32/20, .32 Ideal, .32/40.

STEVENS Single Shot Rifles
"IDEAL"

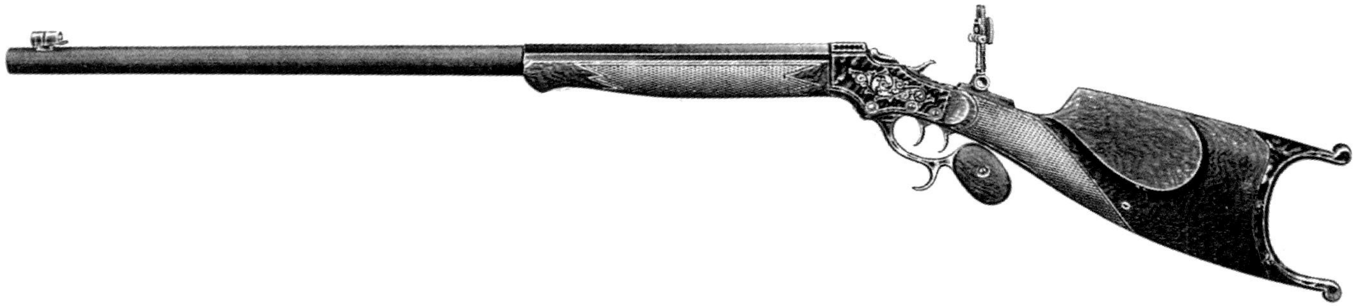

No. 49 "WALNUT HILL" Take-Down List Price $46.25

Barrel No. 3; half octagon; length 28 inches.

Frame Drop forged; casehardened; engraved.

Action No. 3 loop lever; AUTOMATIC EJECTOR IN REGULAR .22-CALIBER ONLY; plain extractor in other calibers; drop-down breech block.

Stock Walnut, checkered; pistol grip; cheek piece; No. 3 heavy Swiss butt plate; length 13 inches, drop $3\frac{1}{4}$ inches; walnut fore-end, checkered.

Sights No. 210 Globe interchangeable front and No. 104 Wind-Gauge Vernier peep with combination eye cup (H sights).

Weight With No. 3, 28″ barrel about 10 lbs.

Ammunition Rim fire — .22 short, .22 long rifle, .22 W. R. F., .25 Stevens, .32 long

Center fire — .22/15/60, .25/20 Stevens, .25/21, .25/25, .28/30/120, .32 long, .32/20, .32 Ideal, .32/40.

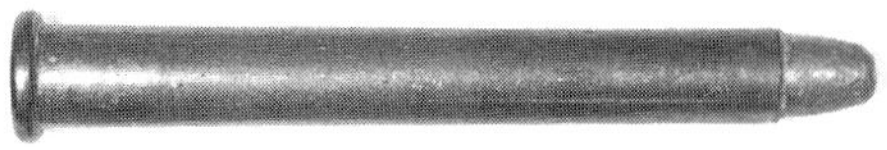

.22-15-60, 2.01" case, brass primer, headstamp "U. M. C. 22-15-60." For the Stevens Ideal and other single shot rifles.

.25-21-86, 2.039". Same as above, single groove in the bullet. H. s. " W R A Co. 25-21."

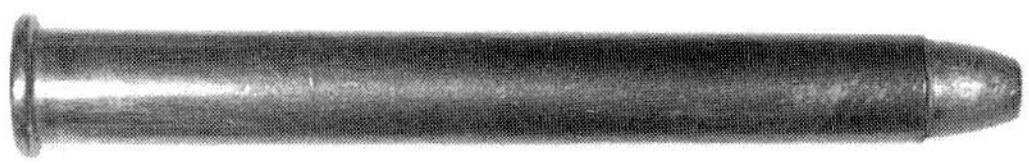

.25-25-86, 2.37". Same as above.

.28-30-120, 2.501" case.

.32-1.472"case. This is the .32 Smith & Wesson Rifle case loaded with paper-patched bullet. It is marked on the side as ".32 Stevens Perfection." The head is unmarked with nickeled boxer primer. The lower case has the bullet flush with the case mouth, h.s. "U M C .32 S & W R."

.32-25-150, 1.88" Ideal.

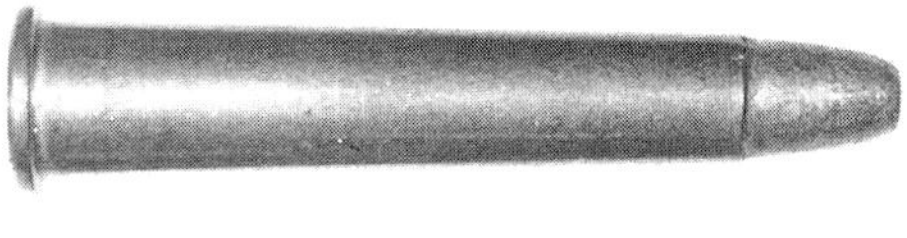

.32-35-165, 1.879", the same case as used in the Model 82 Maynard rifles but the Maynard case is 1.84" long.

.38-35, 1.62". The large brass primer is early and the iron anvil within takes a magnet. Later cases have the small brass primer, no iron anvil. No. h.s.

.38-35, 1.62 shot. Small brass primer, white cardboard wad at the case mouth.

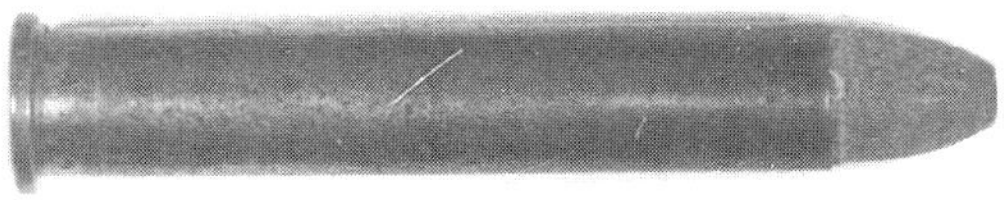

.38-45, 2.2". This is the Hart's patent with two-piece case, iron anvil, large brass primer.

.38-45, 2.19", later case with small brass primer, no iron anvil, no marks.

.38-45, 2.181 shot. Small copper primer with white wad over the shot load.

.44-50, 1.965" Hart's patent case with large brass primer, iron anvil within, flat-nosed bullet.

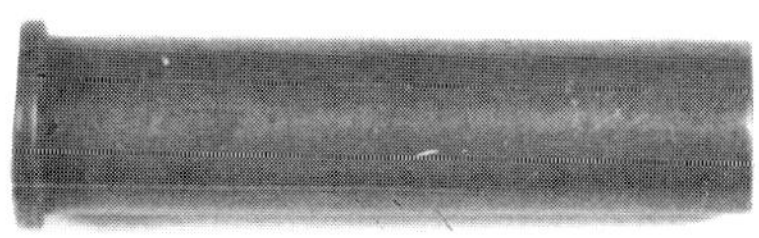

.44-50, 1.932" shot load with large brass primer, iron anvil, tan wad over shot load.

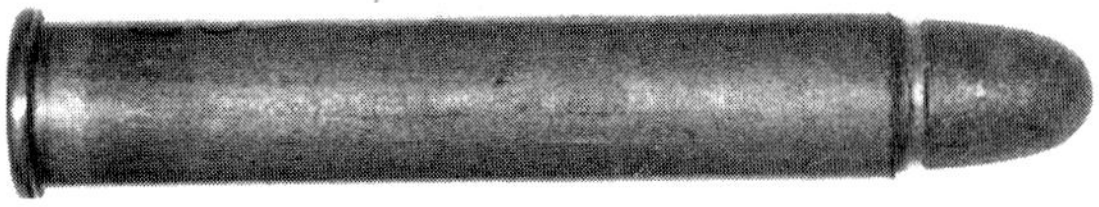

.44-65, 2.451", an early load with large brass primer, iron anvil, single-groove lead bullet, no marks.

.44-65-2.471", a later loading with small brass primer, paper-patch express bullet.

.44-65, 2.451" shot load. Small brass primer with white over-shot wad at case mouth.

BULLETS.

PATCHED.

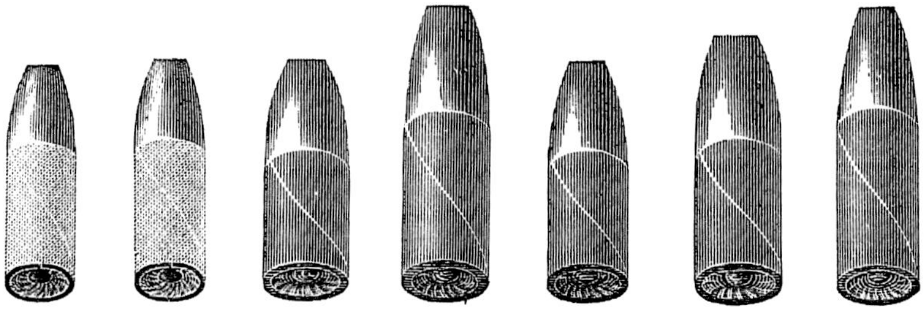

PRICE PER 1,000.

32-165	32-185	38-255	38-330	40-285	40-330	40-370
$10.00	$9.75	$11.00	$11.50	$11.00	$11.50	$12.50

GROOVED.

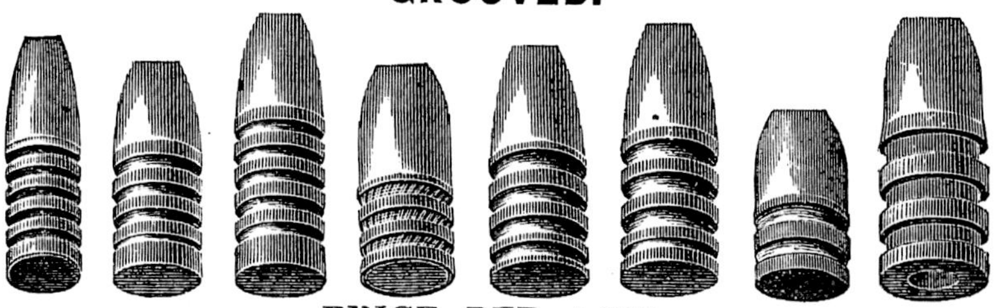

PRICE PER 1,000.

32-185	38-255	38-330	40-260	40-285	40-330	45-285	45-405
$7.75	$9.50	$11.00	$9.00	$9.50	$11.00	$9.50	$12.00

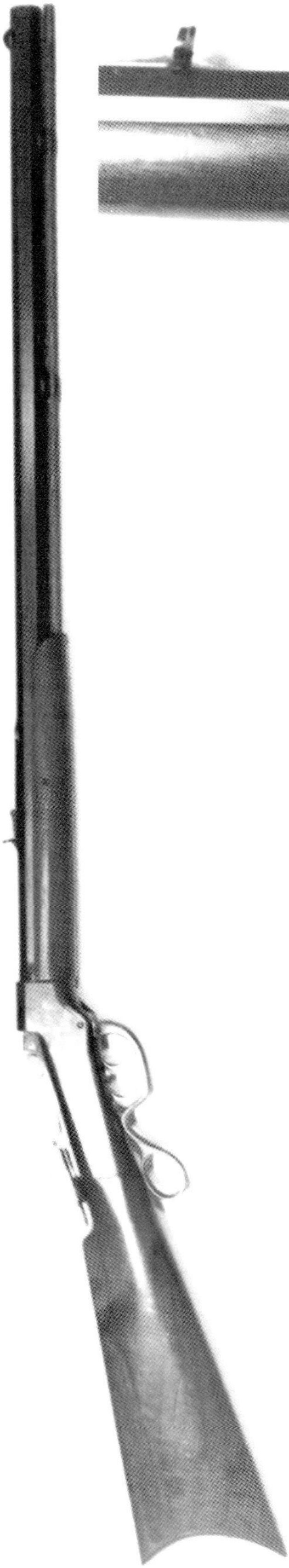

THE BALLARD RIFLE

The rifle shown here is a No. 5 Pacific with 30" round barrel, Rocky Mountain open rear sight on the barrel, and folding peep sight on the tang, fitted with double set trigger and wood cleaning rod under the barrel. The caliber marked on the barrel is .40-63. It weights nine pounds. This is less than the 10 pounds prescribed by the Marlin catalogue description on the next page. Otherwise, it conforms to the description. The Ballard action is a falling block but the block rocks back as it drops and then rocks forward as it closes. The box label below is representative, but is for the 40-85-2 15/16"—a longer case for which the No. 5 was also chambered.

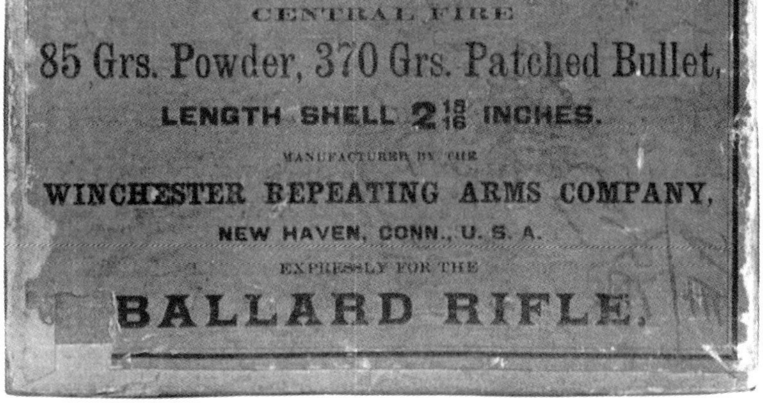

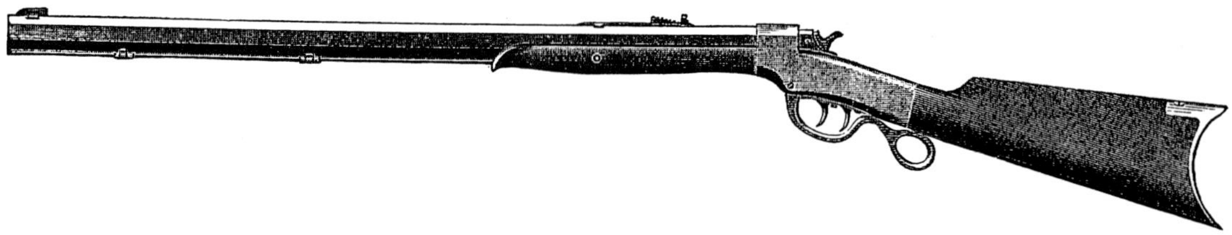

No. 5—PACIFIC RIFLE.

Octagon Barrel, double set Triggers, extra heavy Wrought Frame, Cleaning Rod under the barrel, Rocky Mountain Sights, using either Everlasting Shells or factory ammunition, except 44 W. C. F., which uses factory cartridges only, and 45–100, which uses the Everlasting Shells only.

As its name indicates, the favorite on the Pacific coast. It is THE hunting rifle of the West, and enjoys the best commendations from hunters of buffalo, elk, deer, bear and other large game.

30 inch................ $22.50 | 32 inch........................ $24.00

38–55 are made 30 inch, 10 lbs.; 32 inch, 11 lbs.
40–63 " 30 " 10 lbs.; 32 " 11 lbs.
40–85 " 30 " 10 lbs.; 32 " 11 lbs.

44 W. C. F. are made 30 inch, 10 lbs.
45–70 are made 30 inch, 10½ lbs.; 32 inch, 11½ lbs.
45–100 " 30 " 10½ lbs.; 32 " 11½ lbs.

☞ This style cannot be made in smaller calibre than 38–55.

MARLIN FIRE ARMS CO., NEW HAVEN, CONN.

No. 6 1-2—OFF-HAND RIFLE.

Rigby Barrel, Mid-Range Vernier Peep and Wind-Gauge Sights, fine English Walnut Stock, modified Swiss pattern, Pistol Grip, full checked, Horn Tip, improved "Off-Hand" nickeled Butt Plate, which exactly fits the arm, finely engraved Frame and Barrel, and every part highly finished. Uses either Everlasting Shells or regular factory ammunition.

Stands without a rival for mid-range practice, being used by many of the most celebrated shots of America and Europe.

32–40 and 38–55, 28 and 30 inch, 9¾ lbs., $57.50

This gun can be furnished with Shot Gun Butt, and rubber Heel Plate, if preferred, at same price.

.32-.96" marked "Ballard Short," plain flat head.

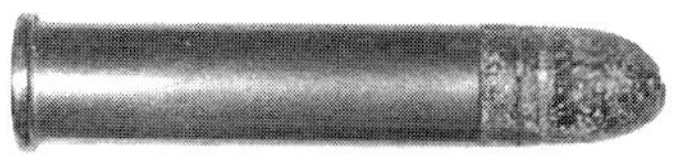

.32-115 Ballard Extra Long, 1.24" case.

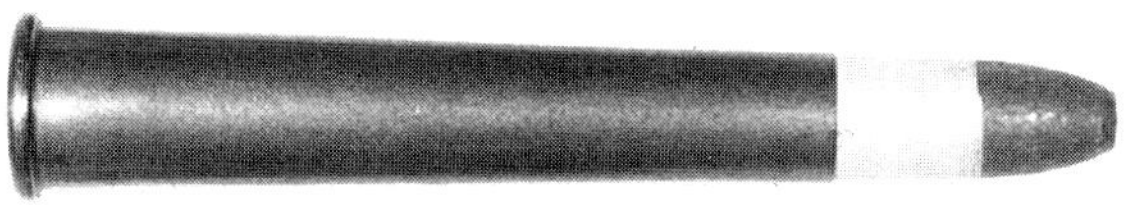

.32-40-165 Ballard, 2.59". Found usually with paper-patched bullet. This same case became the Winchester .32-40, chambered also in Marlin and Savage rifles, and widely used in single shot target rifles.

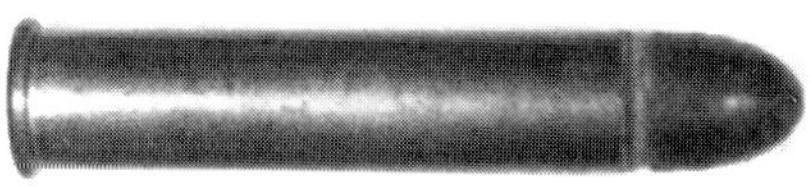

.38-30 Extra Long Ballard, 1.63". This is the center fire version of an earlier rimfire cartridge with a 1.48" case.

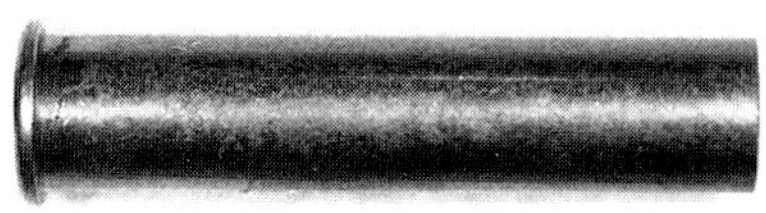

.38-50-250, 2" Ballard Everlasting. Normally found with the paper-patched bullet. A target load used in the Ballard No 4 and Pacific No. 5 rifles.

.38-55-255, 2.05". Paper-patched normally for Ballard rifles, this became the .38-55 for the Winchester and other lever-action magazine rifles.

.40-63-330, 2.38" case. This was also available loaded with 70 grains of powder.

.40-65,2.38" Ballard Everlasting. This is a heavier case and slightly larger in dimensions than that above, so it will not chamber in the rifles made for .40-63 or.40-70. Bullet weights varied according to wishes of the shooter.

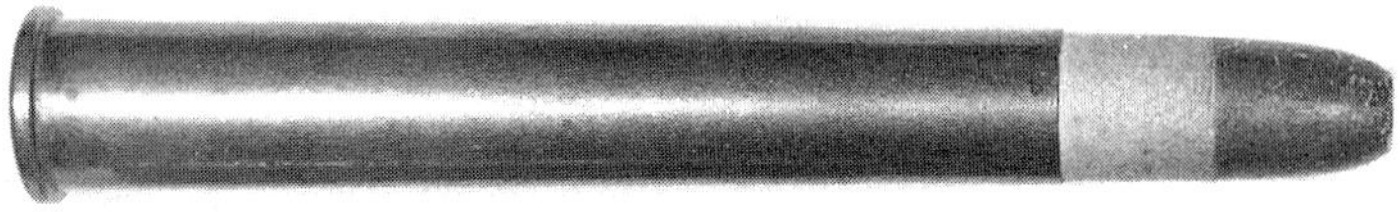

.40-85 and 40-90-370, 2.95" Ballard. The only difference between the two cartridges is the powder charge. The .40-90 may have the bullet set out farther to take make room for the larger charge.

.40-90, 3 1/6" Everlasting. Like the .40-65, this is a heavier case and slightly longer than the .40-85/90, and intended for hand loading.

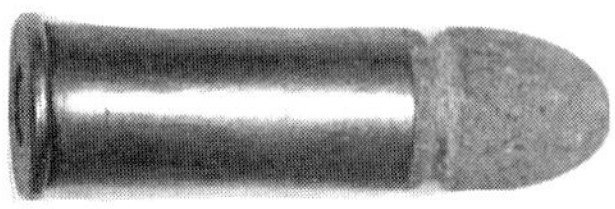

44-35-227 Long, 1.09" case.

.44-50-265 Extra Long, 2.10" case.

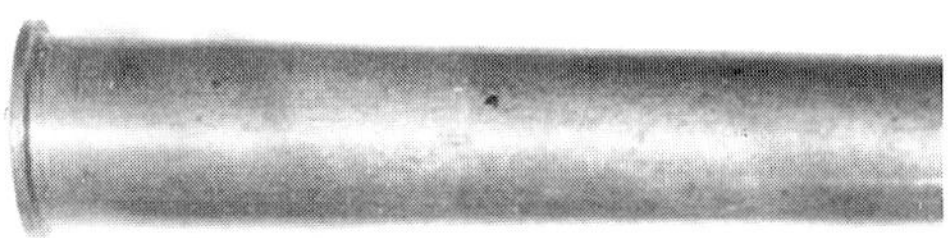

.44-75, 2.5" Everlasting, nickeled case.

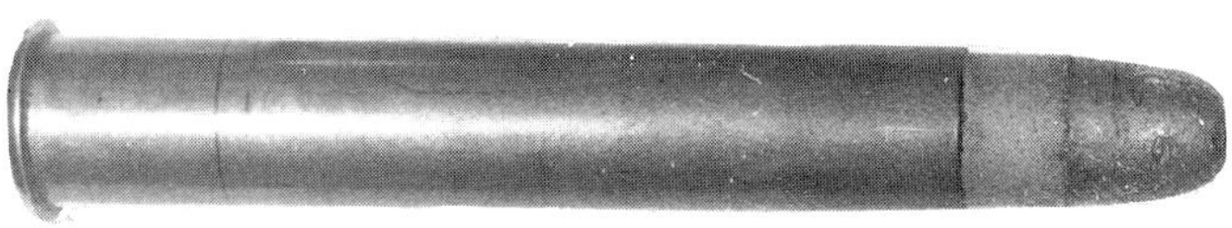

.44-100, 2.80" case.

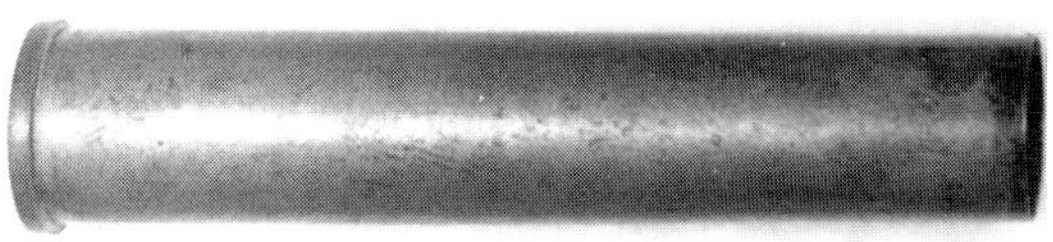

.45-100, 2.80". The same case as for the .44-100 but reamed out at the mouth to accept the larger bullet. Any weight of bullet up to 535 grains may be found in this case as in the .44-100.

20
WINCHESTER
Trade Marks Reg. U. S. Pat. Off. and Throughout the World
.50-70 MUSKET CARTRIDG
CENTER FIRE
WINCHESTER MAKE
Manufactured by the WINCHESTER REPEATING ARMS CO., New Haven, Conn., U. S. A.

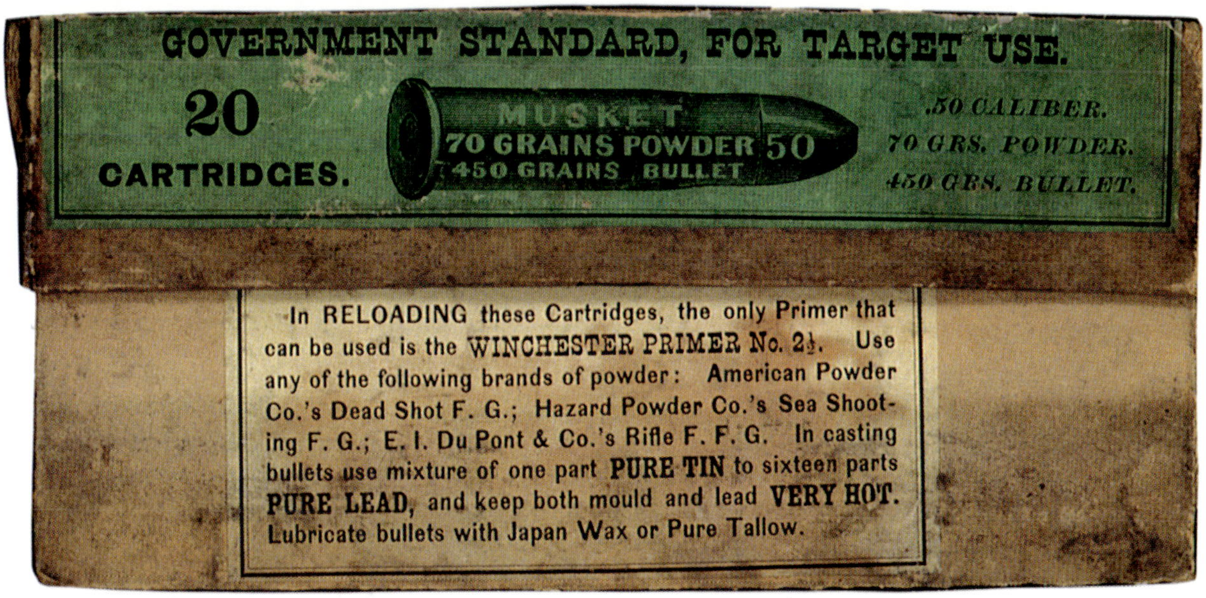
GOVERNMENT STANDARD, FOR TARGET USE.
20
CARTRIDGES.
MUSKET
70 GRAINS POWDER 50
450 GRAINS BULLET
.50 CALIBER.
70 GRS. POWDER.
450 GRS. BULLET.
In RELOADING these Cartridges, the only Primer that can be used is the WINCHESTER PRIMER No. 2½. Use any of the following brands of powder: American Powder Co.'s Dead Shot F. G.; Hazard Powder Co.'s Sea Shooting F. G.; E. I. Du Pont & Co.'s Rifle F. F. G. In casting bullets use mixture of one part PURE TIN to sixteen parts PURE LEAD, and keep both mould and lead VERY HOT. Lubricate bullets with Japan Wax or Pure Tallow.

10 CARTRIDGES,
.50 CALIBRE.
2½ inch Shell, 90 Grains Powder,
1 1/16 inch Patched Bullet, 473 Grains
FOR
SHARPS RIFLE.
MANUFACTURED BY THE
Union Metallic Cartridge Company,
BRIDGEPORT, CONN., U. S. A.

10 CARTRIDGES,
45 CALIBRE.
2 1/10 inch Shell, 70 Grains Powder,
1 1/10 inch Patched Bullet, 420 Grains
FOR
SHARPS RIFLE.
Union Metallic Cartridge Co.,
BRIDGEPORT, CONN., U. S. A.

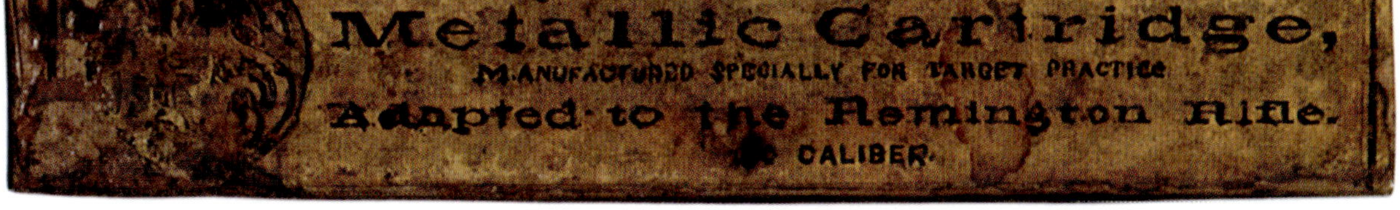
REMINGTONS'
Metallic Cartridge,
MANUFACTURED SPECIALLY FOR TARGET PRACTICE
Adapted to the Remington Rifle.
CALIBER.

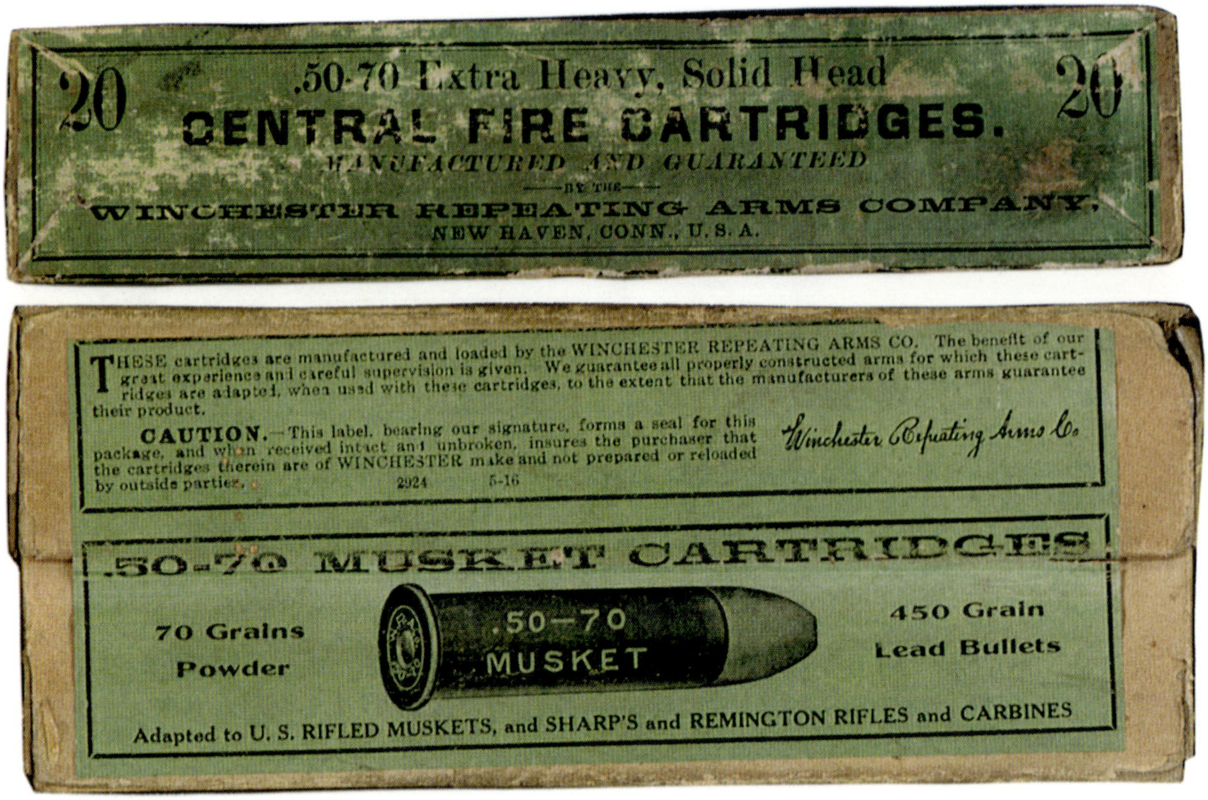
20 .50-70 Extra Heavy, Solid Head 20
CENTRAL FIRE CARTRIDGES.
MANUFACTURED AND GUARANTEED
BY THE
WINCHESTER REPEATING ARMS COMPANY,
NEW HAVEN, CONN., U. S. A.
THESE cartridges are manufactured and loaded by the WINCHESTER REPEATING ARMS CO. The benefit of our great experience and careful supervision is given. We guarantee all properly constructed arms for which these cartridges are adapted, when used with these cartridges, to the extent that the manufacturers of these arms guarantee their product.
CAUTION.—This label, bearing our signature, forms a seal for this package, and when received intact and unbroken, insures the purchaser that the cartridges therein are of WINCHESTER make and not prepared or reloaded by outside parties.
2924 5-16
Winchester Repeating Arms Co.
.50-70 MUSKET CARTRIDGES
70 Grains Powder
.50—70 MUSKET
450 Grain Lead Bullets
Adapted to U. S. RIFLED MUSKETS, and SHARP'S and REMINGTON RIFLES and CARBINES

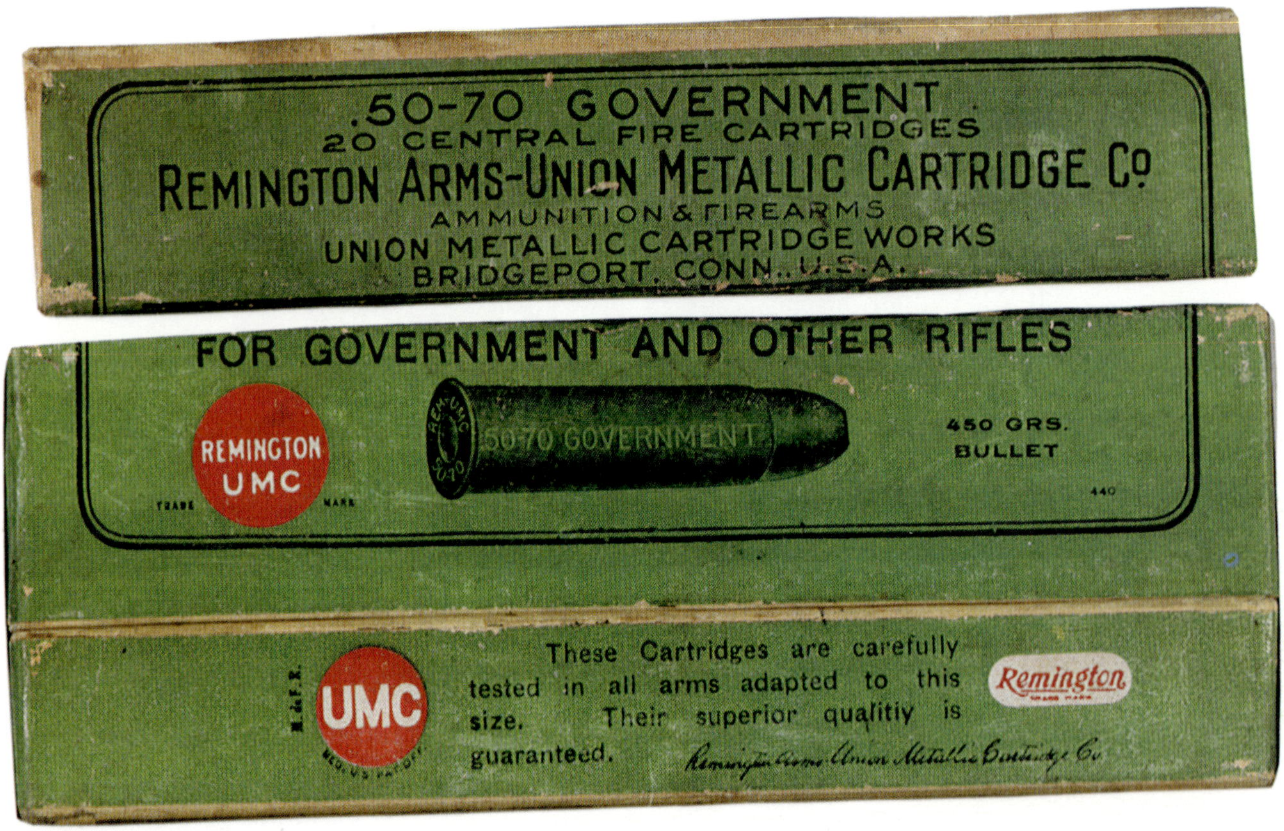
.50-70 GOVERNMENT
20 CENTRAL FIRE CARTRIDGES
REMINGTON ARMS-UNION METALLIC CARTRIDGE Co
AMMUNITION & FIREARMS
UNION METALLIC CARTRIDGE WORKS
BRIDGEPORT, CONN., U.S.A.
FOR GOVERNMENT AND OTHER RIFLES
REMINGTON UMC
TRADE MARK
50-70 GOVERNMENT
450 GRS. BULLET
440
UMC
These Cartridges are carefully tested in all arms adapted to this size. Their superior quality is guaranteed.
Remington Arms-Union Metallic Cartridge Co.
Remington

5 CARTRIDGES,
45 CALIBRE.
inch Shell, 100 Grains Powder,
inch Patched Bullet, 550 Grains,
Alloy 1 to 14
FROM
Sharps Rifle Co.,
BRIDGEPORT, CONN.

10 CARTRIDGES,
.45 CALIBRE.
inch Shell. 90 Grains Powder,
inch Patched Bullet. 500 Grains.
FOR
SHARPS RIFLE
MANUFACTURED BY THE
Union Metallic Cartridge Company,
BRIDGEPORT, CONN., U. S. A.

10 CARTRIDGES,
.45 CALIBRE.
inch Shell. 90 Grains Powder.
inch Patched Bullet. 550 Grains.
FOR
SHARPS RIFLE
MANUFACTURED BY THE
Union Metallic Cartridge Company,
BRIDGEPORT CONN., U. S. A.

10
10
.45 Caliber
CENTRAL FIRE
CARTRIDGES
inch Shell, Grains Powder
500 GRAINS PATCHED BULLET
MANUFACTURED BY THE
WINCHESTER REPEATING ARMS CO.
NEW HAVEN, CONN., U. S. A.
Expressly for SHARP'S RIFLE.

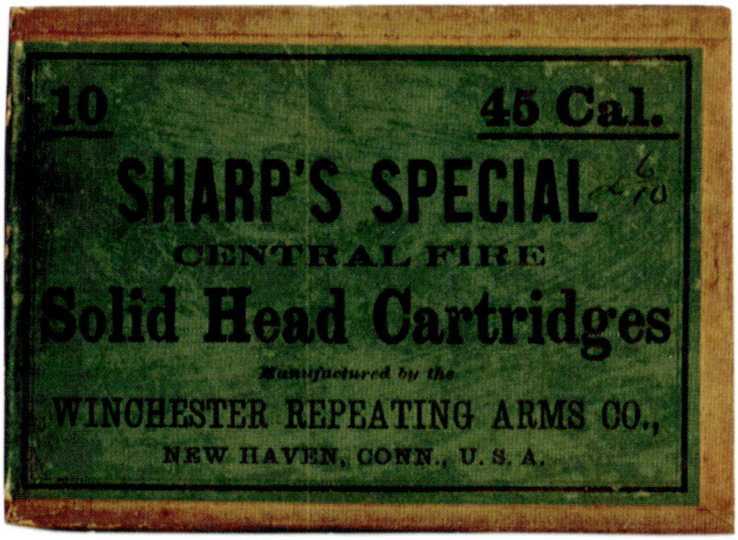
10
45 Cal.
SHARP'S SPECIAL
CENTRAL FIRE
Solid Head Cartridges
Manufactured by the
WINCHESTER REPEATING ARMS CO.,
NEW HAVEN, CONN., U. S. A.

10 CARTRIDGES,
45 CALIBRE.
2 inch Shell, 100 Grains Powder,
PATCHED BULLET.
PUT UP EXPRESSLY FOR
SHARPS RIFLE,
BY
Sharps Rifle Co.,
BRIDGEPORT, CONN.

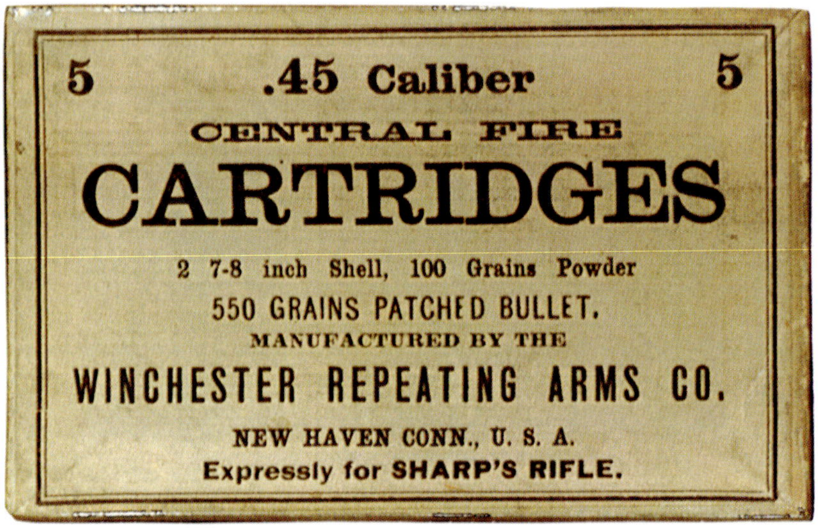
5 .45 Caliber 5
CENTRAL FIRE
CARTRIDGES
2 7-8 inch Shell, 100 Grains Powder
550 GRAINS PATCHED BULLET.
MANUFACTURED BY THE
WINCHESTER REPEATING ARMS CO.
NEW HAVEN CONN., U. S. A.
Expressly for SHARP'S RIFLE.

5 CARTRIDGES,
.45 CALIBRE.
2 7/8 inch Shell, 105 Grains Powder,
1 inch Patched Bullet, 550 Grains
MANUFACTURED EXPRESSLY FOR
SHARPS RIFLES
BY
UNION METALLIC CARTRIDGE CO.
BRIDGEPORT, CONN.
Under Berdan's Patent, March 20th, 1866, and September 29th, 1868, and S. W. Wood's Patents April 1st, 1867, and April 2nd, 1872.

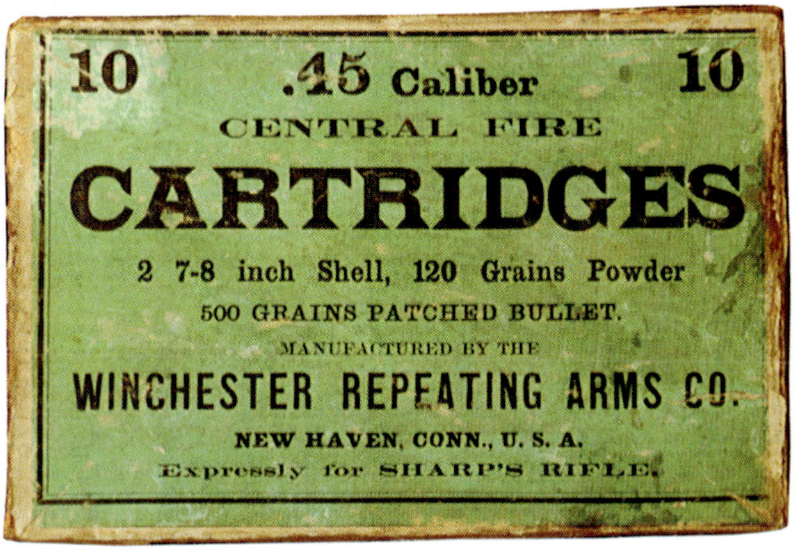
10 .45 Caliber 10
CENTRAL FIRE
CARTRIDGES
2 7-8 inch Shell, 120 Grains Powder
500 GRAINS PATCHED BULLET.
MANUFACTURED BY THE
WINCHESTER REPEATING ARMS CO.
NEW HAVEN, CONN., U. S. A.
Expressly for SHARP'S RIFLE.

.45-105 SHARPS
5 CENTRAL FIRE CARTRIDGES
2⅞ inch Shell
TRADE REMINGTON UMC MARK
550 GRS. PATCHED BULLET
FOR SHARPS RIFLE
REMINGTON ARMS-UNION METALLIC CARTRIDGE Co
AMMUNITION & FIREARMS
UNION METALLIC CARTRIDGE WORKS
BRIDGEPORT, CONN., U.S.A.

10 .50-90
CENTRAL FIRE
SOLID HEAD CARTRIDGES.
FOR SHARP'S RIFLE.
MANUFACTURED BY THE
Winchester Repeating Arms Company,
NEW HAVEN, CONN., U. S. A.

20
SOLID HEAD RE-LOADING CARTRIDGES,
Calibre: .44.–77 Grs. Powder; 405 Grs. Ball.
MANUFACTURED BY THE
UNITED STATES CARTRIDGE CO., LOWELL, MASS., U. S. A.

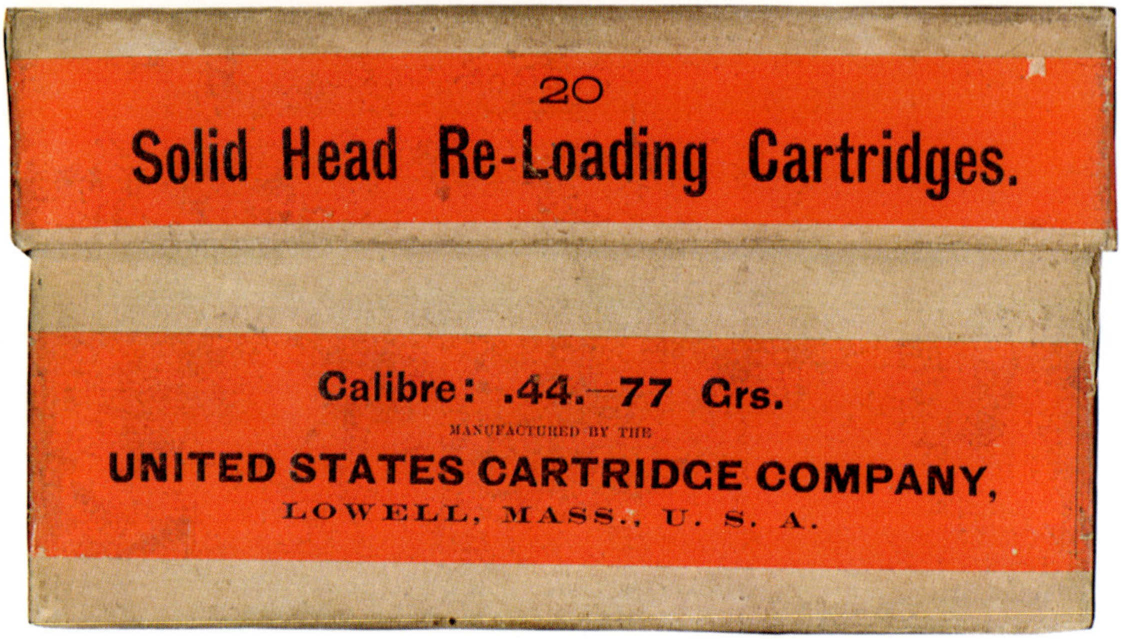
20
Solid Head Re-Loading Cartridges.
Calibre: .44.–77 Grs.
MANUFACTURED BY THE
UNITED STATES CARTRIDGE COMPANY,
LOWELL, MASS., U. S. A.

FOR REMINGTON RIFLE.
20 CARTRIDGES.
44 CALIBER.
470 Grs. BULLET.
77 Grs. POWDER.
In RELOADING these CARTRIDGES
The only PRIMER that can be used is the
Winchester Primer No. 2½
For convenience in carrying, use ANSON MILLS'
WOVEN CARTRIDGE BELT.

20
SOLID HEAD
CARTRIDGES,
CENTRAL FIRE RELOADING.
44-77 GRs
MANUFACTURED BY THE
WINCHESTER REPEATING
ARMS CO.,
New Haven, Conn.,
U. S. A.

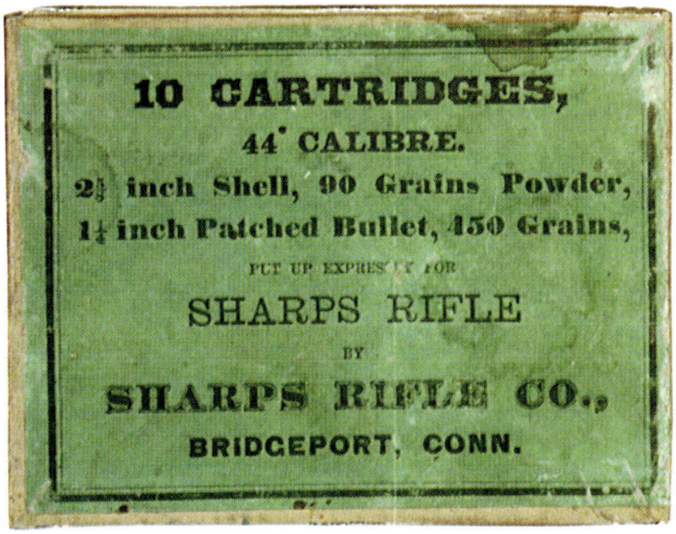
10 CARTRIDGES,
44 CALIBRE.
2½ inch Shell, 90 Grains Powder,
1¼ inch Patched Bullet, 450 Grains,
PUT UP EXPRESSLY FOR
SHARPS RIFLE
BY
SHARPS RIFLE CO.,
BRIDGEPORT, CONN.

20 METALLIC
CENTRE FIRE CARTRIDGES,
FOR
100 CALIBRE,
Put up expressly for Sharps' Improved Rifle,
in Shells susceptible of being reloaded many times,
BY
Sharps Rifle Company,
HARTFORD, CONN.

10 CARTRIDGES,
.44 CALIBRE.
2⅝ inch Shell, 90 Grains Powder,
1⅞ inch Patched Bullet, 500 Grains,
FOR
SHARPS RIFLE,
MANUFACTURED BY THE
Union Metallic Cartridge Company,
BRIDGEPORT, CONN. U. S. A.

10
44-90
CENTRAL FIRE
SOLID HEAD CARTRIDGES
FOR SHARP'S RIFLE.
MANUFACTURED BY THE
Winchester Repeating Arms Co.,
NEW HAVEN, CONN., U. S. A.

20 CENTER FIRE CARTRIDGES. 20
Manufactured by the
Winchester Repeating Arms Company,
NEW HAVEN, CONN., U. S. A.
FOR SHARPS RIFLE.
20
CARTRIDGES.
40-50 GRS
40 Caliber.
50 Grs. Powder.
In RELOADING these CARTRIDGES
the only PRIMER that can be used
IS THE
Winchester Primer No. 2½

20 CENTER FIRE CARTRIDGES. 20
Manufactured by the
WINCHESTER REPEATING ARMS CO.
New Haven, Conn., U. S. A.
FOR SHARP'S RIFLE.
20
CARTRIDGES.
44
CALIBER.
470 Grs.
BULLET.
77 Grs.
POWDER.

20 CENTRAL FIRE CARTRIDGES. 20
MANUFACTURED AND GUARANTEED BY THE
WINCHESTER REPEATING AR CO.,
NEW HAVEN, CONN., U. S. A.

FOR SHARP'S RIFLE.
20 Cartridges.
.40 Caliber.
1⅞ Inch. Shell.
45 Grs. Powder.
265 Grs. Bullet.
These cartridges are loaded by the WINCHESTER REPEATING ARMS COMPANY.
The benefit of our great experience and careful supervision is given; beyond this we assume no responsibility.
Winchester Repeating Arms Co.

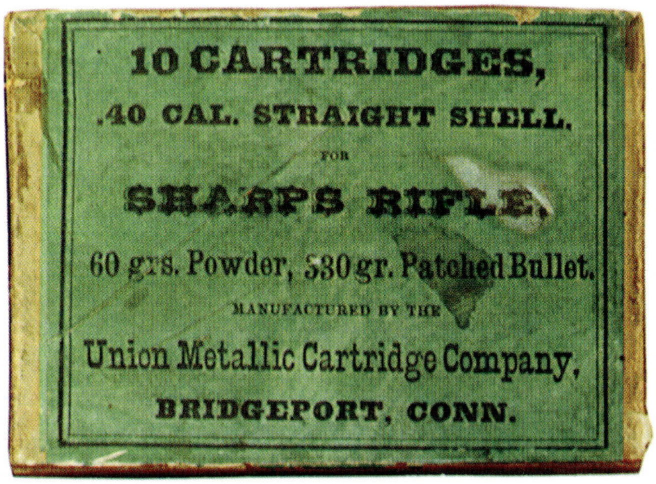
10 CARTRIDGES,
.40 CAL. STRAIGHT SHELL.
FOR
SHARPS RIFLE.
60 grs. Powder, 530 gr. Patched Bullet.
MANUFACTURED BY THE
Union Metallic Cartridge Company,
BRIDGEPORT, CONN.

10 CARTRIDGES,
.40 CAL. STRAIGHT SHELL.
FOR
SHARPS RIFLE.
65 grs. Powder, 330 gr. Patched Bullet.
MANUFACTURED BY THE
Union Metallic Cartridge Company,
BRIDGEPORT, CONN.

10 CARTRIDGES,
.40 CAL. STRAIGHT SHELL.
FOR
SHARPS RIFLE.
70 grs. Powder, 330 gr. Patched Bullet.
MANUFACTURED BY THE
Union Metallic Cartridge Company,
BRIDGEPORT, CONN.

10 CARTRIDGES,
40 CALIBRE.
2⅝ inch Shell, 90 Grains Powder,
1¼ inch Patched Bullet 370 Grains.
PUT UP EXPRESSLY FOR
SHARPS RIFLE
BY THE
Union Metallic Cartridge Company,
BRIDGEPORT, CONN., U. S. A.

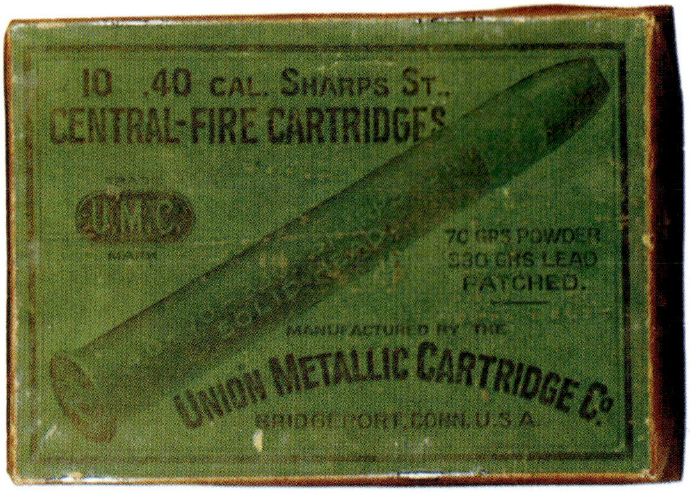
10 .40 CAL. SHARPS ST.
CENTRAL-FIRE CARTRIDGES
UMC
70 GRS POWDER
330 GRS LEAD
PATCHED.
MANUFACTURED BY THE
UNION METALLIC CARTRIDGE Co
BRIDGEPORT, CONN. U.S.A.

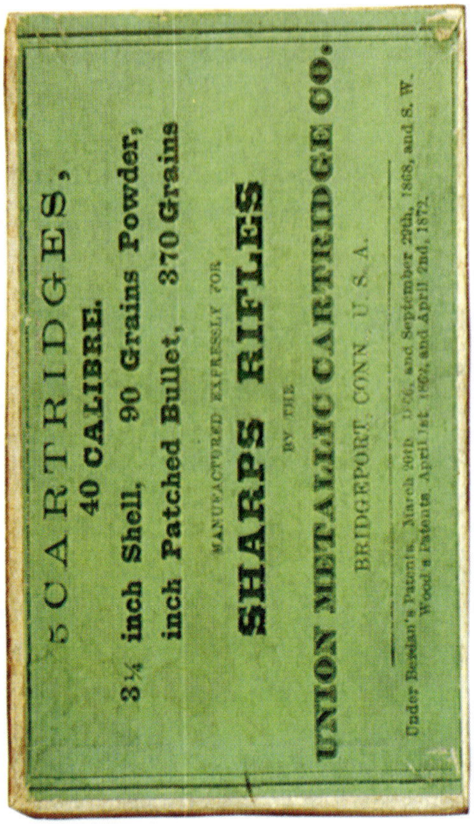
5 CARTRIDGES,
40 CALIBRE.
3¼ inch Shell, 90 Grains Powder,
inch Patched Bullet, 370 Grains
MANUFACTURED EXPRESSLY FOR
SHARPS RIFLES
BY THE
UNION METALLIC CARTRIDGE CO.
BRIDGEPORT, CONN. U. S. A.
Under Berdan's Patents, March 20th, 1866, and September 29th, 1868, and S. W. Wood's Patents, April 1st, 1862, and April 2nd, 1872.

10 CARTRIDGES,
40 CALIBRE.
2⅝ inch Shell, 90 Grains Powder,
1¼ inch Patched Bullet 370 Grains.
PUT UP EXPRESSLY FOR
SHARPS RIFLE
BY THE
Union Metallic Cartridge Company,
BRIDGEPORT, CONN., U. S. A.

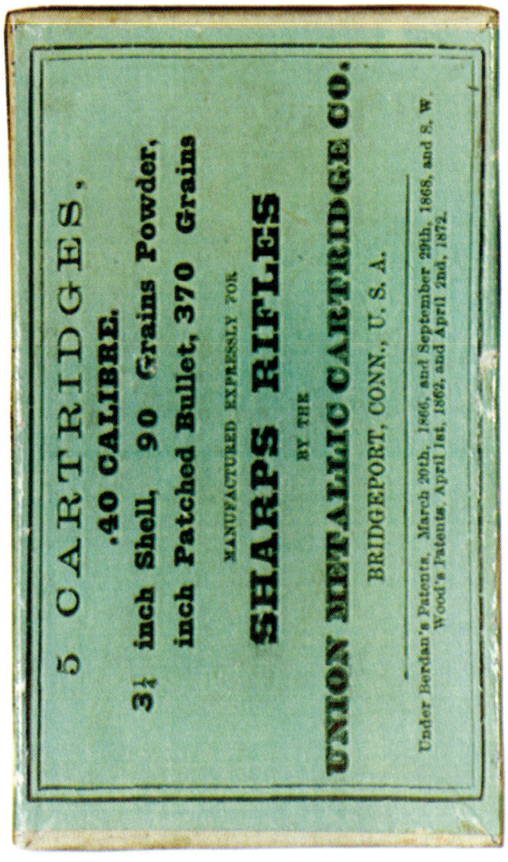
5 CARTRIDGES,
.40 CALIBRE.
3¼ inch Shell, 90 Grains Powder,
inch Patched Bullet, 370 Grains
MANUFACTURED EXPRESSLY FOR
SHARPS RIFLES
BY THE
UNION METALLIC CARTRIDGE CO.
BRIDGEPORT, CONN., U. S. A.
Under Berdan's Patents, March 20th, 1866, and September 29th, 1868, and S. W. Wood's Patents, April 1st, 1862, and April 2nd, 1872.

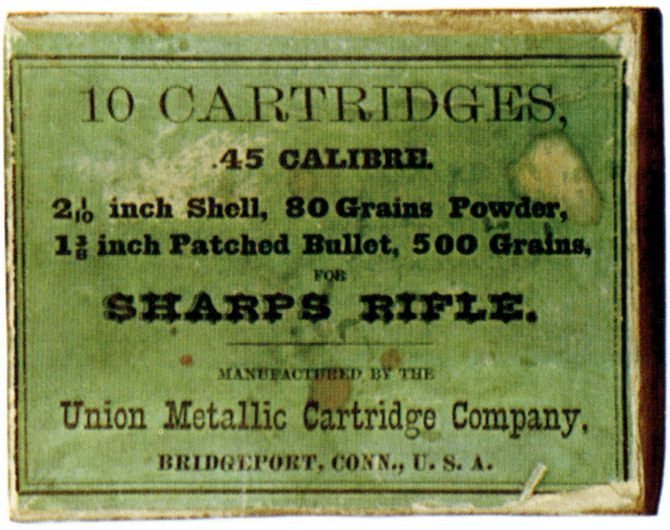
10 CARTRIDGES,
.45 CALIBRE.
2$\frac{1}{10}$ inch Shell, 80 Grains Powder,
1⅜ inch Patched Bullet, 500 Grains,
FOR
SHARPS RIFLE.
MANUFACTURED BY THE
Union Metallic Cartridge Company.
BRIDGEPORT, CONN., U. S. A.

SPECIAL
FROM 200 TO 500 YARDS RANGE.
TWENTY
CARTRIDGES.
40 Calibre, 70 Grains.

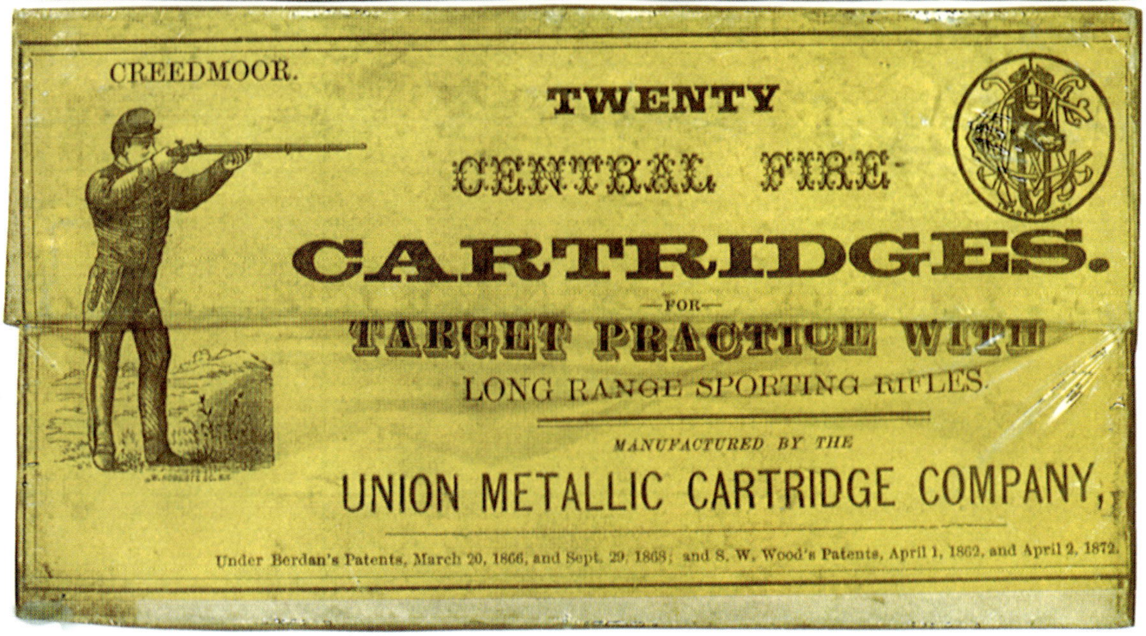
CREEDMOOR.
TWENTY
CENTRAL FIRE
CARTRIDGES.
—FOR—
TARGET PRACTICE WITH
LONG RANGE SPORTING RIFLES.
MANUFACTURED BY THE
UNION METALLIC CARTRIDGE COMPANY,
Under Berdan's Patents, March 20, 1866, and Sept. 29, 1868; and S. W. Wood's Patents, April 1, 1862, and April 2, 1872.

TWENTY
CARTRIDGES.
.44 Calibre, 77 Grains.

TWENTY
CENTRAL FIRE
Cartridges.
—FOR—
LONG RANGE SPORTING RIFLES
MANUFACTURED BY THE
UNION METALLIC CARTRIDGE COMPANY

SPECIAL.
FROM 200 TO 500 YARDS RANGE.
TWENTY
CARTRIDGES.
40 Calibre, 50 Grains.

CREEDMOOR.
TWENTY
CENTRAL FIRE CARTRIDGES
—FOR—
E. REMINGTON & SON'S
LONG RANGE SPORTING RIFLES.
MANUFACTURED BY THE
Union Metallic Cartridge Company.
Under Berdan's Patents, Mch. 20, 1866, and Sept. 29, 1868; and S.W. Wood's Patents, Apr. 1, 1862 and Apr. 2, 1872.

SPECIAL
FROM 500 TO 1000 YARDS RANGE.
TWENTY
CARTRIDGES.
44 Calibre, 90 Grains.

CREEDMOOR.
TWENTY
CENTRAL FIRE
Cartridges.
—FOR—
LONG RANGE SPORTING
MANUFACTURED BY THE
UNION METALLIC CARTRIDGE COMPANY,

25
SOLID HEAD
Metallic Shells,
Manufactured by
E. REMINGTON & SONS, ILION, N. Y.
ADAPTED TO THE
Remington ~~Military~~ Creedmoor Rifle.
~~SPECIAL~~ .44 Cal. ~~STRAIGHT~~ SHELL—2 ~~6-10~~ 1/4
In reloading use the Remington Primer No 2½.

20 CARTRIDGES,
40 CALIBRE.
1 7/8 inch Straight Shell, 50 Grains Powder,
15/16 inch Patched Bullet, 265 Grains.
FOR
REMINGTON RIFLE.
MANUFACTURED BY THE
Union Metallic Cartridge Co.
BRIDGEPORT, CONN., U. S. A.

E. REMINGTON & SONS,
"SPECIAL"
Metallic Cartridges.
Patented October 26, 1858; July 15, 1862
Extended October 26, 1873.
90 FOR R
REMINGTON MATCH RIFLE
CALIBRE .44.

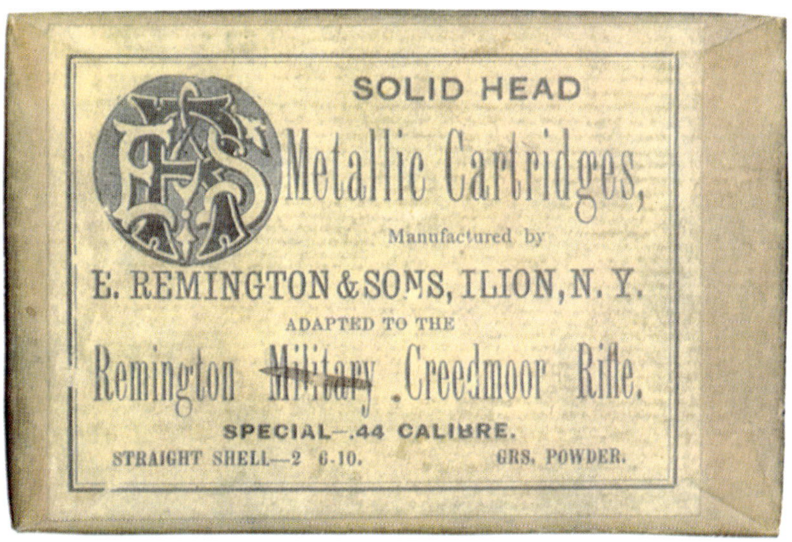
SOLID HEAD
Metallic Cartridges,
Manufactured by
E. REMINGTON & SONS, ILION, N. Y.
ADAPTED TO THE
Remington ~~Military~~ Creedmoor Rifle.
SPECIAL—.44 CALIBRE.
STRAIGHT SHELL—2 6-10. GRS. POWDER.

Solid Head Metallic Cartridges,
MANUFACTURED BY
E. Remington & Sons
ILION, N. Y.
ADAPTED TO THE REMINGTON CREEDMOOR RIFLE.
Special---44 Calibre---Straight Shell
Grs. Powder.

25
SOLID HEAD
Metallic Shells
Manufactured by
E. Remington & Sons, Ilion, N. Y.
ADAPTED TO THE
REMINGTON MID-RANGE RIFLE
40 Cal.—45 Grs. Powder. Straight Shell
In reloading use the Remington
Primer No. 3½.

10 CARTRIDGES,
.45 CALIBRE
500 grs. Patched Bullet.
FOR
REMINGTON'S
LONG RANGE RIFLE.
MANUFACTURED BY THE
Union Metallic Cartridge Company,
BRIDGEPORT CONN., U. S. A

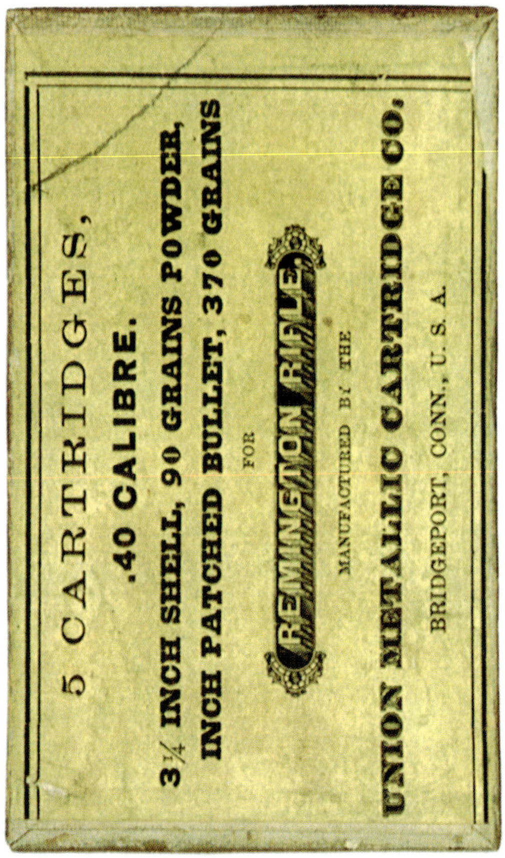
5 CARTRIDGES,
.40 CALIBRE.
3¼ INCH SHELL, 90 GRAINS POWDER,
INCH PATCHED BULLET, 370 GRAINS
FOR
REMINGTON RIFLE
MANUFACTURED BY THE
UNION METALLIC CARTRIDGE CO.
BRIDGEPORT, CONN., U. S. A.

10 CARTRIDGES,
.45 CALIBRE
REMINGTON'S
LONG RANGE RIFLE.
Union Metallic Cartridge Company,
BRIDGEPORT CONN., U. S. A

For BALLARD & MARLIN RIFLES.
20
Cartridges.
.38 Calibre.
55 Grs. Powder.
255 Grs. Bullet.
BALLARD & MARLIN
Also for Winchester Single Shot Rifle.
GUARANTEE.
Colt, Winchester, Marlin, Bullard, Whitney, Smith & Wesson,
Ballard, Stevens, Wesson, Remington and others of first-class manufacture.
Union Metallic Cartridge Co.

10 CARTRIDGES .40 Cal.
CENTRAL FIRE
85 Grs. Powder, 370 Grs. Patched Bullet,
LENGTH SHELL 2 15/16 INCHES.
MANUFACTURED BY THE
WINCHESTER REPEATING ARMS COMPANY,
NEW HAVEN, CONN., U. S. A.
EXPRESSLY FOR THE
BALLARD RIFLE,

U.M.C. TRADE MARK
TEN CARTRIDGES. .40 CAL.
2 15/16 INCH SHELL, 90 GRAINS POWDER.
370 GRAINS PATCHED BULLET.
MANUFACTURED EXPRESSLY FOR THE
BALLARD RIFLE.
(These Cartridges are adapted to Ballard Rifle, using 40-85 Everlasting Shells)
BY THE
UNION METALLIC CARTRIDGE CO.,
BRIDGEPORT, CONN., U. S. A.
In Reloading these Shells use U. M. C. No. 1 1-2 Primer.

For BALLARD & BULLARD RIFLES.
20
Cartridges
32-40 BALLARD
32 Calibre.
40 Grs. Powder.
165 Grs. Lead
PATCHED BULLET.
GUARANTEE.
We hereby guarantee the Ammunition manufactured and sold by The Union Metallic Cartridge Co.;
also the following named Arms when being used with the U. M. C. Cartridges and Shells, which are
made expressly for them, viz.: Colt, Winchester, Marlin, Bullard, Whitney, Smith
& Wesson, Ballard, Stevens, Wesson, Remington and others of first-class manu-
facture.
Union Metallic Cartridge Co.

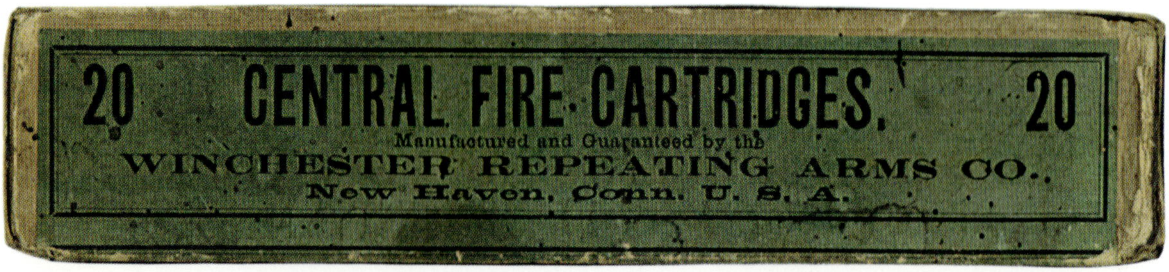
20 CENTRAL FIRE CARTRIDGES. 20
Manufactured and Guaranteed by the
WINCHESTER REPEATING ARMS CO.,
New Haven, Conn. U. S. A.

FOR REMINGTON RIFLE.
20
44 Calibre.
In RELOADING these CARTRIDGES
the only PRIMER that can be used is the
Winchester Primer No. 2½
For convenience in carrying, use ANSON MILLS'
WOVEN CARTRIDGE BELT.

.50-70 GOVERNMENT
50 CENTRAL FIRE
PRIMED METALLIC SHELLS
WITH No. 2½ PRIMER
REMINGTON
UMC
TRADE
MARK
For Government and other Rifles
REMINGTON ARMS-UNION METALLIC CARTRIDGE Co.

PROVIDENCE TOOL COMPANY, PROVIDENCE RHODE ISLAND

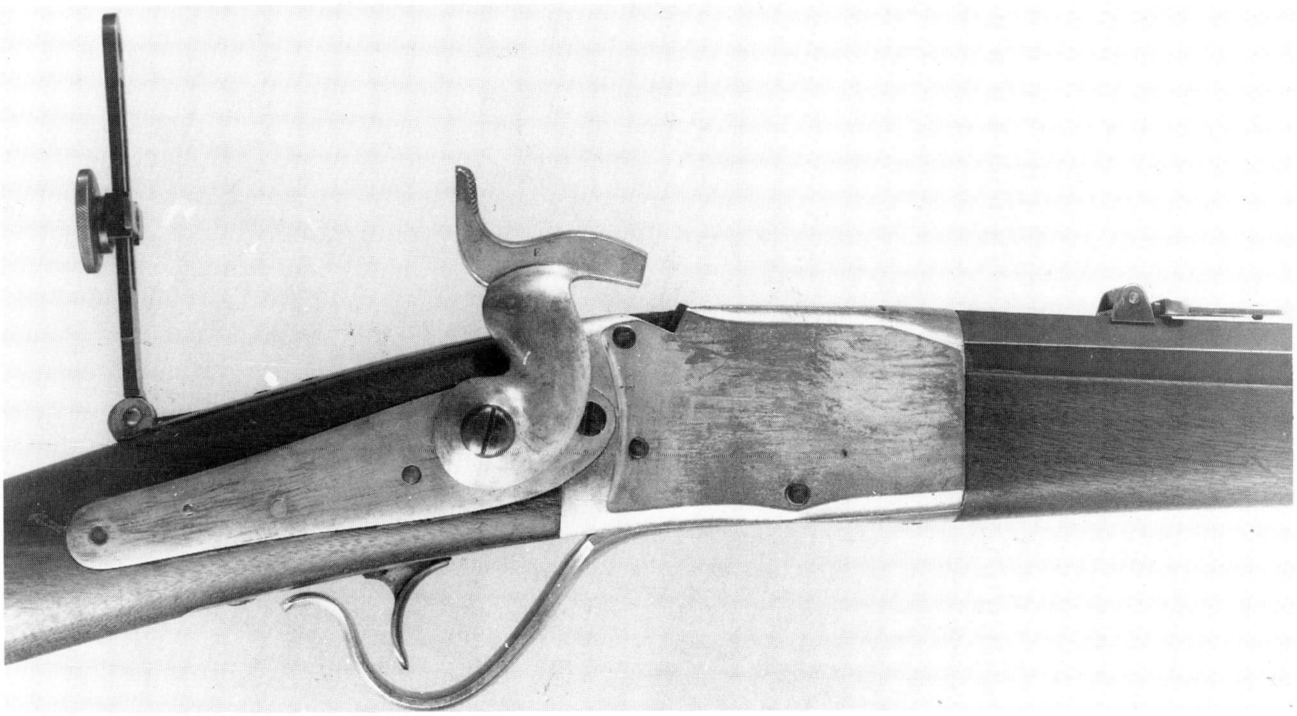

1092. Bousfield, G. T., [*Peabody, H. O.*]. April 19. *Disclaimer.*

FIG. 2.

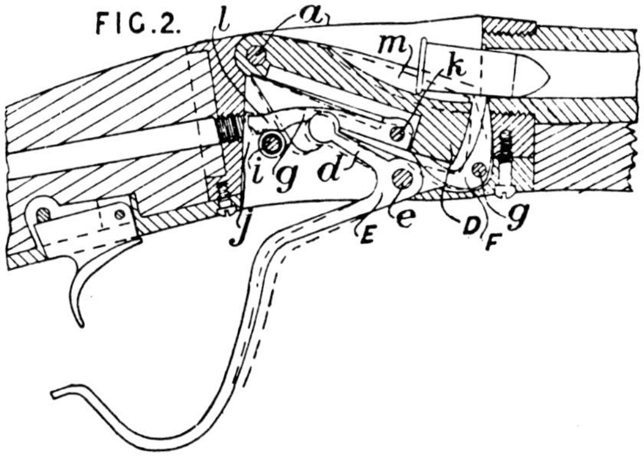

Breech actions, hinged breech-block. — The breech-block D, hinged at *a*, is turned down by the short arm *d* of the trigger-guard lever E pivoted at *e*. The breech-block is depressed a little more than is necessary to open the breech, so as to operate the cartridge-extractor F, pivoted at *g*. The breech-block is lifted again to bring the groove *m* on the upper face of the breech-block into line with the bore by the spring-controlled lever *g* pivoted at *k*, the fixed roller *i* engaging in the depression *l* or the notch *j*, according as the breech is open or closed. By a memorandum of disclaimer, filed July 10, 1871, by J. Batley, the use of the groove *m* in the breech-block is disclaimed.

Providence Tool Company manufactured rifles based on a rear-hinged breechblock invented by Henry O. Peabody that tipped down at the front to expose the chamber, operated by the under lever that also served as the trigger guard. It was one of the strongest of the early breech-loading, single shot systems, and was used to produce military rimfire, and later centerfire, rifles for Canada, Roumania, Spain, France, and the state of Connecticutt. The .50-70 specimen above is one of approximately 300 to 500 sporting rifles believed to have been made on the original outside hammer design. The tinned action surface is original and the rifle has the correct tang and barrel sights. Some recent "conversions" from military weapons have been palmed off as sporters, so it is best to ensure originality before investing.

F. von Martini, a Swiss inventor, modified the Peabody action to a concealed hammer design in 1868 and this action formed the basis for 600,000 Peabody-Martini military rifles made by Providence Tool Company for Turkey and 7,000 for the Japanese navy. It was also used in the Martini-Henry rifle adopted by Great Britain, and the Martini-Henry action target rifles that the company marketed as the "Creedmoor" and "What Cheer" models after famous rifle ranges of the day.

Sources: **Providence Tool Company Military Arms**, Edward A. Hull, 1978. **Patents For Inventions (Small Arms), 1855-1876**, British Patent Office.

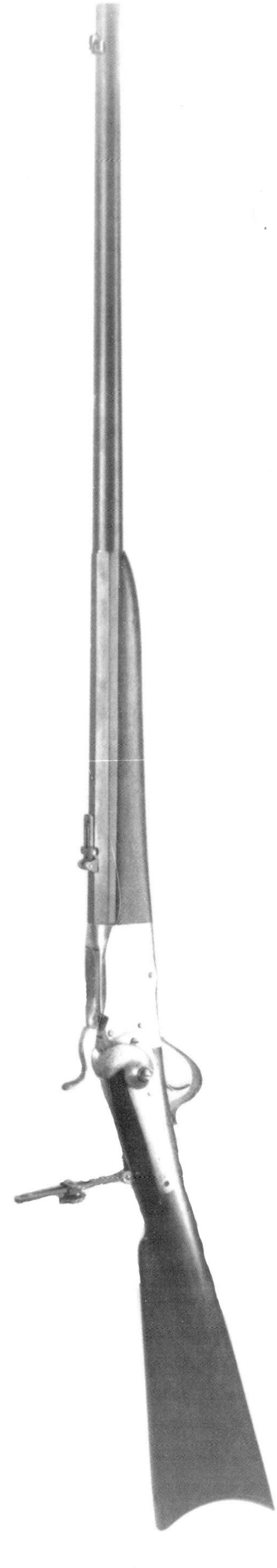

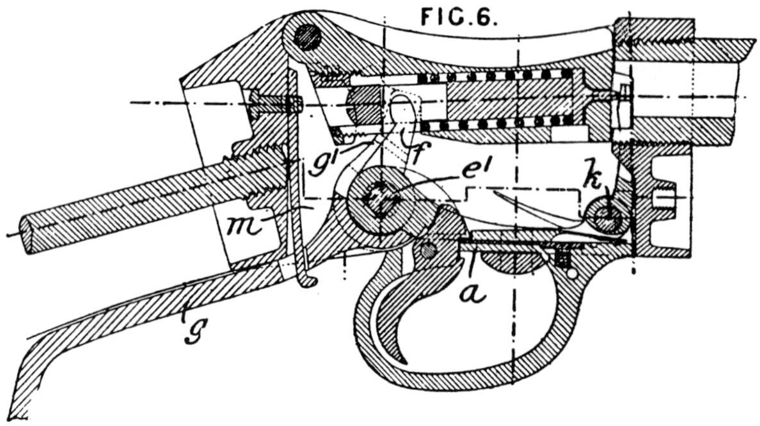

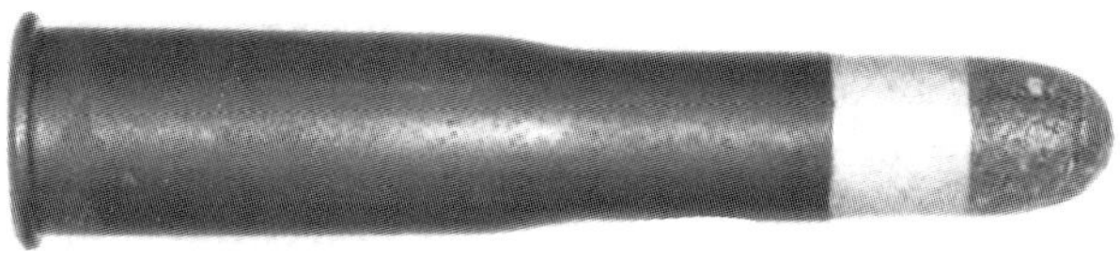

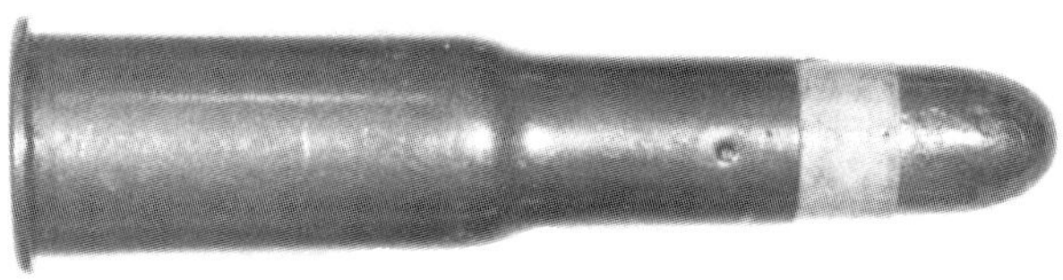

Shown at top is one of F. von Martini's patent drawings for the concealed hammer modification of the Peabody rifle used under license by Providence Tool Company to produce the 11.43x55R Turkish military rifle. Below it is the action of the carbine model, this one chambered for the American .50-70 cartridge. The rim and head dimensions of the three "What Cheer" target cartridges shown following are the same as the Turkish military cartridge. The lower cartridge, the .45 Martin-Henry, is larger in rim and head dimensions than the Turkish and Peabody target and sporting loads. At left is a full-length view of the Peabody sporter chambered for the .50-70 military cartridge.

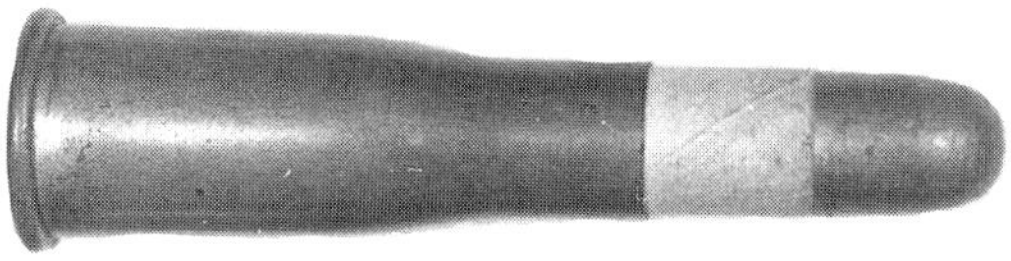

.40-70-330 Peabody "What Cheer" for the Peabody-Martini target rifles named after the rifle range near Providence, Rhode Island. Berdan primed, plain head.

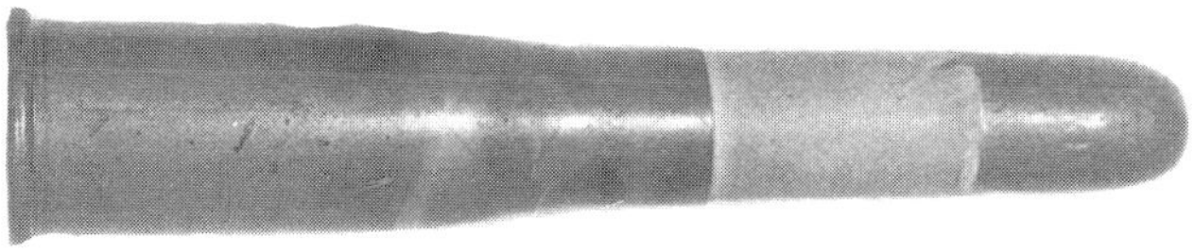

.40-90-500 "What Cheer". This cartridge has the longest of the .40 caliber paper-patched target bullets used in the black powder target rifles.

.44-95-550 "What Cheer." Also was loaded with 100 grains of powder. This cartridge is referred in an 1878 Providence Tool Company catalogue as for the "Creedmoor" rifle.

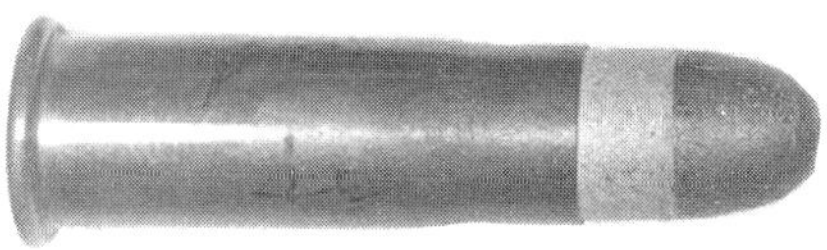

.45-50-290 Sporting. Chambered in a hunting model of the Peabody sporting rifle.

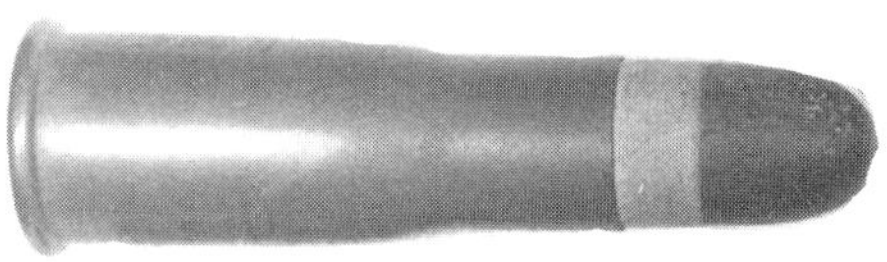

.45 Peabody Carbine. This was produced for a .22" barrel carbine. Turkey purchased several hundred of the carbine through Schuyler, Hartley & Graham. The case capacity is larger than the .45-50. The carbine was also available chambered for .50-70 or .45-70. This cartridge is included to complete the Peabody family of centerfire cartridges.

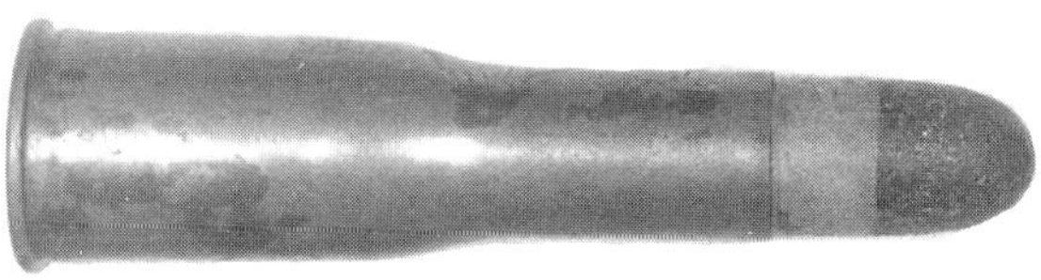

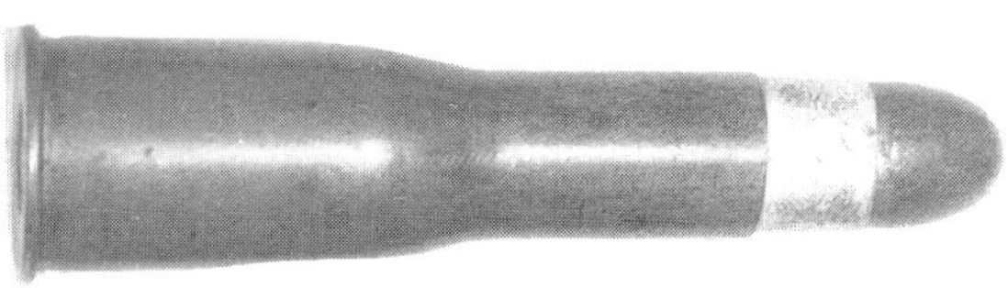

.45-2.2" Peabody based on the Martini-Henry cartridge used by Great Britain in the military rifle, but with shoulder farther back and longer neck. The rifle for which this was made remains unidentified. The lower specimen is the same cartridge but with a slightly different rim and head configuration. The .45 Martini-Henry cartridge in drawn brass form was over capacity for its 85-grain charge so one view is that an attempt was made to modify the cartridge by shortening the powder chamber.

W. MILTON FARROW

W. Milton Farrow patented this falling block, single shot rifle action on October 14, 1884 and followed it with improvements in 1887 (U. S. 306,391 and 372,213). The hammer on this specimen can be seen just barely clearing the frame at top. The reversed front trigger is pressed forward to set the rear one for hair letoff. With its full octagon 14-groove, 30" barrel, chambered for .32-40 Ballard, the rifle weighs 11 lb. The globe front and tang sights are Farrow's own design, as is the rosewood operating lever grip and Schuetzen buttplate.

Milton Farrow's rifle plants were located variously at Holyoke, Massachusetts, Morgantown, West Virginia, Washington, D. C. and at West Palm Beach, Florida where he lost his business to a hurricane in 1928. He was well known in the United States and Europe as a competitive marksman. He lived to 1934, age 86.

Source: **More Single Shot Rifles**, James J. Grant, 1959.

WINCHESTER REPEATING ARMS COMPANY

The group of Winchester experimentals below barely scratches the surface. You could devote a lifetime to this pursuit and still never acquire them all. The basic source of information is Paul Foster's **Winchester's Forgotten Cartridges** published in the 1952 **GUN DIGEST** and re-issued by Bob and Beverly Strauss several years ago. Paul didn't get them all, either. You will find the .31-62 W. C. F. and .46 W. C. F. in the following pages, and one in this group below that are newcomers.

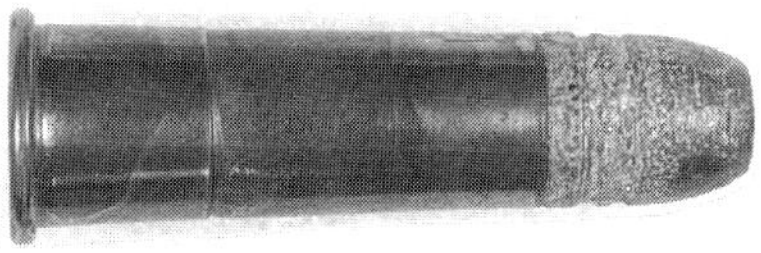

.46 O. F. Winchester patent 60,814, January 1, 1867. This is a two-piece copper case with the rimmed cup over the case body. The primer is within the cup and the case, the head dished out to thin the metal. The charge is compressed black powder with a hole running through its length. Cody Firearms Museum Curator, Howard Madaus, tried this specimen in the early experimental rifles and reported that it chambers perfectly in the 1867 Swiss trial lever action rifle but slides somewhat too far into the chamber of the English trial series lever action, both of which were based on the Model 1866 rifle. (See Vol Two, p. 28 for two others, one of which is necked with a larger case body and probably was made for the English trial rifle).

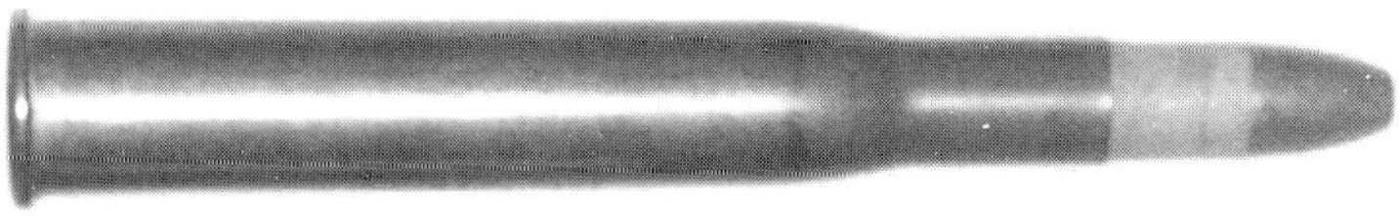

.32-80-165. This is necked down from the .40-90 Sharps Straight, so headstamped. Case length is 3".

.32-70 U. S. Navy rimless. Headstamped ".32 U.S.N". With full metal jacket bullet, this was undoubtedly a military experimental only. Paul Foster reports a rimmed version but had none to show in his article.

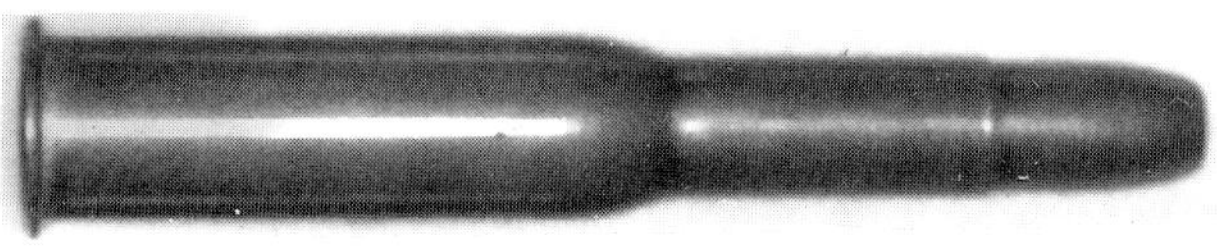

.35-2.619, full-copper jacketed, flat-nosed bullet, plain Winchester head. The head is .605", head .504", neck .377", bullet .353. This came from the collection of J. P. Gibault, Williams Lake, British Columbia back in the 1970s. Mr. Gibault had been a Winchester employee. The specimen is similar to the two necked down Murata specimens on p. 216, Volume Two, but the rim and head are much smaller than those. Oddly enough, it has the same rim and head size as .31-62 shown under the Model 86 cartridges following.

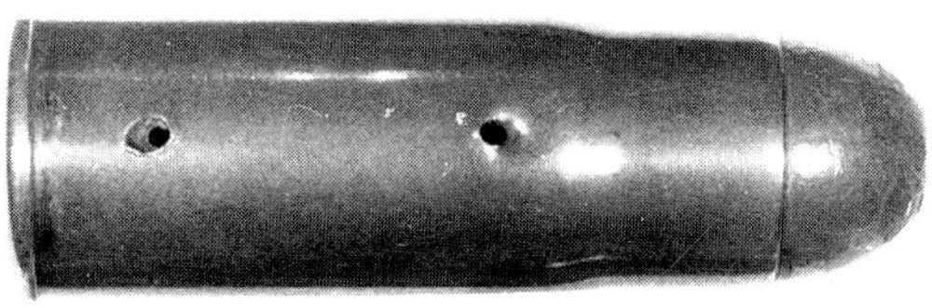

.70-150. The is a dummy from the Big "W" Winchester display board. It is the copper-tube hollow point with raised "X" on the nose. No headstamp, this thick-rimmed specimen, either dummy or loaded, is the only one recognized as authentic. Thin-rimmed versions made up later are often seen, even with the proper raised "X" copper tube in the bullet.

WINCHESTER REPEATING
FIRE ARMS
AND
TRADE H MARK
METALLIC
CARTRIDGES
TRADE H MARK
MANUFACTURED BY
THE
WINCHESTER
REPEATING ARMS
COMPANY
NEW HAVEN
CONNECTICUT
1875

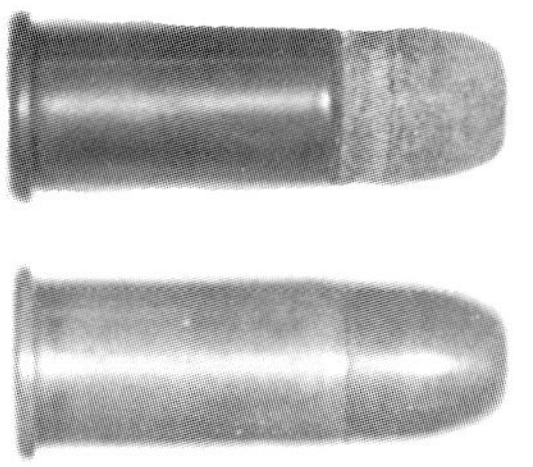

.44-200, .875", no headstamp. This is the center fire version of the .44 Henry rim fire, reportedly for rifles made for Brazil. The cartridge was also loaded by Societe Francais Munitions (S.F.M.). The second specimen is a hollow jacketed dummy.

Oliver Winchester's first "cartridge" was included in his U. S. patent no. 57808, September 4, 1866 below describing changes to the Henry rifle from which came the first Winchester, the Model 66. This was just a chamber piece, or "stopper", with percussion nipple to allow loading from the muzzle with loose powder and bullet.

It is sometimes desirable to use this arm as a muzzle-loader. For this purpose I construct a steel case, S^5, as denoted in Figs. 9 and 10, formed to fit the rear end of the barrel like the shell of a metallic cartridge, placing in the rear end of the said tube a cone-seat to receive an ordinary percussion-cap. The arm thus arranged may be loaded from the muzzle in the usual manner, while the breech-pin is held against the said case. When thus loaded (or, if preferred, before the loading is commenced) place a percussion-cap upon the cone, then cock the arm, as before described, and release the hammer, which, on striking the piston M, will explode the percussion-cap and ignite the powder within the arm. In order that the latch and points on the breech-pin and spindle may not interfere with this operation I cut recesses in the flange of the said case, as seen in Figs. 9 and 10.

O. F. WINCHESTER.

Witnesses:
JOHN E. EARLE,
M. A. HINE.

Archaeologist Charles Adkins, Bureau of Land Management, reports that re-primed rimfire .44 Henry cartridge cases were found in the Eskimo village of Aniganigaruk, Central Brooks Range, Alaska. The report states that a T-shaped slit was cut across the head of the shell into the powder chamber, and one wing of a "G. D." (top hat) percussion cap with two wings removed was inserted into this slit. The remaining wing was folded over the rim to hold it in place. The cartridge had to be positioned carefully in the chamber so that the cap was struck by the Henry or Model 66 rifle's rim firing pin. The cartridge cases are now part of the artifact collection of the University of Alaska at Fairbanks. (Illustrations courtesy Mr. Adkins).

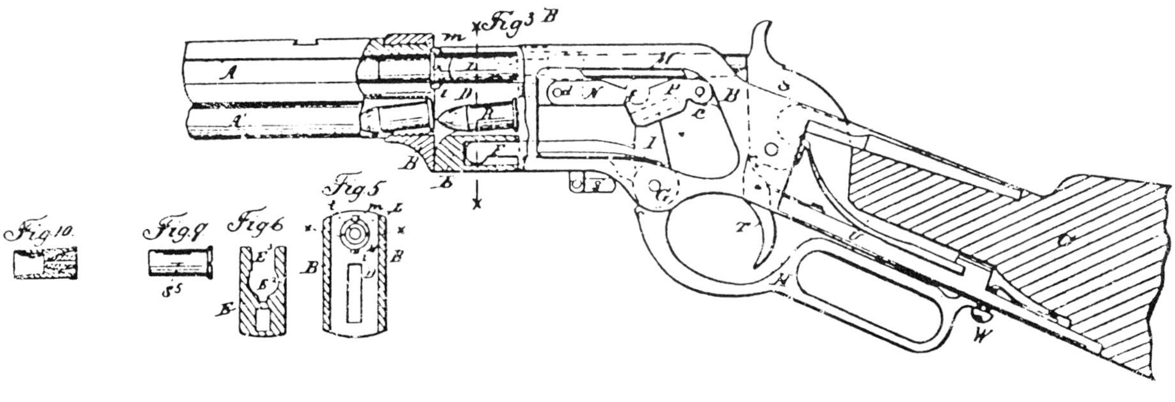

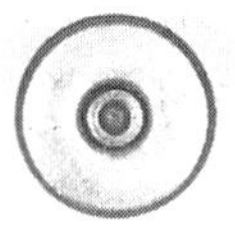

44-40-200. Originally announced by Winchester as the the ".44 Winchester Model 1873," it soon became known as the the .44-40 and has remained so. The first cartridges were loaded with the Milbank primer that looked like it had been struck by the firing pin. (See patents appendix). This was soon dropped as being unreliable and replaced by a rounded boxer-type primer. The concern that centerfire cartridges would be accidentally set off in the magazine by recoil was solved by the flat-nosed bullet.

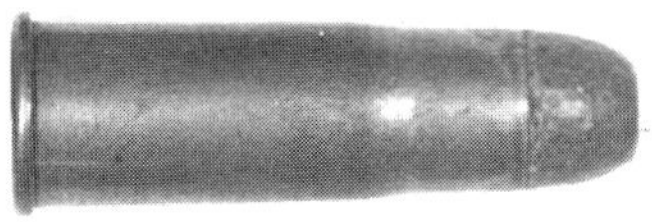

.38-40-180. Necked down version of the .44-40 which gave higher velocity.

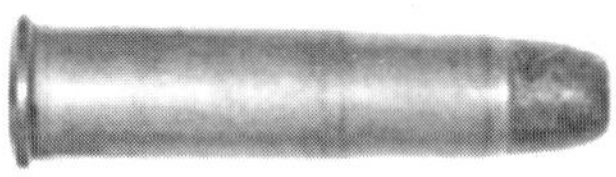

.32-20-80 or 100. This little cartridge gives still higher velocity than the two above, but is far less suitable for any but small game because of the light bullets.

MUSKET AND CARBINE.

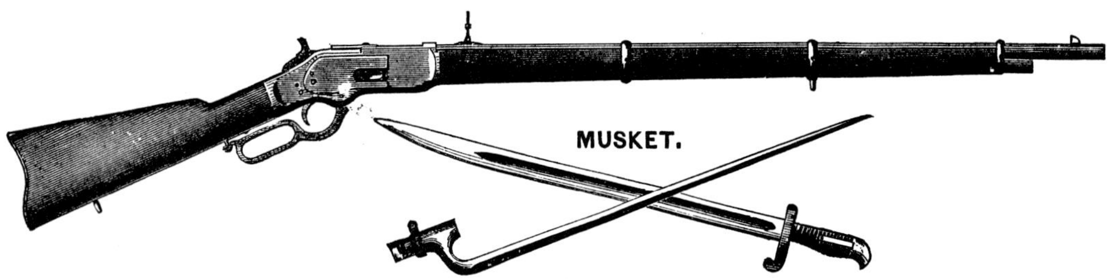

MUSKET.

Price	$26.00	Weight	9½ lbs.
Length of Barrel	30 inch.	Angular Bayonet	$3.00
Caliber	.44	Saber	4.00
Number of Shots	17		

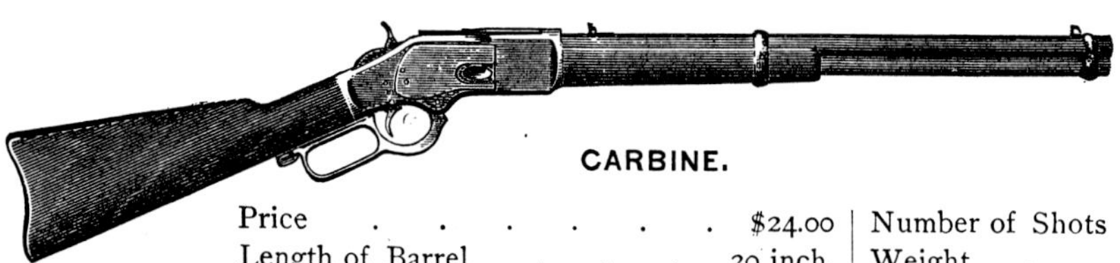

CARBINE.

Price	$24.00	Number of Shots	12
Length of Barrel	20 inch.	Weight	7¼ lbs.
Caliber	.44		

This Model can only be used with Central Fire Cartridge .44, W. C. F.

Powder, 40 grains. Bullet, 200 grains. Price per 1,000, $20.00. Case contains 2,000.

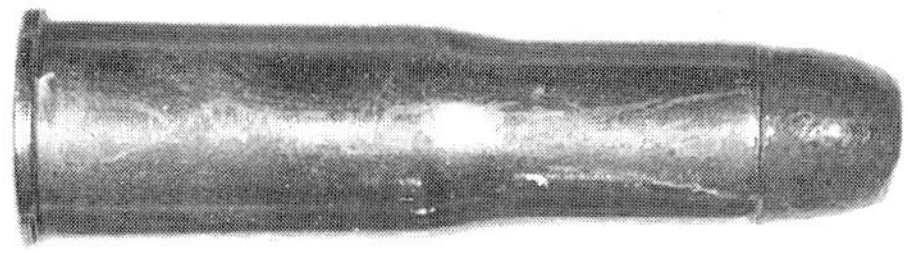

45-75-350. Introduced along with the rifle in 1876.

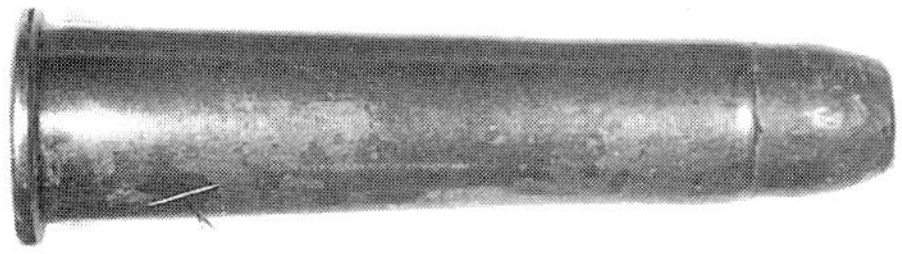

.40-60-210. Also introduced with the rifle in 1876.

.45-90-500 Winchester Special. An experimental cartridge for the Model 1876 rifle announced in the 1878 catalogue but not produced in any quantity. It was apparently meant for use as a long range match use. Because of its length it would have had to be loaded singly. Paul Foster shows a specimen with blunt nose bullet.

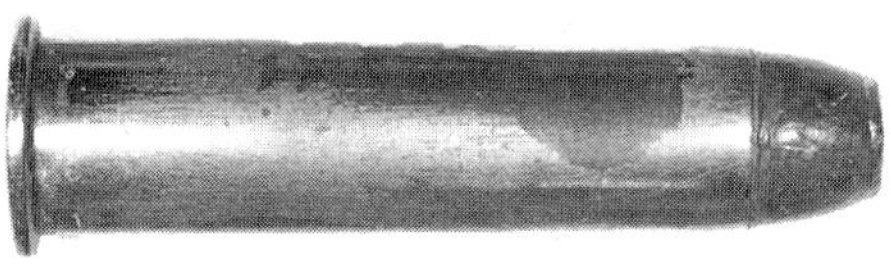

.45-60-300. This cartridge was added to the Model 76 line in 1879.

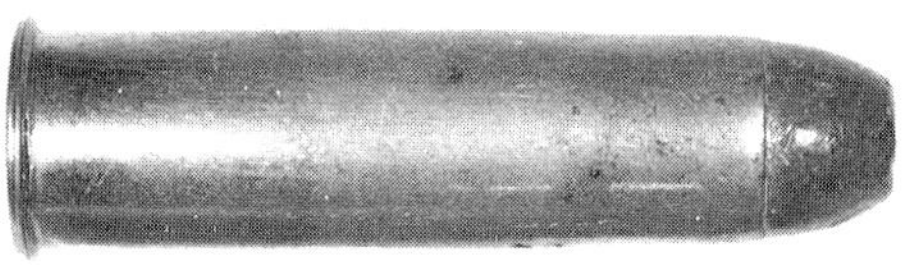

.50-95-300. The most powerful of the four standard cartridges for the Model 76, also introduced in 1879.

CENTENNIAL, OR MODEL 1876.

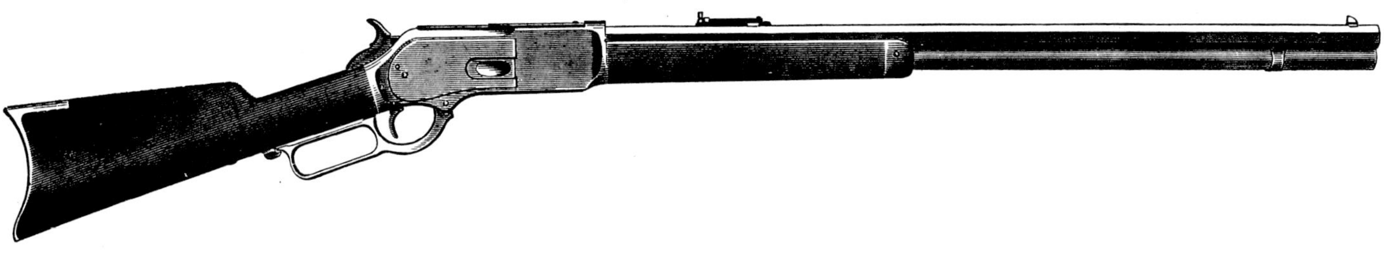

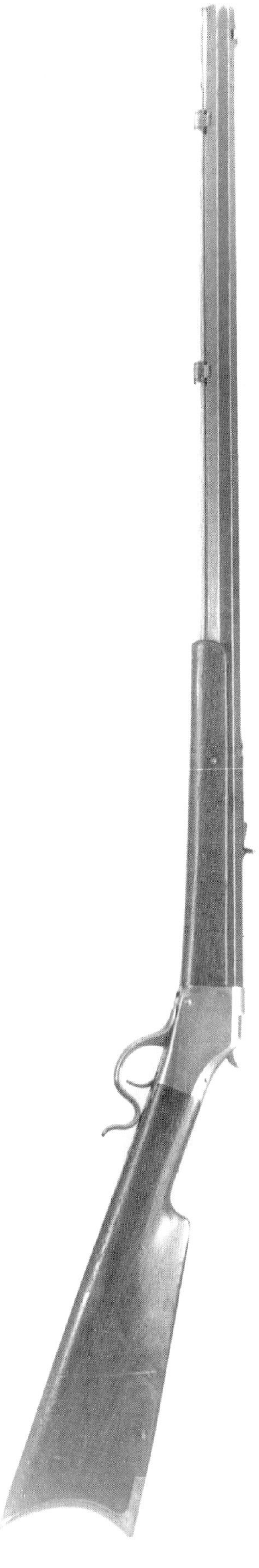

FROM BROWNING TO WINCHESTER

At left is a Pacific Model Browning Brothers, Ogden, Utah rifle chambered for the .40-70 Sharps bottle necked case. The rifle was developed and patented by John Moses Browning and produced by him and his brother, Mathew. Above is another with double-set triggers chambered for the .40-90 Sharps bottle necked case. Both rifles were supplied with rarely-seen wood cleaning rods. Browning also produced and sold an assortment of reloading tools and supplies—bullet mould, ladle, powder measure, primer punch, screwdriver, even strips of lead.

Winchester purchased the single shot patent from Browning Brothers in 1883 and created their Model 1885 rifle, little changed from the Browning design. The Model 1885 Single Shot with high wall or low wall action was made for over 60 cartridges from .22 Short to .50-110--more than for any other Winchester rifle. The Winchester Single Shot was discontinued not long after WWI but remained a popular target rifle, particularly in .32-40, and many are still in service with single shot enthusiasts.

WINCHESTER SINGLE SHOT RIFLE.

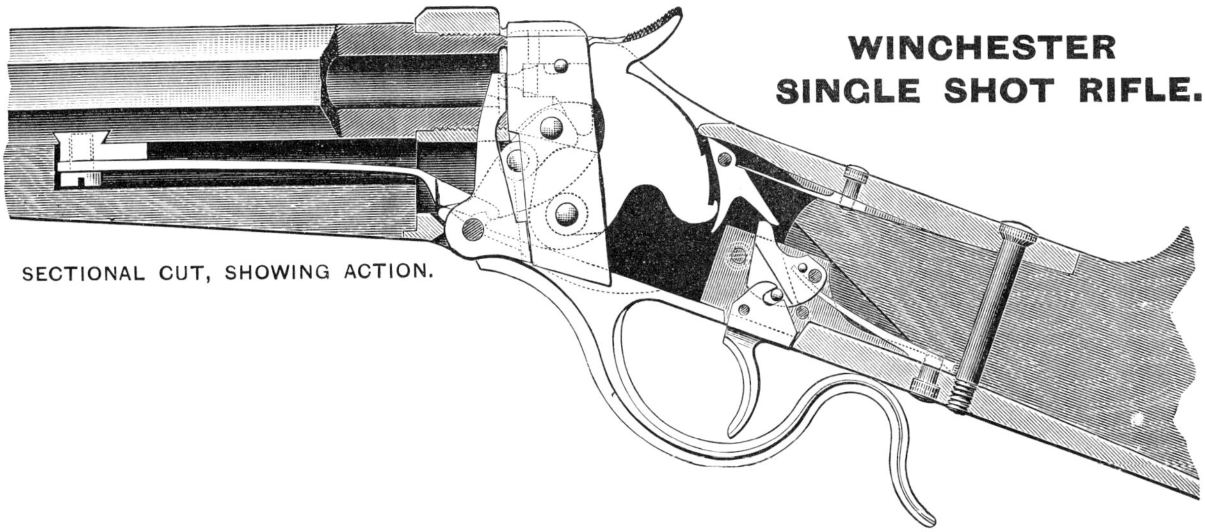

SECTIONAL CUT, SHOWING ACTION.

Lightweight Single Shot Carbine.

Fifteen Inch Round Barrel, Plain Trigger, Straight Grip Stock and Forearm of Plain Wood, Sling Ring on Frame, Caliber .44 W. C. F., Weight 4¼ pounds, ... $14.50

Plain Sporting Rifle.

Octagon Barrel, Plain Trigger, Plain Walnut Stock, Rifle or Shotgun Butt Stock, Blued Frame, ... $15.00

Special Sporting Rifle.

Octagon Barrel, Plain Trigger, Fancy Walnut Checked Pistol Grip Rifle or Shotgun Butt Stock, ... $30.00
Shotgun Butt Stock, with either metal or rubber butt plate, same price as Rifle Butt Stock.

SINGLE SHOT

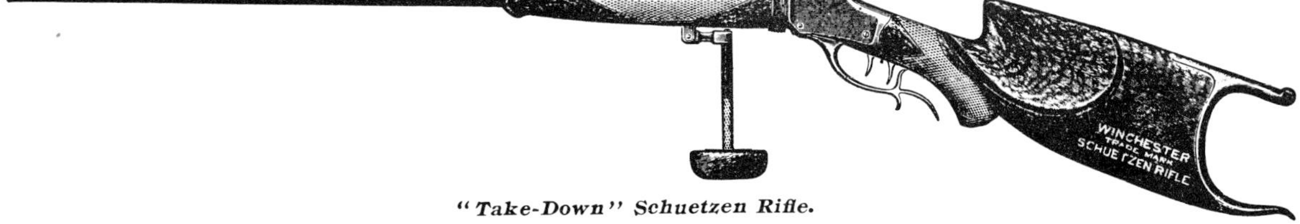

"Take-Down" Schuetzen Rifle.

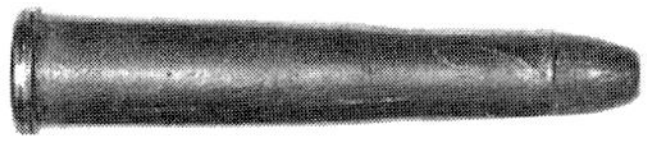

.22-13-45 Winchester Single Shot. Introduced with the Model 1885 single shot rifle. It was among 31 centerfire and 10 rimfire chambers offered in the October 1886 catalogue.

.38-90-217, 3 ¼". First offered in the October 1886 catalogue, Winchester always referred to this cartridge and its companion loads below as Winchester Express Single Shot.

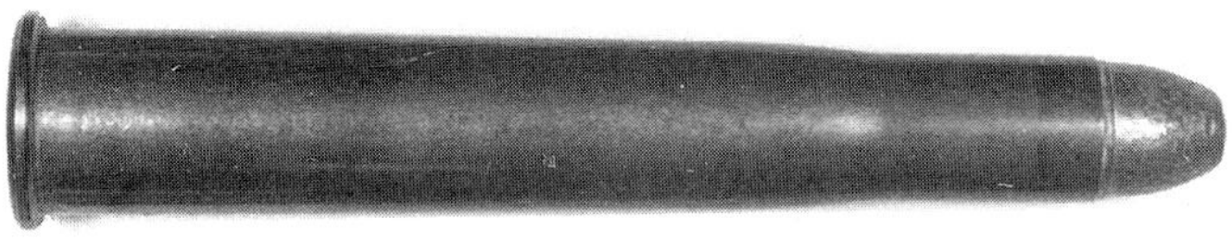

.40-110-260, 3 ¼". The same case as above but necked up to .40.

.45-125-300, 3 ¼". The Model 1885 rifle was offered chambered for all three of these loads through 1909. They do not appear in the 1910 catalogue.

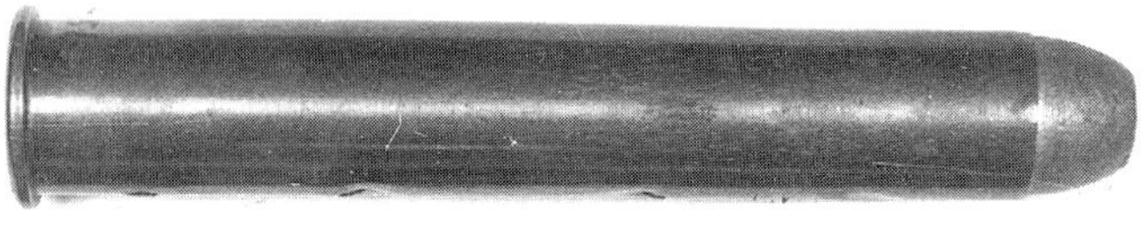

.50-140-473. This was an experimental and neither rifles no cartridges were listed or sold. Most surviving cartridges are the four-holed Winchester cartridge board dummy. However, loaded specimens exist.

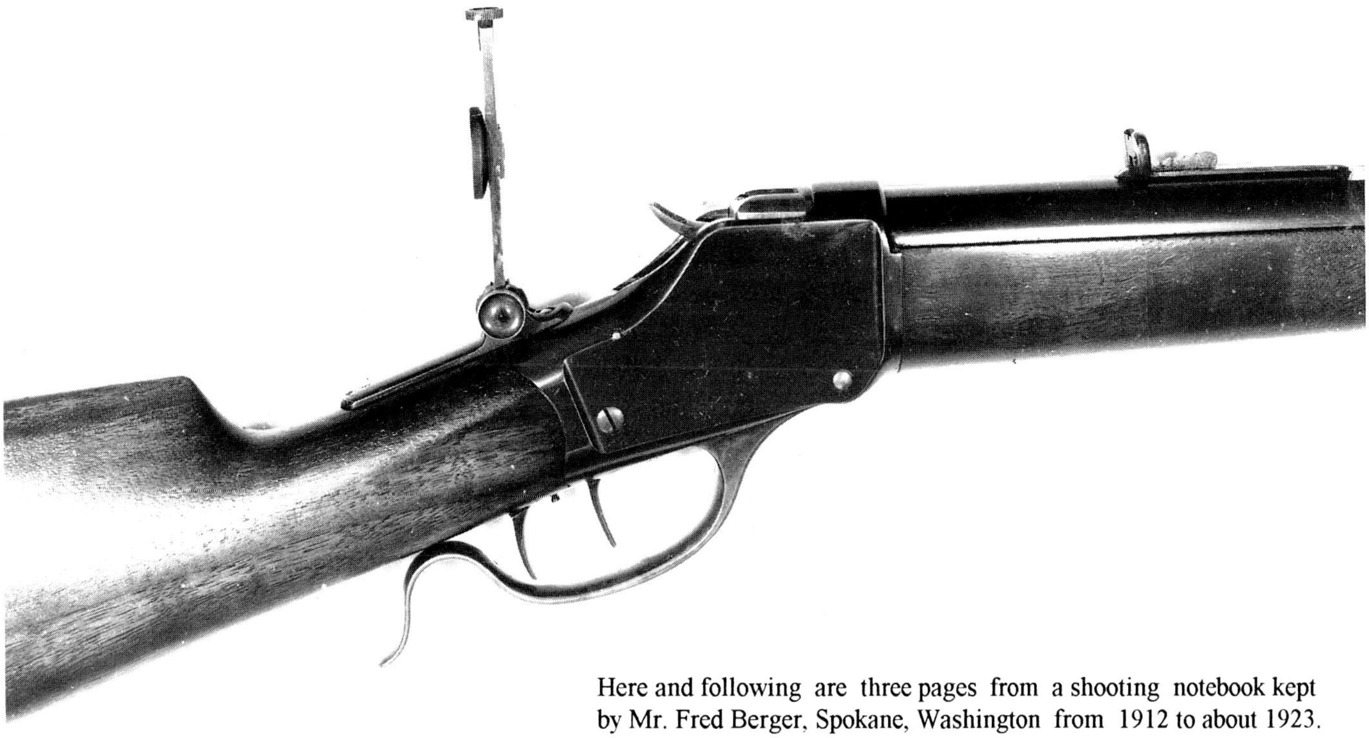

Here and following are three pages from a shooting notebook kept by Mr. Fred Berger, Spokane, Washington from 1912 to about 1923. The targets were shot with his Model 1885 Winchester single shot .32-40. The 6"x9"40-page notebook is leather bound with a Peters Cartridge Company logo on the cover. The rifle is a Model 1885 with 30" No. 4 barrel chambered for .38-55, with double set triggers. The tang sight is original equipment.

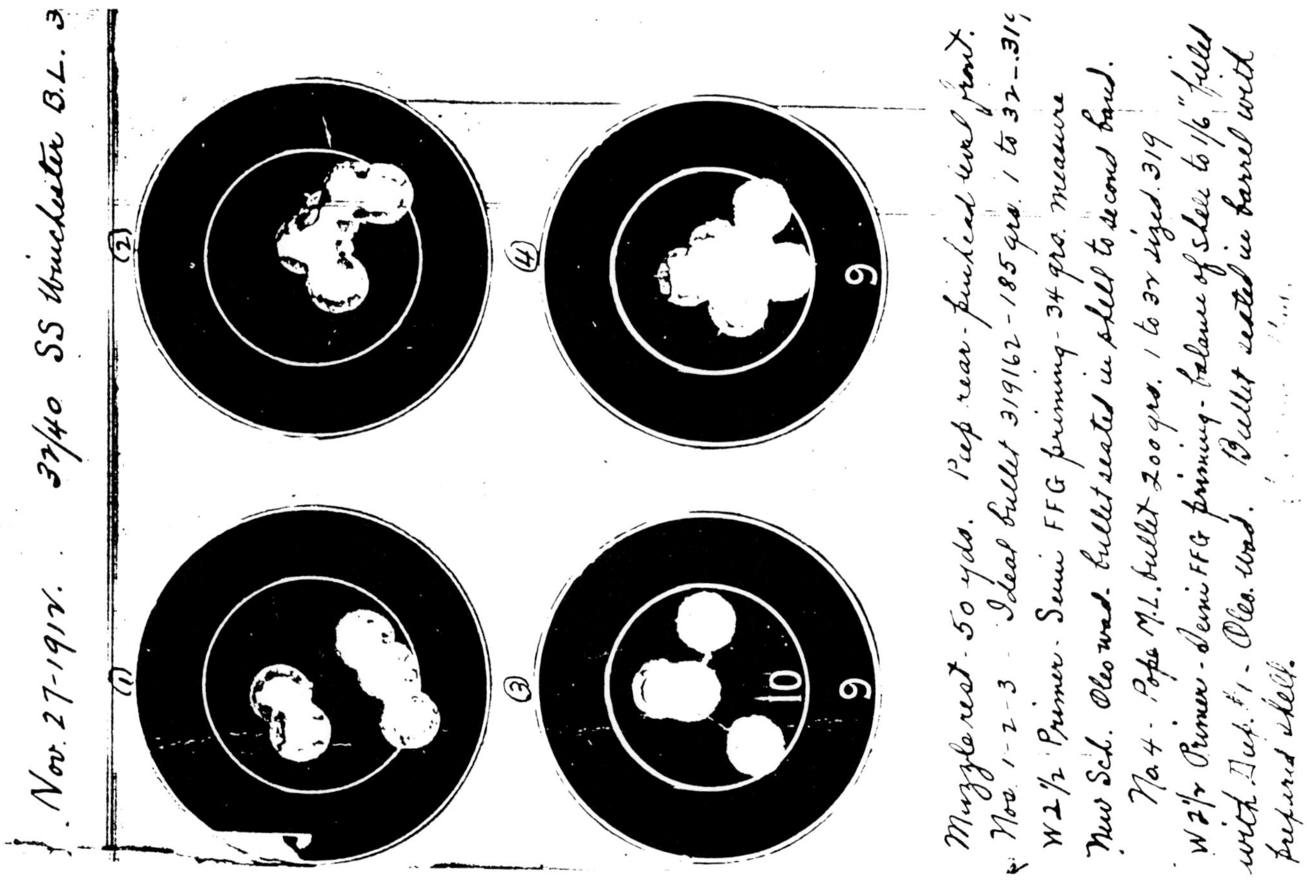

32/40 Win. S.S. 30" #4   Nov. 18-1914.

Weather cold and damp.
Light changeable.

Lyman fine bead front. Lyman W.S. rear. Watson eye cup

50 yards - rest.
5 shots each target.
Pope bullet 1-36 - sized 31?
Home made lub. No wad
seated with prepared shell
Primer W 2½
Priming SS FFG 5 gr. m
Load New Sch. 32 gr. m
Ideal measure.
Off shot was poor hold
not being able to see well

32/40 Win S.S. 30" #4.   Aug. 15-1920.
The ranch.

Weather warm - Sunny - Light clear.
Distance 75 yds. muzzle & elbow rest
Win. A5 + scope E 76 W57½ Focus 1-5½

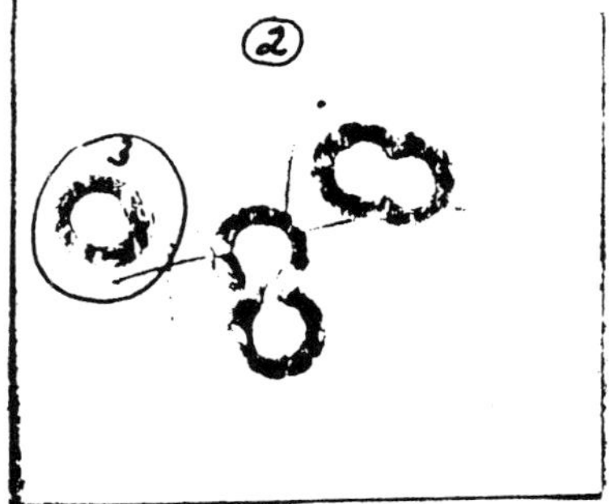

Fired 1 shot for fouling and sighting. Load for targets 1 and 2 the same, - UMC #2½ prime 5 grs. meas #80 - 32 grs. meas. FFG Semi. no wad. Pope bullet 1-32 sized 321 - White lub. Seated with prepared shell. Used W shell

Shot #3 target #2 was a left hold. It struck where aimed in spite of the fact that it was a tipper

Gun was not cleaned between shots. Same shell used for all shots. Used Ideal measure.

THE WINCHESTER MODEL 1886 CARTRIDGES

The Winchester Model 1886 first appeared in October 1886. Four cartridges were illustrated with it—.40-82-260, .45-70-500, .45-70-405, and 45-90-300.

In November 1887 the following were added--.38-56-255, .40-65-260, .50-110-300.

In April 1894 the following appeared in the catalogue with the Model 83--.40-70-330, .38-70-255, .38-56-255.

In 1889 Winchester first listed the .45-82-405, the .45-85 with 300-grain hollow point express and 350-grain solid bullets in the .45-90 case. These loads bear the .45-82 and 45-85 headstamps but are not shown in the catalogues. In August 1895 Winchester added the .50-100-450. Then the .33 Winchester was added in 1902.

By October 1911 Winchester had trimmed the cartridges available in the Model 86 to .33 Winchester, .45-70 with four bullet weights, .45-90 and .50-110.

The 1918 catalogue lists only .33 W.C.F. and .45-70.

The Model 86 was made to 1936 when it was replaced by the Model 71 in .348 caliber. Ammunition for the Model 86 chamberings was continued for many years.

SPORTING RIFLE, MODEL 1886.

Twenty-six Inch Octagon Barrel, Full Magazine, Plain Trigger, Case-hardened Trimmings, ... $21.00

FANCY SPORTING RIFLE, MODEL 1886.

Twenty-six Inch Octagon Barrel, Full Magazine, Plain Trigger, Case-hardened Trimmings, Fancy Walnut Checked Pistol Grip Stock, ... $36.00

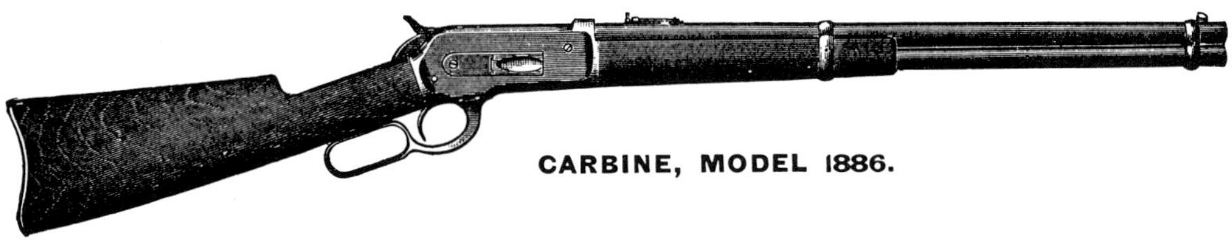

CARBINE, MODEL 1886.

Twenty-two Inch Barrel, Full Magazine, Plain Trigger, Case-hardened Trimmings, weight about 8 pounds, ... $19.00

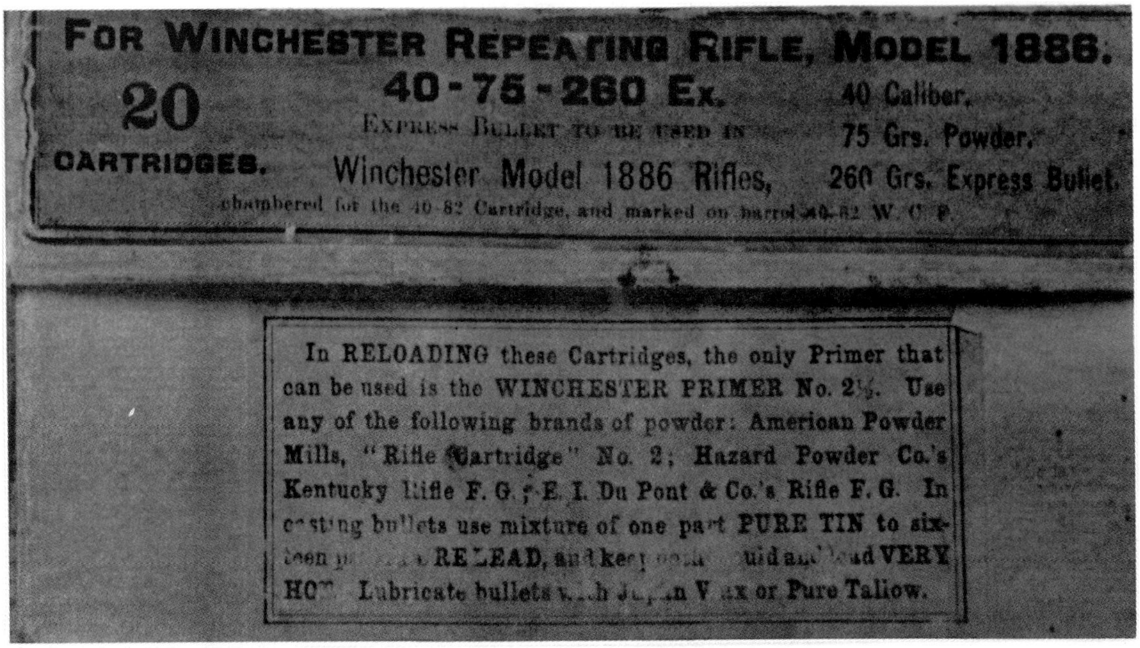
For Winchester Repeating Rifle, Model 1886.
20
Cartridges.
40-75-260 Ex.
Winchester Model 1886 Rifles,
40 Caliber.
75 Grs. Powder.
260 Grs. Express Bullet
In RELOADING these Cartridges, the only Primer that can be used is the WINCHESTER PRIMER No. 2½. Use any of the following brands of powder: American Powder Mills, "Rifle Cartridge" No. 2; Hazard Powder Co.'s Kentucky Rifle F. G.; E. I. Du Pont & Co.'s Rifle F. G.

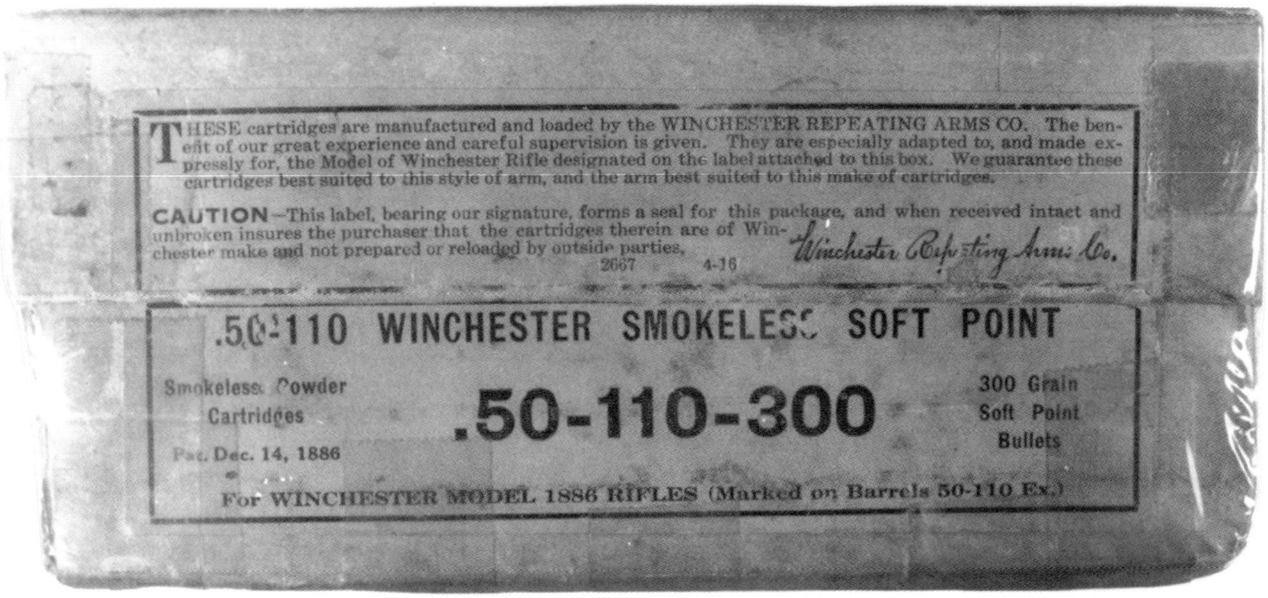
THESE cartridges are manufactured and loaded by the WINCHESTER REPEATING ARMS CO. The benefit of our great experience and careful supervision is given. They are especially adapted to, and made expressly for, the Model of Winchester Rifle designated on the label attached to this box. We guarantee these cartridges best suited to this style of arm, and the arm best suited to this make of cartridges.
CAUTION—This label, bearing our signature, forms a seal for this package, and when received intact and unbroken insures the purchaser that the cartridges therein are of Winchester make and not prepared or reloaded by outside parties.
Winchester Repeating Arms Co.
2667 4-16
.50-110 WINCHESTER SMOKELESS SOFT POINT
Smokeless Powder Cartridges
Pat. Dec. 14, 1886
.50-110-300
300 Grain Soft Point Bullets
For WINCHESTER MODEL 1886 RIFLES (Marked on Barrels 50-110 Ex.)

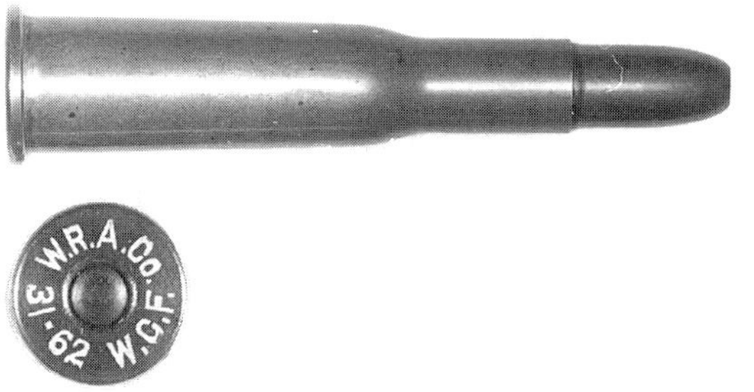

.31-62-200 W. C. F. This is a smokeless powder, copper-jacketed bullet experiment for a Model 1886 carbine in the Winchester Arms Collection at Cody, Wyoming. Records show that this carbine, serial no. 62657, was delivered to the Winchester warehouse in October 1891. Herbert Houze, then curator of the museum, described the rifle in response to an inquiry as having rapid-tapered barrel, full-length magazine tube, and a forend thinner than normal design for the Model 1886. Dimensions of the cartridge are: rim, .605", case 2.102", overall 2.752". (Source on the carbine--The GUN REPORT, Oct. 1994).

.33-200. This smokeless powder cartridge was added to the Model 86 calibers in 1902 and retired in 1936 when the Model 71 in .348 Winchester replaced the Model 86.

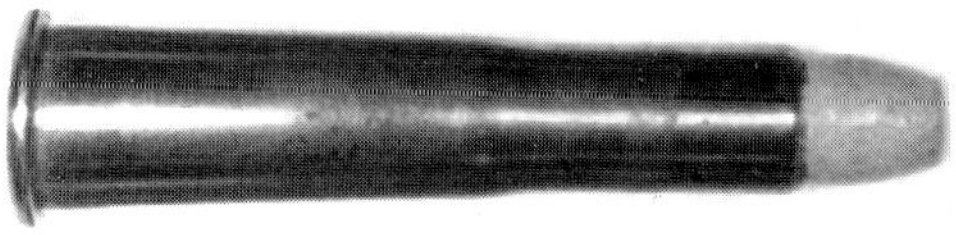

.38-56-255. A case type added in 1887 and rifles were chambered for it until 1911.

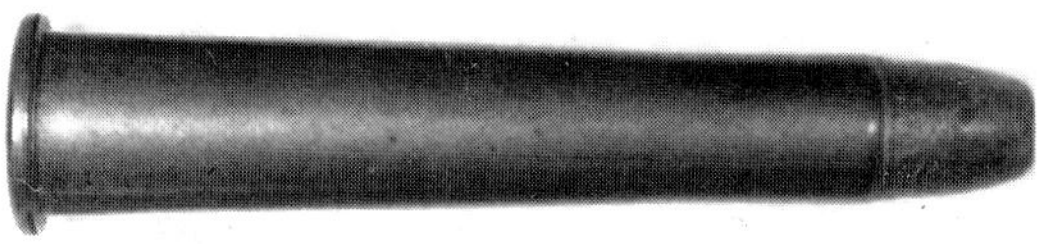

.38-70-255. This cartridge was added in 1894 and discontinued in 1911.

.40-65-260. Added to the Model 1886 in 1887 and available until 1911.

.40-70-330. Another chambering for the Model 1886 added in 1894, discontinued in 1911.

.40-75-260 Express. This is the .40-82 case head-stamped .40-82 W. C. F. with 75 grains of powder and a hollow point, cupper-tubed lead bullet.

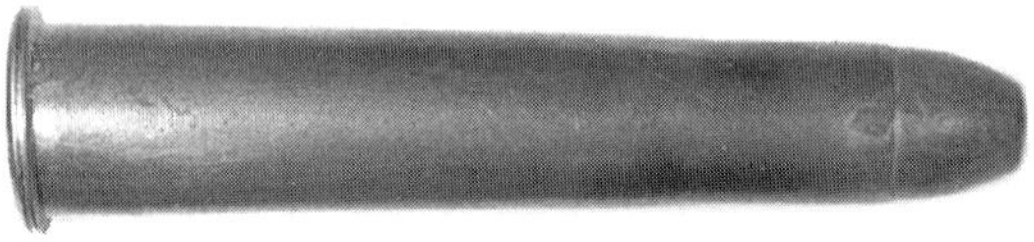

.40-82-260. This was one of the first chamberings for the Model 86 announced in October that year. Rifles chambered for it were made until 1911.

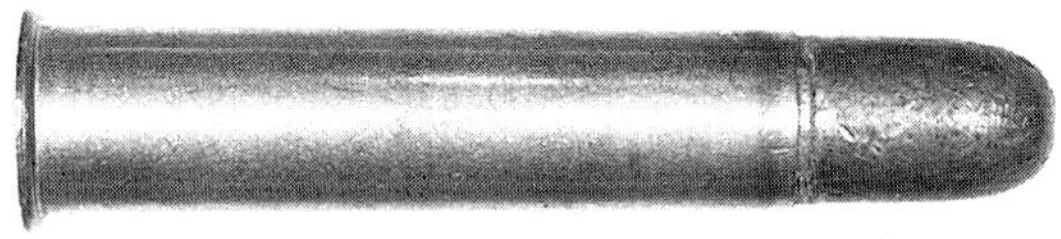

.45-70-405 or 500. In the October 1911 catalogue the .45-70-350 and a .45-70-330 hollow point loads were available. In 1918 the smokeless powder load, marked W. H. V. (Winchester High Velocity) with jacket softpoint bullet was made available.

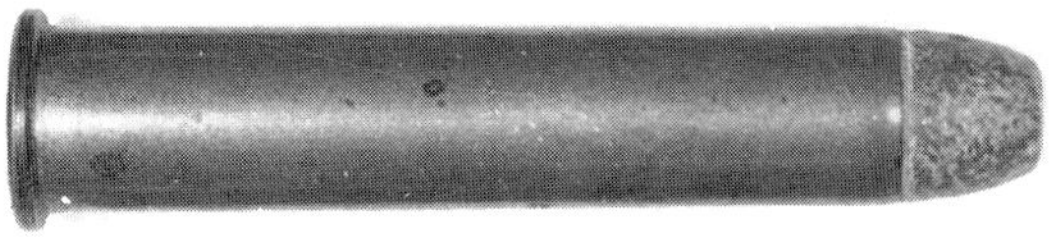

.45-82-405. Not illustrated in the catalogues but was offered from 1889 to 1911 as an optional bullet and charge variation under .45-90.

.45-85-300 gr. hollow point express or 350-gr. solid lead loading in the .45-90 case.

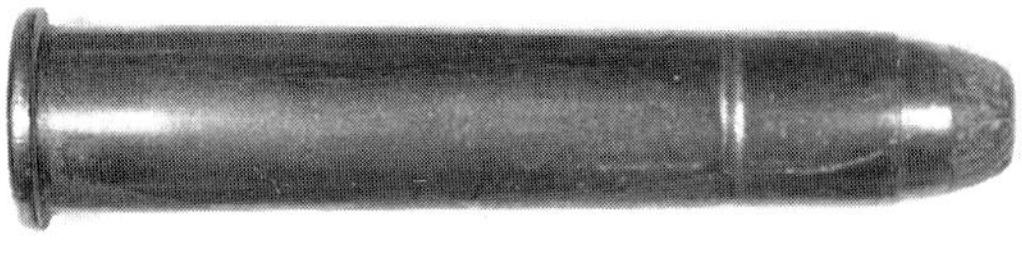

.45-90-300. One of the first cartridges offered in October 1886 and rifles available until 1911. It was made in the W. H. V. smokeless loads and loaded for many years after the rifle was discontinued.

.46-50-400 W. C.F. An experimental from 1912 for a Model 1886 rifle that was never marketed. Shown are three variations--at top with copper-jacketed hollow-nosed bullet closed with a raised "X" copper tube, and copper protected primer; the second and third with slightly shorter case and lead bullets. The lower one, originally in the H. P. White collection, has the plain head. Case length of the final version is 2.406". The other two are 2.3".

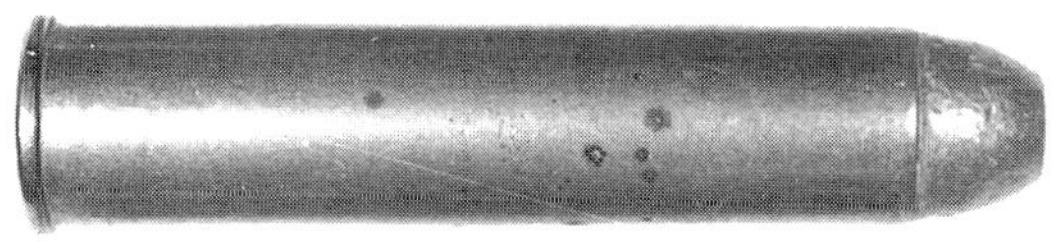

.50-100-450. Same case as the .50-110-200 shown below was first illustrated in the August 1895 catalogue.

.50-105-300 Express, hollow-point copper-tube bullet with the raised "X" on the nose. An experimental based on the .50-110 case. Above is the two-hole display board dummy with wood spacer within. Below is the loaded round.

.50-110-300. Introduced in 1887, this cartridge was soon offered in a smokeless load with jacketed, softnose bullet as Winchester High Velocity. It lasted until 1911 when it and several other cartridges for the Model 86 were discontinued. From then on the rifles were not made in this caliber but the cartridges were produced for many years. Shown below is a shot load with case mouth crimped over a card wad.

.25-20-60 or 86 W. C. F. (Winchester Center Fire)

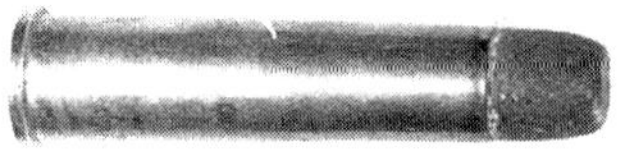

.32-20-80 or 100 W. C. F.

.38-40-180 W. C. F.

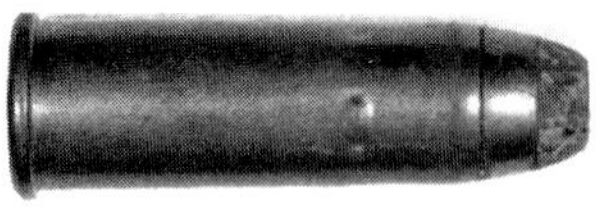

.44-40-200 W. C. F.

MODEL 1892. .25, .32, .38, AND .44 CALIBERS.

CARBINE, MODEL 1892.

20 inch Barrel, Full Magazine, Plain Trigger, weight about 5¾ lbs., number of shots 12, $17.50

SPORTING RIFLE, MODEL 1892.

24 inch Octagon Barrel, Full Magazine, Plain Trigger, weight about 6¾ lbs., number of shots 15, $19.50
24 inch Round Barrel, Full Magazine, Plain Trigger, weight about 6¾ lbs., number of shots 15, 18.00

FANCY SPORTING RIFLE, MODEL 1892.

24 inch Octagon Barrel, Full Magazine, Plain Trigger, Fancy Walnut Checked Pistol Grip Stock, weight about 7½ lbs., number of shots 15, $34.50
24 inch Round Barrel, same style, weight about 7¼ lbs., number of shots 15, 33.00

All of the above can be furnished with Half Magazine at same list price.

For Extras, see page 57.

Shot Gun Butt Stocks, with either metal or rubber butt plate, same price as Rifle Butt Stocks.

Send for our handsomely illustrated 24 page catalogue showing the engraving and checking on fancy guns.

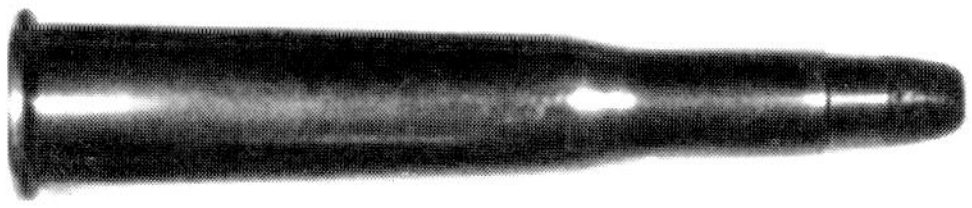

.25-35-117.

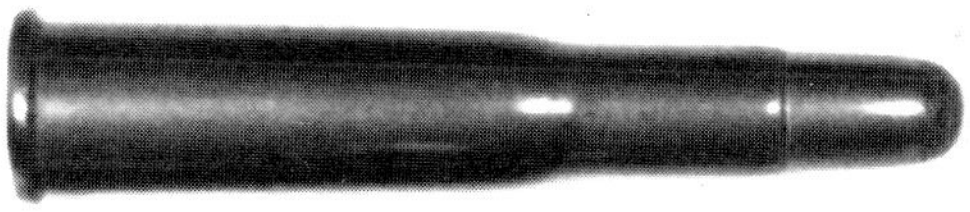

.30-30-150 or 170.

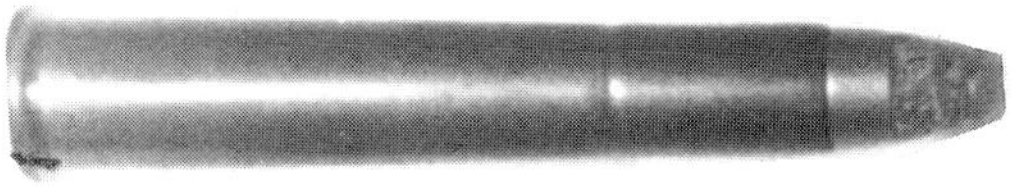

.32-40-165.

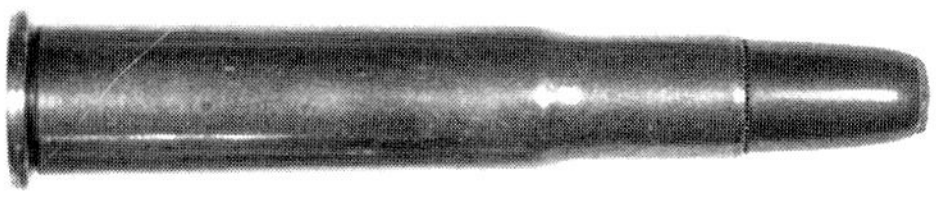

.32-170 Winchester Special.

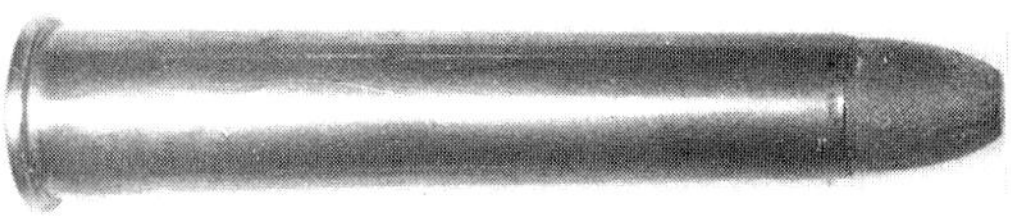

.38-55-255.

THE WINCHESTER REPEATING RIFLE, MODEL 1894.

.25-35 Smokeless.
.30 W. C. F.
.32-40.
.38-55.

CARBINE, MODEL 1894, .32-40 AND .38-55.
20 inch Barrel, Full Magazine, 7 shots, weight 5¾ lbs., $17.50

SPORTING RIFLE, MODEL 1894, .32-40 AND .38-55.
26 inch Octagon Barrel, Full Magazine, 10 shots, weight 7½ lbs., $19.50
26 inch Round Barrel, Full Magazine, 10 shots, weight 7¼ lbs., 18.00

SPORTING RIFLE, MODEL 1894, .25-35 SMOKELESS AND .30 WINCHESTER SMOKELESS.
26 inch Round Nickel Steel Barrel, Full Magazine, 10 shots, weight 7¼ lbs., $23.00

THE WINCHESTER MODEL 1895.

5 SHOTS IN MAGAZINE.

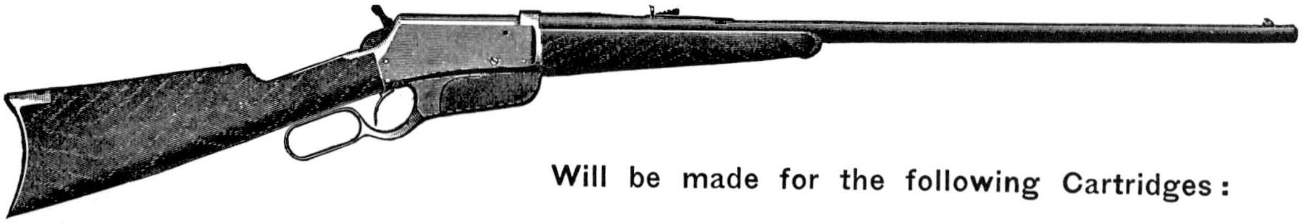

Will be made for the following Cartridges:

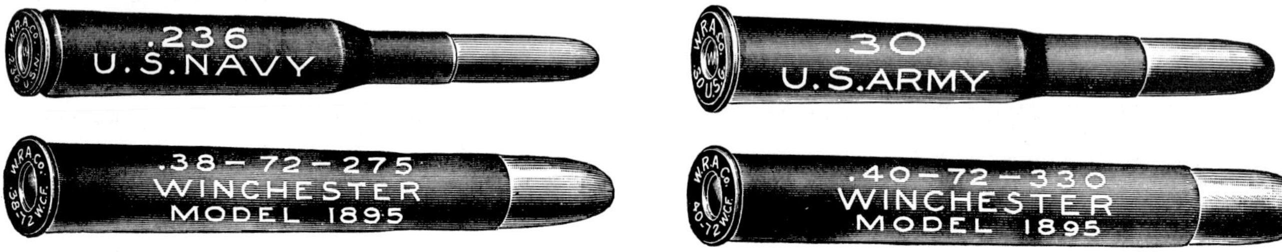

PRICE LIST.

.236 Navy, 28 inch, Round Barrel, ... $25.00
.30 Army, 28 inch, Round Barrel, weight 8¼ pounds, .. 25.00
.38-72 Winchester, 26 inch, Round Barrel, ... 19.50
.38-72 Winchester, 26 inch, Octagon Barrel, ... $21.00
.40-72 Winchester, 26 inch, Round Barrel, ... 19.50
.40-72 Winchester, 26 inch, Octagon Barrel, ... 21.00

The barrels of the .236 Navy and .30 Army will be of nickel steel, of the standard length of 28 inches, and only round barrels can be supplied.

The .38-72 and .40-72 Rifles will be made in both round and octagon styles, with a standard length of 26 inches.

Extras same as on other Repeating Rifles. See page 56.

Set Triggers cannot be furnished for this gun.

Winchester Repeating Rifle, Model 1895.

Made For .30 Caliber Army, .30 Caliber Gov't M. '03 And '06, .303 British, .35 And .405 Cartridges.

Sporting Rifle, Model 1895, .30 Caliber Army And .303 British.

Twenty-eight Inch Round Nickel Steel Barrel, Weight about 8¼ pounds, Magazine Capacity 5, ... $30.00

Fancy Sporting Rifle, Model 1895, .30 Caliber Army And .303 British.

Twenty-eight Inch Round Nickel Steel Barrel, Fancy Walnut Checked Stock and Forearm, Shotgun or Rifle Butt Stock, Weight about 8½ pounds, Magazine Capacity 5, ... $45.00

Shotgun Butt Stock, with either metal or rubber Butt Plate, same price as Rifle Butt Stock.

Winchester announced the Model 95 box magazine rifle in the June 1896 catalogue chambered for the .236 U. S. Navy rimless cartridge. The cartridge was missing from the next and following catalogues. No doubt it was made experimentally but no authentic .236 Model 95 rifles are known to exist. Two rifles with .236, or 6mm Lee, barrels were discovered by a well-known Winchester authority in Montana many years ago, but it turned out that someone had used Lee Navy bolt action rifle barrels to make up the guns. Collector beware!

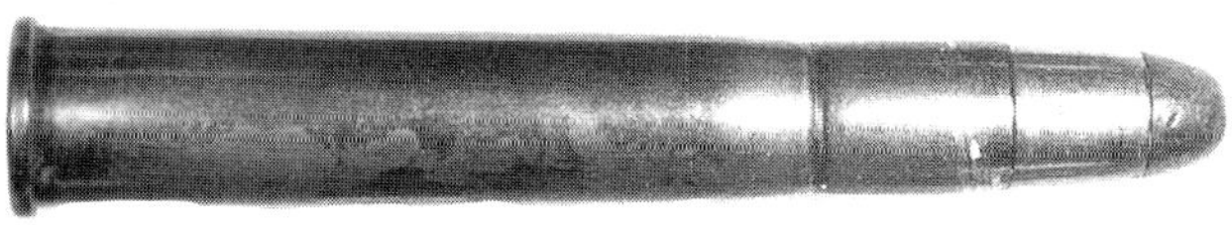

.38-72-275. This cartridge was introduced with the new rifle in the June 1896 catalogue. The first loads were lead bullet with black powder but jacketed bullet smoke less loads were available by 1897. The Model 95 was no longer offered in .38-72 after 1910, though ammunition was available for existing rifles for many years.

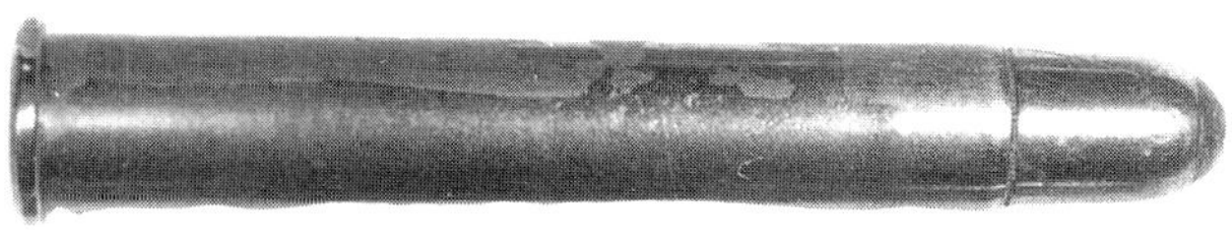

.40-72-330. The same history applies to this cartridge as the .38-72.

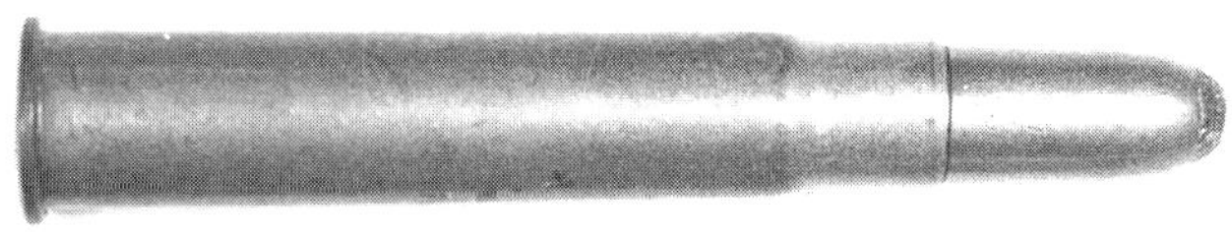

.35-250 Winchester. This cartridge was added to Model 95 loads in 1903 and the rifle was available chambered for it until 1936 when it was discontinued along with the Model 86, and replaced by the .348 Model 71.

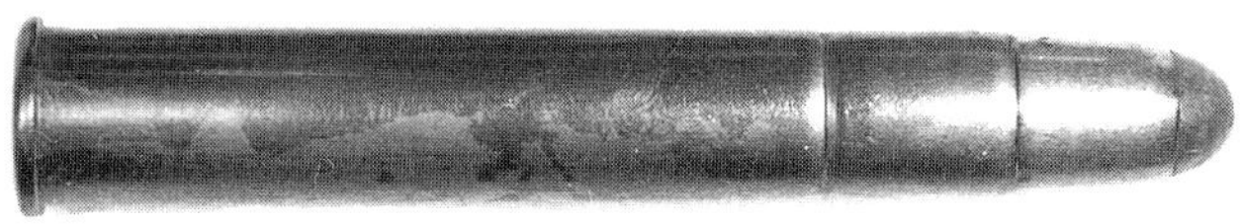

.405-300. Winchester introduced this load in 1904 and made the Model 95 a grizzly and lion rifle. Although the rifle was dropped by Winchester in 1936, the Model 95 retained its popularity with big game hunters around the world. Many double rifles were chambered for .405.

Other modern smokeless powder cartridges for which the Model 1895 was chambered are: .30-40, .30-03, .30-06, .303 British, and 7.62 x 54R (7.62 Russian).

This is the Oliver F. Winchester commemorative medallion struck in 1980 with a facsimile of his signature and dates of birth and death—1810-1890—below his portrait in front of the factory in New Haven, Connecticutt. The medallion is about 1 ½" in diameter. It was inlet into the right hand side of the stock on 19,999 Model 94 rifles chambered for .38-55.

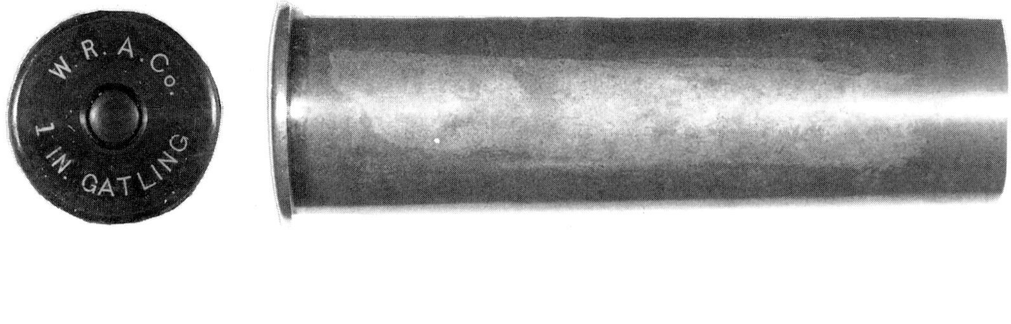

BULLARD REPEATING ARMS COMPANY

The rifle illustrated here was designed and patented by James J. Bullard (U. S. patent no. 245,700, August 16, 1881). Bullard built single shot rifles and two frame sizes of this lever action at Springfield, Massachusetts from 1880 until about 1890. The rack and pinion action was smoother operating than competing lever actions.

Bullard developed his own line of cartridges, only the .45-85 using the same case as the .45-90 Winchester. The .45-70 Springfield was also an option. The .32-40 and .38-45 were made for the small-framed rifle; all of the others for the large frame. Bullard evidently chambered his single shot rifle for the .32-40 Ballard (see label among the boxes illustrated with Sharps and Remington).

Bullard improved on the Winchester Model 1876 but then John Browning designed what became the Winchester Model 1886. Marlin joined in the competition and the Bullard company passed out of the picture, along with the short-lived Burgess and Kennedy rifles produced by the Whitney Arms Company, New Haven, Connecticutt, both of which were chambered for .45-70. With the great variety of cartridges made available by Winchester, the Model 86 dominated this field. (Photo and drawing courtesy of R. Bruce McDowell, author of "Evolution of the Winchester," Armory Publications, 1985 and "Colt Conversions," Krause Publications, 1997).

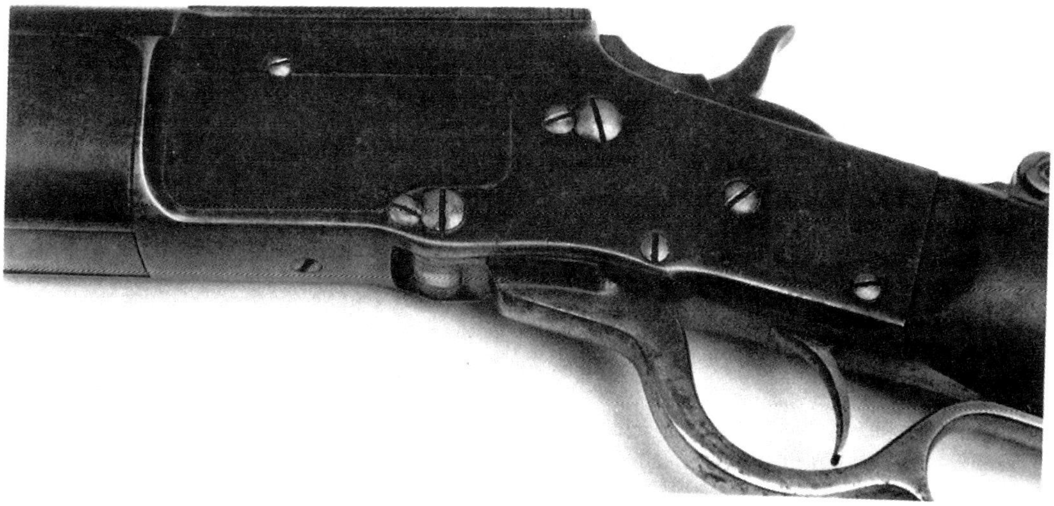

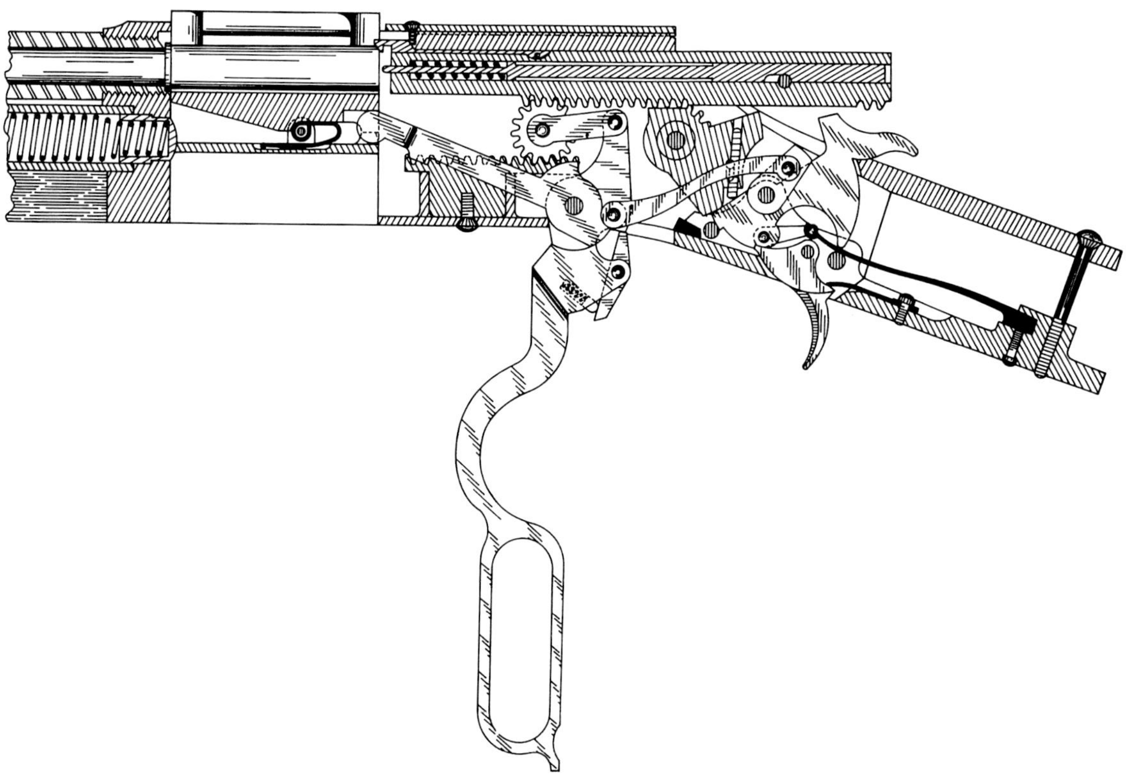

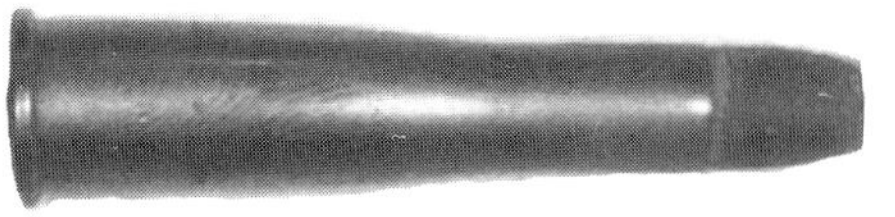

.32-40-150, 1.85" case. This cartridge appears in the Winchester catalogue in January 1885, and was last offered in January 1913.

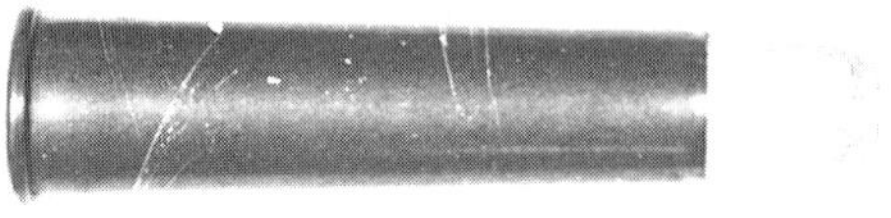

.38-45-190, 1.80" case. Also appeared in the Winchester catalogue in 1885, and also gone by 1914.

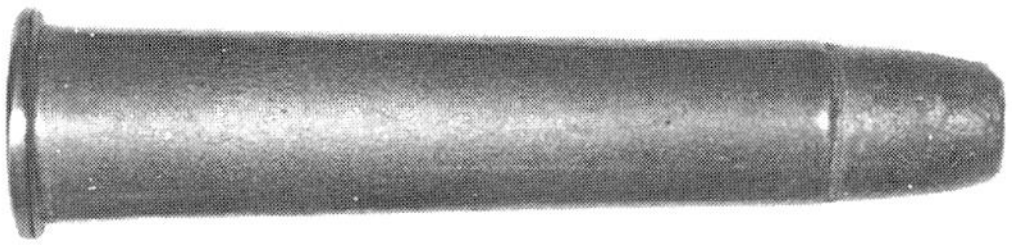

.40-70-232, 2.1" case. Appears in the September 1884 Winchester catalogue. The .40-60 Marlin, and .40-65 Winchester cases are the same. The most likely scenario is that Marlin originated the case, Bullard changed powder charge and bullet weight to make his own cartridge, then Winchester used the case to make it their own .40-65 for the Model 86 which came after the Marlin and Bullard rifles. Finally, Colt chambered the Lightning slide action rifle for this case as the .40-60 Colt.

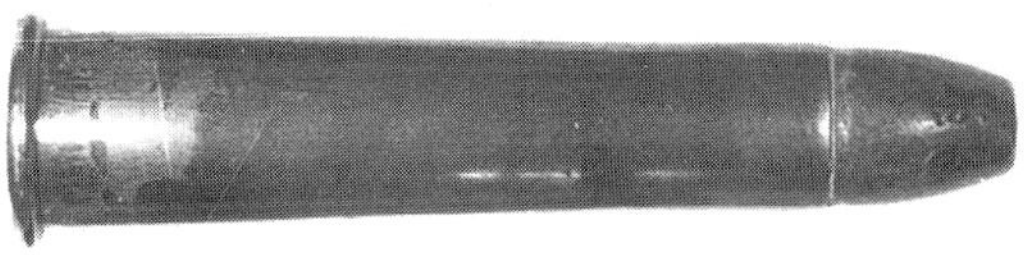

.40-75-258, 2.1" case. This is also identical in dimensions to the .40-70 Bullard, loaded with heavier bullet and five grains more powder. It appeared about 1887.

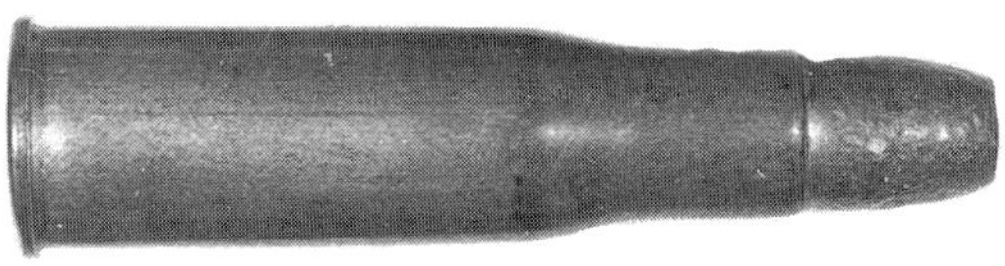

.40-90-300, 2.04" case. This cartridge appeared in 1884 and was no longer listed after 1910.

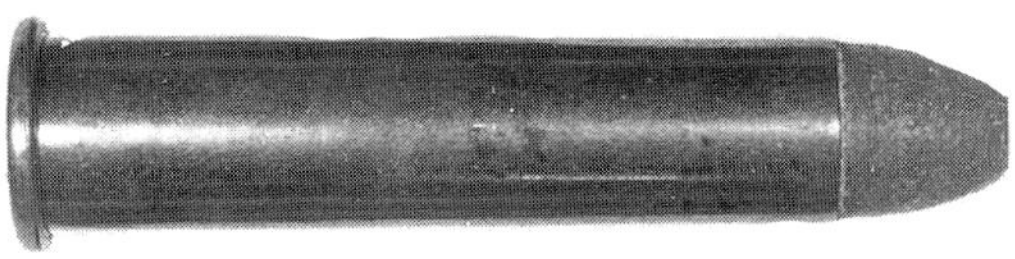

.45-85-295, 2.15" case. Using the .45-70 military case, this was the longest lived of the Bullard cartridges. It was still on Winchester's list in 1916 but gone in 1918.

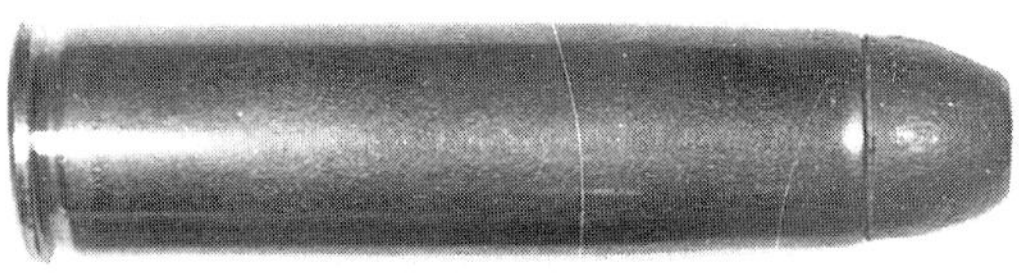

.50-115-300, 2.19" case. Appeared in 1886. This is the first semi-rimmed, grooved cartridge case for rifle. Date of disappearance not available, but most likely well before WWI. The U. M. C. 1889 list following shows it with solid or express bullet (copper-tubed hollow point). This cartridge is not found in any Winchester catalogue.

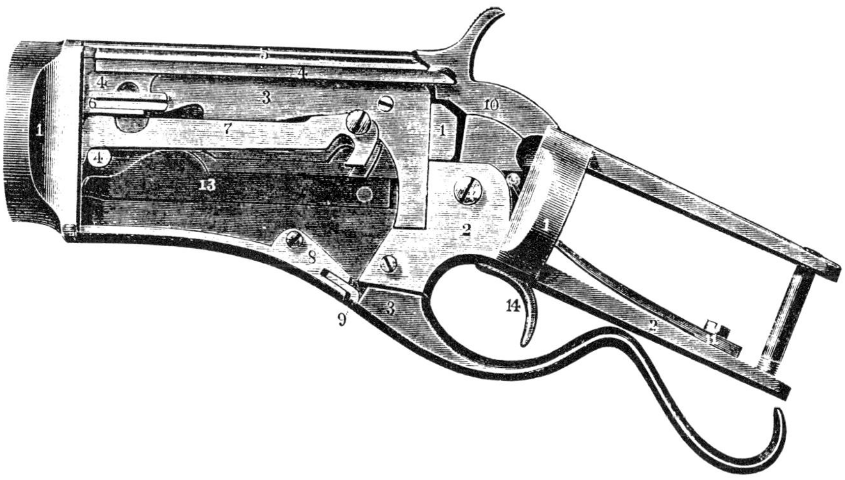

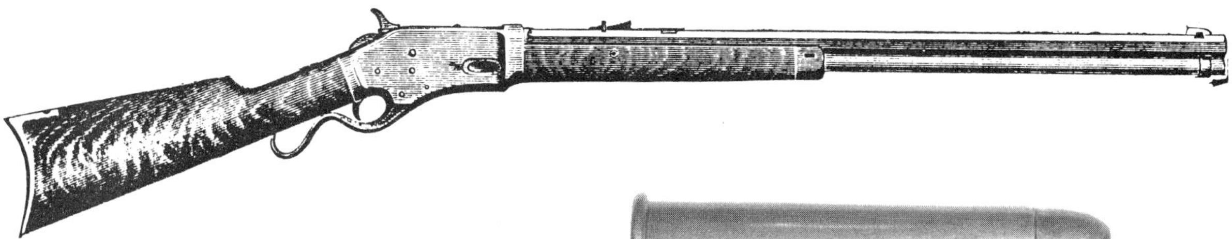

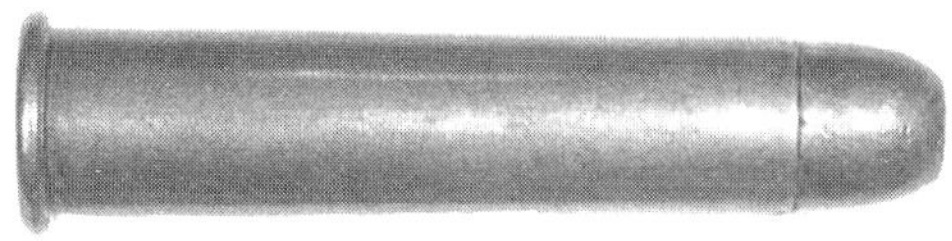

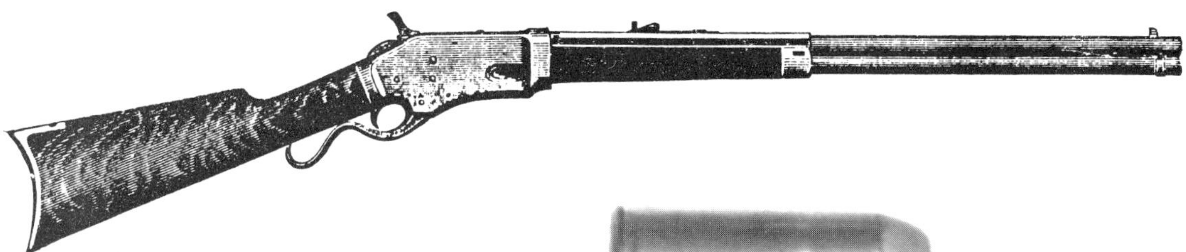

The action of the rifles shown above were patented by Andrew Burgess on July 12, 1887 (U. S. 366,558). Operation was similar to the Bullard but without the teeth of the rack and pinion design of the Bullard. It and the Kennedy below were made briefly by the Whitney Arms Company, New Haven, Connecticutt. The Burgess was chambered for the .45-70 military cartridge, the Kennedy for the .44-40 Winchester. (Illustrations from a reprint of the Whitney Arms Company catalogue ca 1960 by Fortress Publications, Stoney Creek, Ontario, Canada. Original from the collection of John H. Hintlian).

THE MARLIN FIREARMS COMPANY

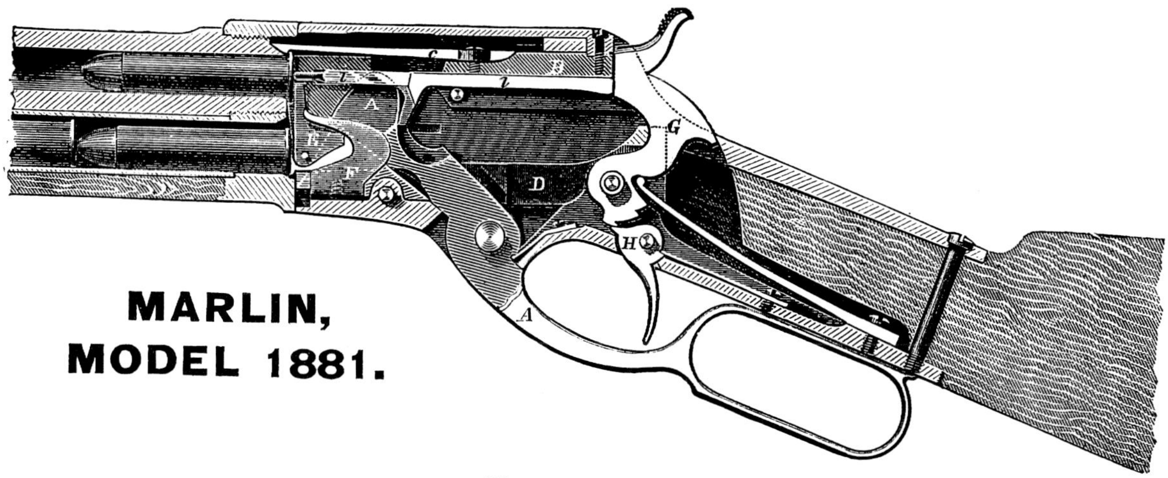

Figure No. 1.

Figure No. 1 shows the Arm in a closed position. A represents the Lever; B the Bolt; C the Extractor; D the Carrier Block; E the Ejector; F the Carrier Block Spring; G the Hammer; H the Trigger; I the Firing Pin.

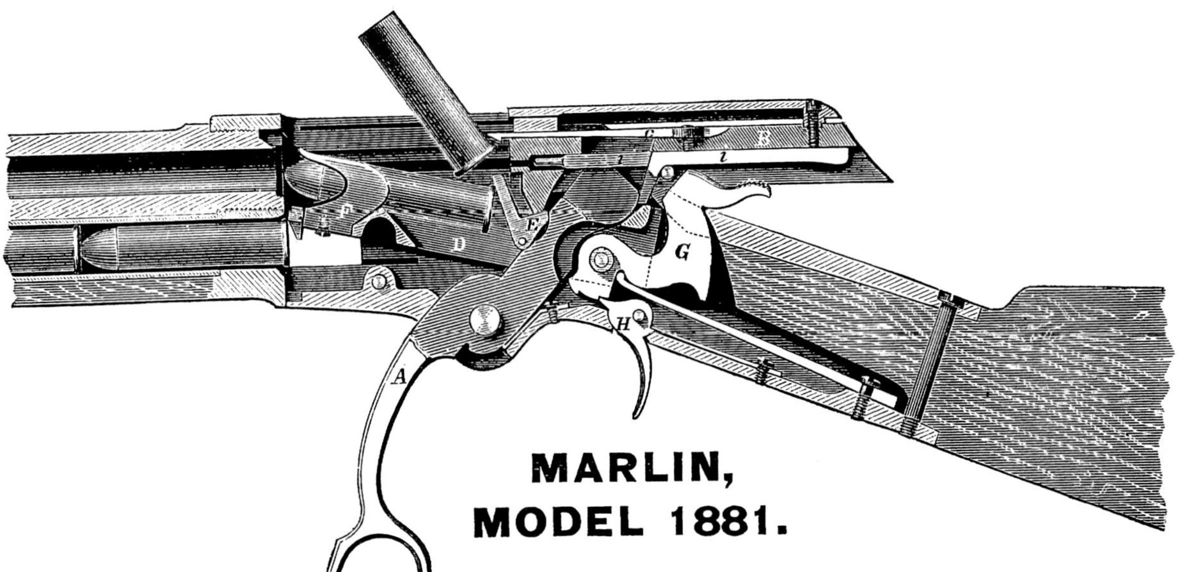

Figure No. 2.

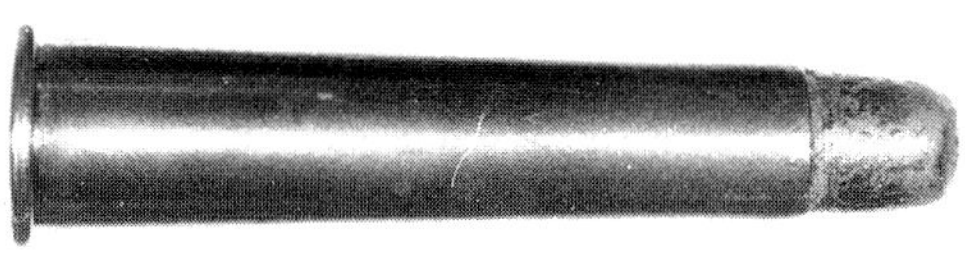

.40-60-260 Marlin, 2.1" case. This cartridge first appears in the June 1883 Winchester catalogue. It was later used by Winchester in their Model 1886 rifle as the .40-65-260. Colt also used this cartridge in their Lightning slide action, tubular magazine rifle in the Marlin loading.

The .40-60 Marlin was the company's only original cartridge. The following pages from the 1905-06 Marlin catalogue illustrate the various cartridges that may be found with the Marlin headstamp, some with different bullet weight and powder charge than the Winchester loads. In this catalogue the .40-60-260 that they developed in the early 1880s is no longer listed as such but is still there as the .40-65-260 Marlin Safety. Perhaps there had been confusion between the Marlin cartridge and the shorter .40-60 Winchester for the Model 76.

Same as .32 W. C. F.

Same as .38 W. C. F.

Same as .44 W. C. F.

CARBINES

Made for .25-20, .32-20, .38-40 and .44-40 cartridges.

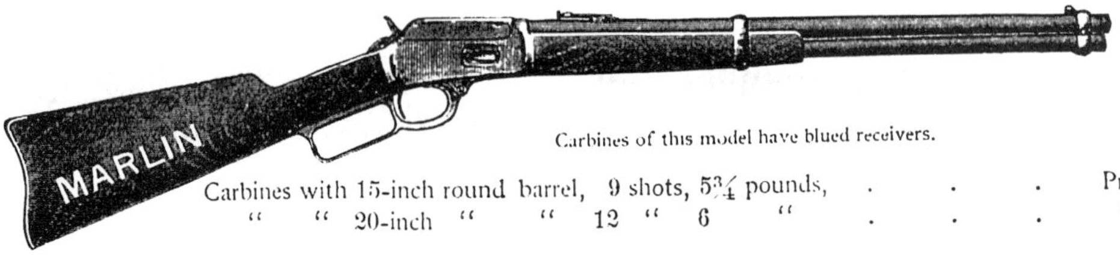

Carbines of this model have blued receivers.

Carbines with 15-inch round barrel, 9 shots, $5\frac{3}{4}$ pounds,	. . .	Price	$12.95
" " 20-inch " " 12 " 6 "	. . .	"	12.95

The standard carbine has a sling ring on left side of receiver. This will be omitted if so ordered. When a carbine is ordered with butt and forearm swivels for sling strap, the ring is omitted unless ordered to be retained.

MUSKETS

Made for .38-40 and .44-40 cartridges only.

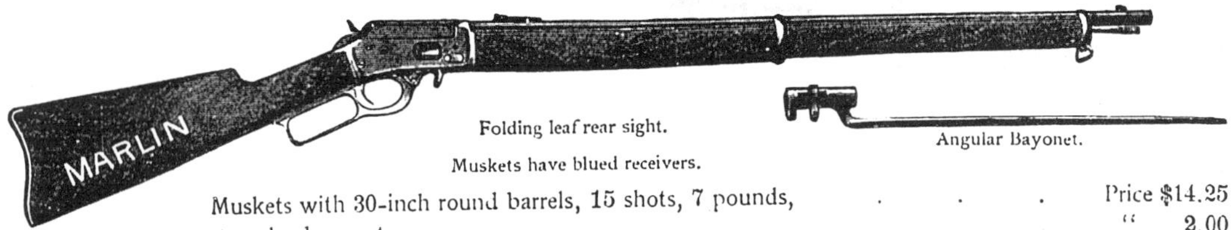

Folding leaf rear sight.

Muskets have blued receivers.

Angular Bayonet.

Muskets with 30-inch round barrels, 15 shots, 7 pounds,	. . . .	Price	$14.25
Angular bayonets,	. . . .	"	2.00

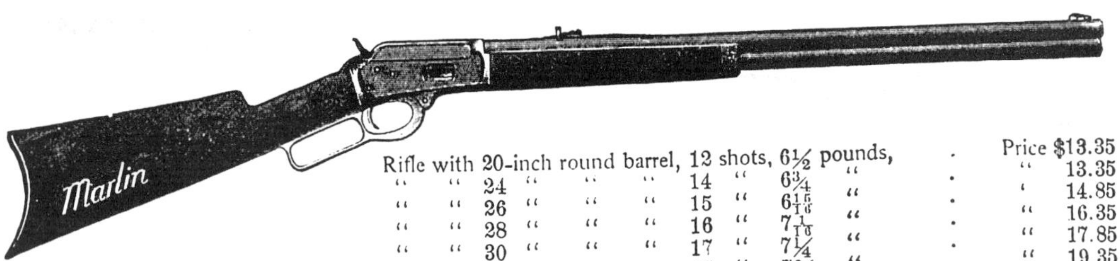

Rifle with	20-inch round barrel,		12 shots,		$6\frac{1}{2}$ pounds,	.	Price	$13.35
" "	24 " " "		14 "		$6\frac{3}{4}$ "	.	"	13.35
" "	26 " " "		15 "		$6\frac{15}{16}$ "	.	"	14.85
" "	28 " " "		16 "		$7\frac{1}{16}$ "	.	"	16.35
" "	30 " " "		17 "		$7\frac{1}{4}$ "	.	"	17.85
" "	32 " " "		17 "		$7\frac{3}{8}$ "	.	"	19.35

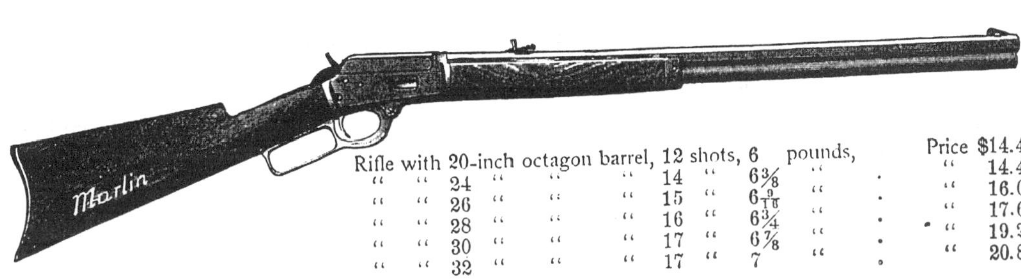

Rifle with	20-inch octagon barrel,		12 shots,		6 pounds,	.	Price	$14.40
" "	24 " " "		14 "		$6\frac{3}{8}$ "	.	"	14.40
" "	26 " " "		15 "		$6\frac{9}{16}$ "	.	"	16.00
" "	28 " " "		16 "		$6\frac{3}{4}$ "	.	"	17.60
" "	30 " " "		17 "		$6\frac{7}{8}$ "	.	"	19.20
" "	32 " " "		17 "		7 "	.	"	20.80

Longest magazine made is 30 inches.
Short magazine, to end of forearm, six shots, same price as full magazine.
Half magazine, $3\frac{1}{4}$ inches beyond forearm, **eight** shots, same price as full magazine.
Shotgun rubber buttplate, makes rifle one-eighth pound lighter, same price as rifle buttplate.
Half-octagon barrel same price as octagon.
Rifles of this model have case-hardened frames.
Extras, see page 23.

Made in .25-20, .32-20, .38-40 and .44-40 calibers

Model 1893

Marlin

(See also pages 74 to 94)

Made in .25-36, .30-30, .32-40, .32 Special H. P. S. and .38-55 Calibers

CARBINE

All barrels in this model, except Grade "B," are made of *Special Smokeless Steel*

.25-36, .30-30, .32 Special H. P. S., .32-40 or .38-55.

Carbine with 15-in. round barrel,	5 shots	6¼ lbs.,	Price	$15.50
" " 20-in. "	" 7 "	6¾ "	"	15.50

The standard carbine has a sling ring on left side of receiver. This will be omitted if so ordered. When a carbine is ordered with butt and forearm swivels for sling strap, the ring is omitted unless ordered to be retained.

Marlin

Model 1893

All barrels in this model, except Grade "B," are made of "Special Smokeless Steel"

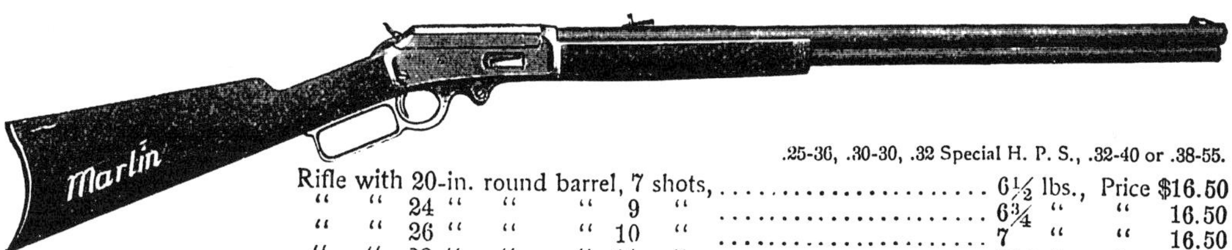

.25-36, .30-30, .32 Special H. P. S., .32-40 or .38-55.

Rifle with 20-in. round barrel,	7 shots,	6½ lbs.,	Price $16.50
" " 24 " "	" 9 "	6¾ "	" 16.50
" " 26 " "	" 10 "	7 "	" 16.50
" " 28 " "	" 11 "	7¼ "	" 18.00
" " 30 " "	" 12 "	7½ "	" 19.50
" " 32 " "	" 12 "	7¾ "	" 21.00

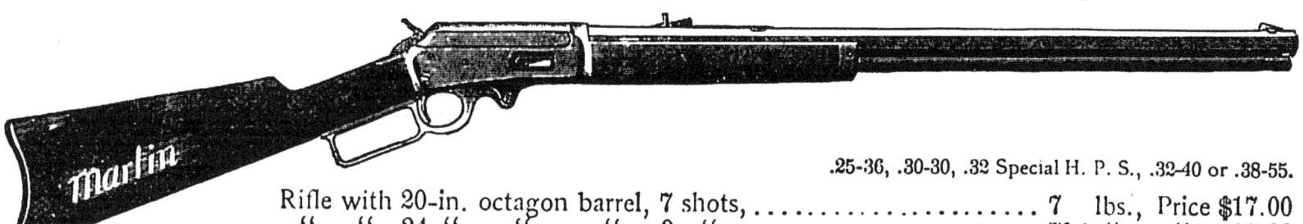

.25-36, .30-30, .32 Special H. P. S., .32-40 or .38-55.

Rifle with 20-in. octagon barrel,	7 shots,	7 lbs.,	Price $17.00
" " 24 " "	" 9 "	7¼ "	" 17.00
" " 26 " "	" 10 "	7½ "	" 17.00
" " 28 " "	" 11 "	7¾ "	" 18.50
" " 30 " "	" 12 "	8 "	" 20.00
" " 32 " "	" 12 "	8¼ "	" 21.50

Longest magazine made is 30 inches.

Rifles can be furnished with full, half or short magazine.

Capacity of half magazine four (4) cartridges.

Capacity of special short magazine two (2) cartridges.

Rifle with half-octagon barrel, same price as with octagon barrel.

The standard rifle is fitted with rifle butt as illustrated; buttstock with rubber buttplate will be furnished at same price. With rubber butt, weight of rifle is an eighth of a pound less.

Carbines and rifles of this model have case-hardened receivers, but if specially ordered can be furnished with blued receivers at same price.

The length of rifle buttstock is 13 inches; the length of stock with rubber buttplate is 13⅛ inches; the drop is 3 inches.

The length over all of a Model 1893 Rifle with 26-inch barrel is 44⅞ inches. With buttstock removed the length of the rifle over all is 35 inches. Take-Downs will pack in the length of the barrel. Extras can be furnished as described on page 23.

18

Model 1895

Marlin

Made in .38-56, .40-65, .40-70, .40-82, .45-70 and .45-90 calibers
All barrels in this model are made of "SPECIAL SMOKELESS STEEL"
(See also pages 101 to 113)

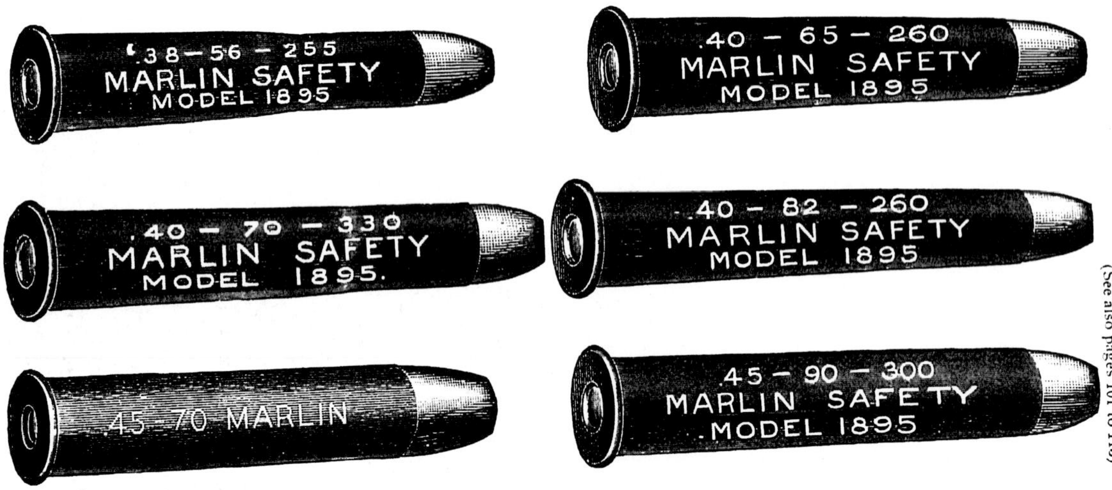

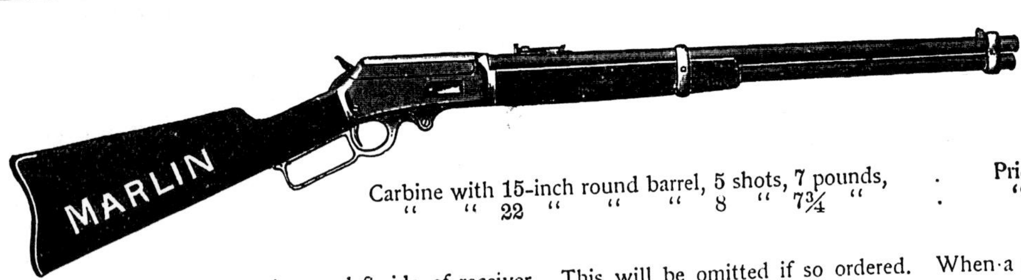

Carbine with 15-inch round barrel,	5 shots,	7 pounds,	.	Price	$15.60
" " 22 " " "	8 "	7¾ "	.	"	15.60

The standard carbine has a sling ring on left side of receiver. This will be omitted if so ordered. When a carbine is ordered with butt and forearm swivels for sling strap, the ring is omitted unless ordered to be retained.

19

Model 1895

Marlin

All barrels in this model are made of "Special Smokeless Steel"

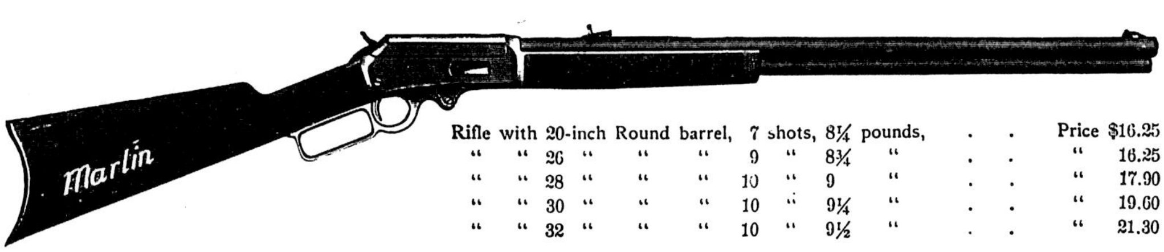

Rifle with	Barrel	Shots	Pounds		Price
Rifle with	20-inch Round barrel,	7 shots,	8¼ pounds,	. .	Price $16.25
" "	26 " " "	9 "	8¾ "	. .	" 16.25
" "	28 " " "	10 "	9 "	. .	" 17.90
" "	30 " " "	10 "	9¼ "	. .	" 19.60
" "	32 " " "	10 "	9½ "	. .	" 21.30

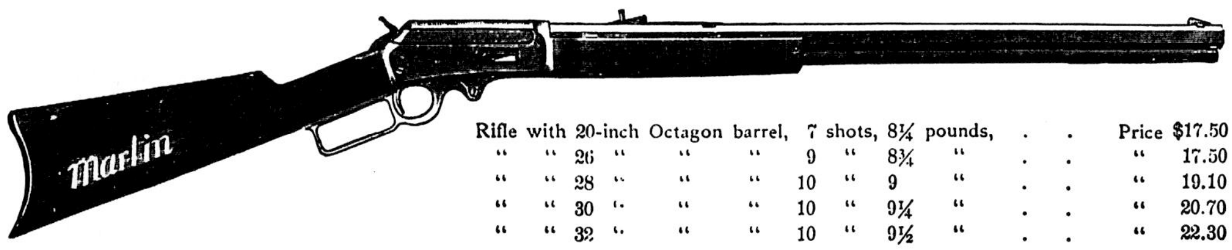

Rifle with	Barrel	Shots	Pounds		Price
Rifle with	20-inch Octagon barrel,	7 shots,	8¼ pounds,	. .	Price $17.50
" "	26 " " "	9 "	8¾ "	. .	" 17.50
" "	28 " " "	10 "	9 "	. .	" 19.10
" "	30 " " "	10 "	9¼ "	. .	" 20.70
" "	32 " " "	10 "	9½ "	. .	" 22.30

Longest magazine made is 30 inches.

The capacity is for .38-56, .40-65 or .45-70 cartridges. A magazine will hold one less of the .40-82, .40-70 and .45-90 cartridges, which have longer shells.

The half-magazine rifle will take 5 of the shorter cartridges, or 4 of the longer cartridges. A short magazine rifle will take 3 of the shorter cartridges and 3 of the longer ones.

Rifle with half-octagon barrel, same price as with octagon barrel.

The standard rifle is fitted with rifle butt as illustrated; buttstock with rubber buttplate will be furnished at same price. With rubber butt, weight of rifle is one-eighth of a pound less.

Carbines and rifles of this model have case-hardened receivers, but if specially ordered, can be furnished with blued receivers at same price.

The length of rifle buttstock is 13 inches; the drop, 3¼ inches. The length of stock with rubber butt is 13⅛ inches; the drop, 3 inches. The length over all of a Model 1895 Rifle with 26-inch barrel is 45⅝ inches. With buttstock removed, the length of the rifle over all is 35½ inches. Take-Downs will pack in the length of the barrel. Extras can be furnished as described on page 23.

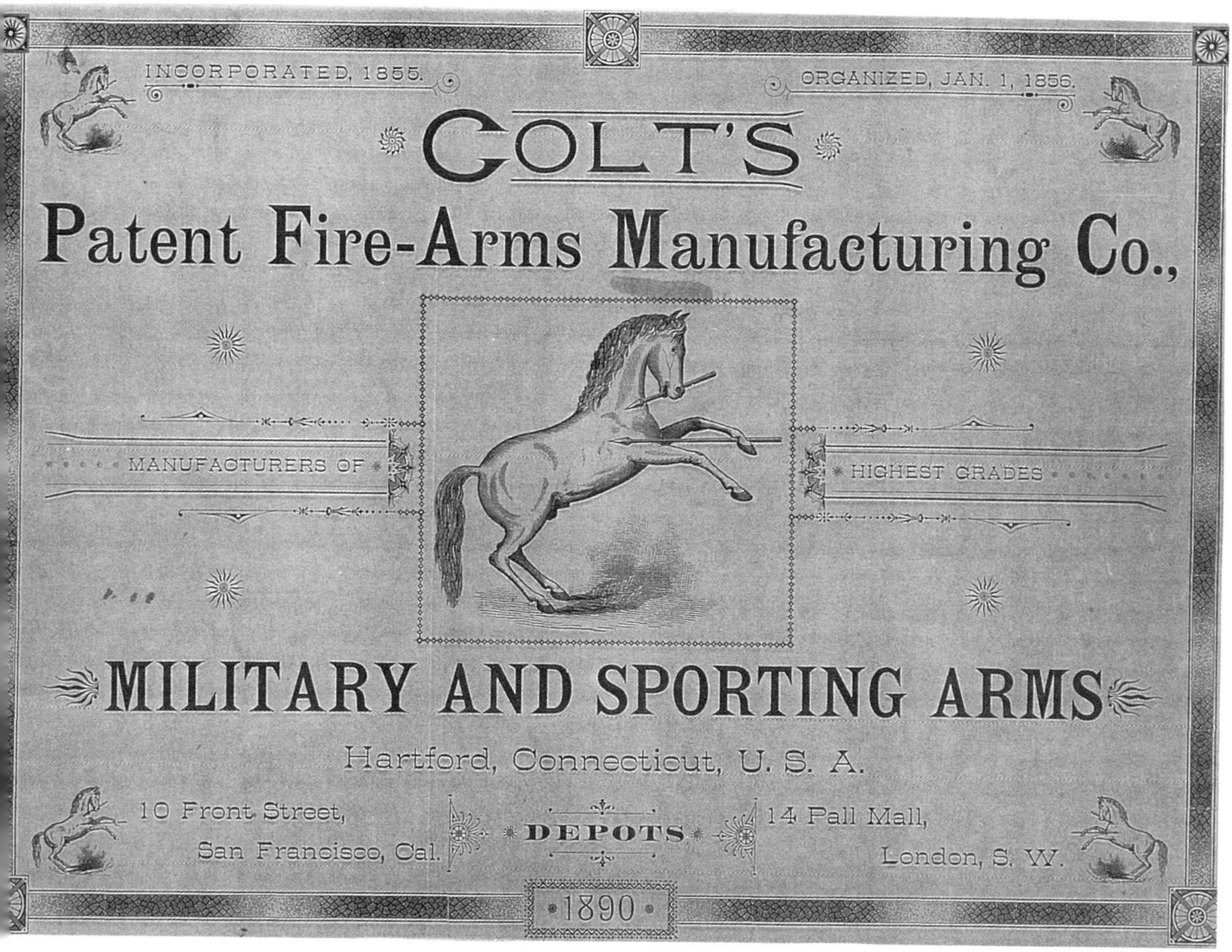

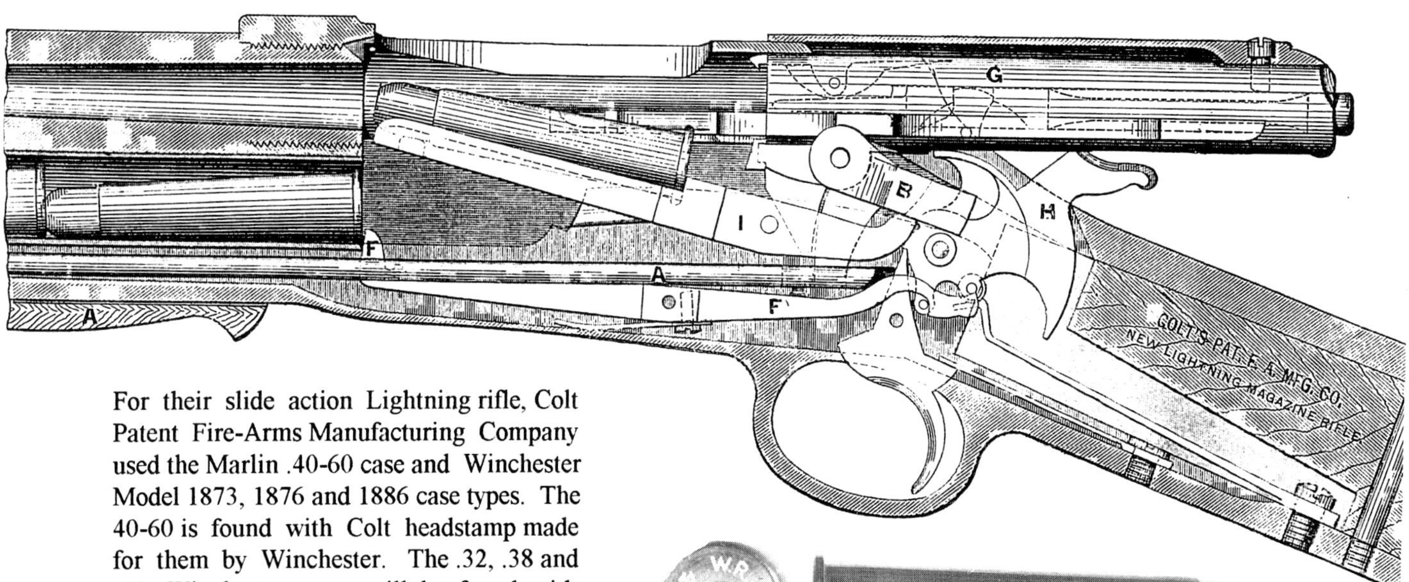

For their slide action Lightning rifle, Colt Patent Fire-Arms Manufacturing Company used the Marlin .40-60 case and Winchester Model 1873, 1876 and 1886 case types. The 40-60 is found with Colt headstamp made for them by Winchester. The .32, .38 and .44 Winchester cases will be found with C. L. M. R. (Colt Lightning Magazine Rifle).

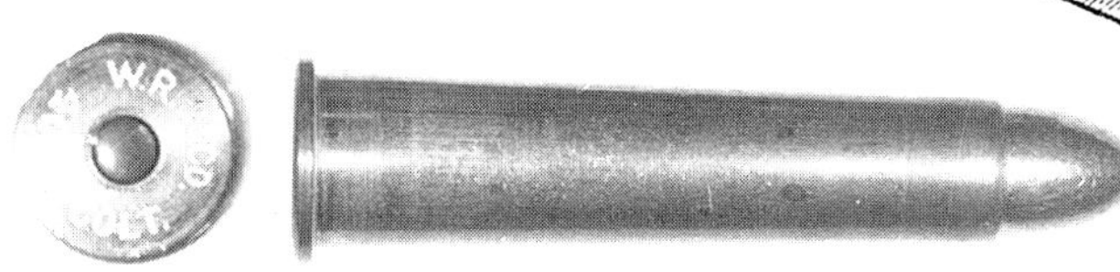

NEW LIGHTNING MAGAZINE RIFLES.

.32, .38, and .44 Calibres. Central Fire.

CARBINE.

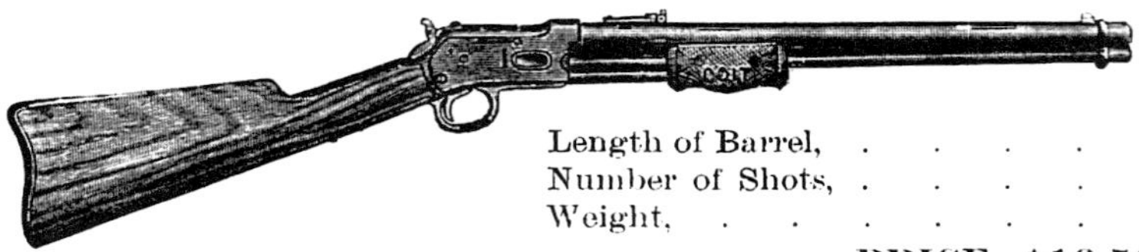

Length of Barrel, ...	20 inches.
Number of Shots, ...	12.
Weight, ...	6¼ lbs.

PRICE, $16.50.

SPORTING RIFLE, ROUND BARREL.

Length of Barrel, ...	26 inches.
Number of Shots, ...	15.
Weight, ...	6¾ lbs.

PRICE, $16.50.

SPORTING RIFLE, OCTAGON BARREL.

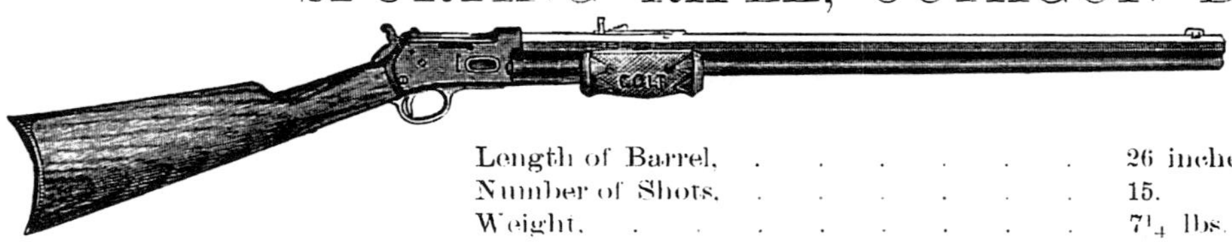

Length of Barrel, ...	26 inches.
Number of Shots, ...	15.
Weight, ...	7¼ lbs.

PRICE, $18.00.

This Rifle possesses advantages over all others now in the market, from the adoption of the left-handed method of handling, which enables the breech-action to be much simplified and lightened, without detracting from its strength. The materials used in its construction are the best of their kind, and the workmanship is of the same high standard as that of the other arms manufactured by this Company.

Attention is called to the positive action of the magazine gate, which, after one cartridge is fed to the carrier, is positively locked, so as to prevent blocking the breech-action, or the escape of a cartridge from the magazine under the carrier.

The cartridge used in the .44 calibre Rifle and Carbine is the same as that of our "Frontier" .44 calibre Army Pistol; it contains 40 grains of powder, and the bullet weighs 200 grains. The cartridge used in the .38 calibre Rifle contains 40 grains of powder, and the bullet weighs 180 grains. The cartridge used in the .32 calibre Rifle contains 20 grains of powder, and bullet weighs 100 grains. All kinds can be obtained from all arms dealers in the United States and Mexico.

The performance of the Rifle is guaranteed only when ammunition made especially for them, and bearing the *fac-simile* of our signature on the box, is used.

The list above gives the details of the various rifles produced, and the prices. Attention is called to the fact that by holding the trigger back, and using the reciprocating motion with the left hand, the rifle can be fired with great rapidity and without further use of the trigger. Use cartridges bearing *fac-simile* of our signature on each box.

We shall also make the above Rifles and Carbines for the New .25-.20-.86 Cartridge.

NEW LIGHTNING MAGAZINE RIFLE.

.38-.56, .40-.60, .45-.60, .45-.85, and .50-.95 Express Calibres.

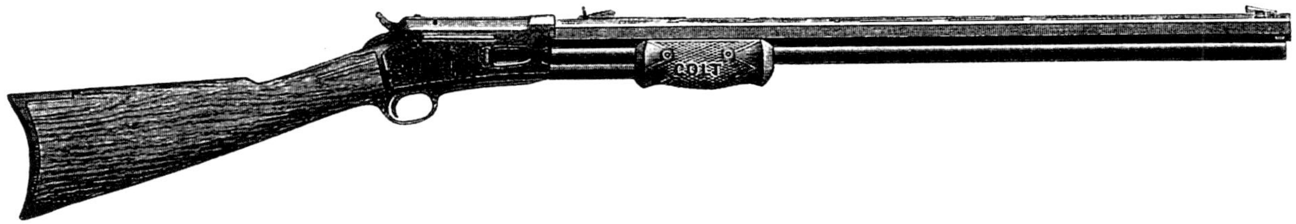

	OCTAGON.	ROUND.	CARBINE.	BABY CARBINE.
Length of Barrel,	28 inches.	28 inches.	22 inches.	22 inches.
Number of Shots,	10.	10.	8.	8.
Weight,	10 lbs.	9¾ lbs.	9 lbs.	8 lbs.
Price,	**$20.50**	**$19.00**	**$19.00**	**$19.00**

The .50-.95 Express Rifle will only be made with a 26 inch Barrel. Prices: Octagon Barrel, $24.00; Round Barrel, $22.50.

This Model is made in the following Calibres:

.38 Calibre,	56 grains powder,	256 grains bullet.
.40 Calibre,	60 grains powder,	260 grains bullet.
.45 Calibre,	60 grains powder,	300 grains bullet.
.45 Calibre,	85 grains powder,	285 grains bullet.
.50 Calibre,	95 grains powder,	300 grains bullet.

Carbines will, if desired, be furnished with Swivels for Strap.

This Rifle possesses peculiar advantages in weight and calibre for sporting purposes, particularly for large game. The materials are the best, and the workmanship of the highest order.

The Cuts and description of this Rifle show that the breech-action is simpler and more perfect than any in the market, making the Rifle absolutely safe against accidental explosion, by a positive automatic lock; and the left-handed method of handling insures the easiest manipulation of the gun.

CARE SHOULD BE TAKEN *in the selection of cartridges, that the 40-60-260 cartridges should be obtained, as the Rifle will not take the cartridge with a bullet weighing 210 grains, the Rifle being* CHAMBERED *for that in which the bullet weighs 260 grains.*

Attention is called to the fact that, by holding the trigger back, and using the reciprocating motion with the left hand, the Rifle can be fired with great rapidity, and without further use of the trigger.

For the *Express Rifle* using the .50 inch calibre cartridge, containing 95 grains of powder, and 300 grains hollow express bullet, a solid bullet can be used if desired, but to obtain the flattest possible trajectory, only the regular hollow express bullet should be used. This arm is especially adapted to those sportsmen who hunt large game, and as an Express Rifle is unexcelled.

We guarantee all of our Rifles and their accuracy, which is unsurpassed, when used with cartridges made expressly for them, and bearing *fac-simile* of our signature on each box.

CARTRIDGES USED IN COLT'S PATENT FIRE-ARMS MANUFACTURING CO'S ARMS.

These Cartridges are manufactured especially for our Arms, and bear a FAC-SIMILE OF OUR SIGNATURE ON EACH BOX.

Price per 1,000, $5.00. Case contains 10,000.

Powder, 4 grains. Lead, 30 grains.

Adapted to the Lightning Magazine Rifles, .22 Calibre.
Old and New Line Revolvers, .22 Calibre.

Price per 1,000, $6.00. Case contains 10,000.

Powder, 5 grains. Lead, 30 grains.

Adapted to the Lightning Magazine Rifles, .22 Calibre,
Old and New Line Revolvers, .22 Calibre.

Price per 1,000, $9.00. Case contains 5,000.

Powder, 6 grains. Lead, 55 grains.

Adapted to New Line Revolvers, 30 Cailbre.

Price per 1,000, $10.50. Case contains 5,000.

Powder, 9 grains. Lead, 55 grains.

Adapted to New Line Revolvers, .30 Calibre.

Price per 1,000, $15.00. Case contains 3,000.

Powder, 13 grains. Lead, 163 grains.

Adapted to New Line Revolvers, .41 Calibre.
Colt's and National Deringer.

Price per 1,000, $17.50. Case contains 3,000.

Powder, 16 grains. Lead, 196 grains.

Adapted to New Line Revolvers, .41 Calibre.

Price per 1,000, $11.00. Case contains 5,000. Primed Shells, per 1,000, $6.00.

Powder, 9 grains. Bullet, 80 grains. Bullets, per 1,000, $3.00.

Adapted to New Line Revolvers, .32 Calibre, using Central Fire Cartridges.

Price per 1,000, $16.00. Case contains 2,000.

Powder, 20 grains. Bullet, 100 grains.

Primed Shells, per 1,000, $9.00. Bullets, per 1,000, $4.00.
Adapted to the Lightning Magazine Rifles, .32 Calibre.
Frontier Revolvers, .32 Calibre.

Price per 1,000, $13.50. Case contains 4,000.

Powder, 15 grains. Bullet, 125 grains. Bullets, per 1,000, $4.00.

Primed Shells, per 1,000, $7.00.
Adapted to New House, New Police, and Double Action Revolvers, .38 Calibre.

Price per 1,000, $14.50. Case contains 2,000.

Powder, 18 grains. Bullet, 150 grains.

Primed Shells, per 1,000, $7.50. Bullets, per 1,000, $4.50.
Adapted to New House, New Police, and Double Action Revolvers, .38 Calibre.

Price per 1,000, $19.00. Case contains 2,000.

Powder, 40 grains. Bullet, 180 grains.

Primed Shells, per 1,000, $10,00. Bullets, per 1,000, $5.00.
Adapted to the Lightning Magazine Rifles, .38 Calibre.
Frontier Revolvers, .38 Calibre.

Price per 1,000, $15.00. Case contains 2,000. Primed Shells, per 1,000, $7.50.

Powder, 15 grains. Bullet, 100 grains. Bullets, per 1,000, $4.00.

Adapted to Army, New House, and Double Action Revolvers, .41 Calibre.

Price per 1,000, $17.50. Case contains, 2,000.

Powder, 22 grains. Bullet, 19 grains.

Primed Shells, per 1,000, $8.00. Bullets, per 1000, $6.00.
Adapted to Army, New House, and Double Action Revolvers, .41 Calibre.

Price per 1,000, $19.00. Case contains 2,000.

Powder, 40 grains. Bullet, 200 grains.

Primed Shells, per 1,000, $10.00. Bullets, per 1,000, $6.00.
Adapted to the New Lightning Magazine Rifles, .44 Calibre.
Frontier Revolvers, Single and Double Action, .44 Calibre.

Price per 1,000, $33.00. Case contains 1,000.
Powder, 56 grains. Bullet, 256 grains.
Primed Shells, per 1,000, $18.00. Bullets, per 1,000, $9.50.
Adapted to the Lightning Magazine Rifles, .38–.56 Calibre.

Price per 1,000, $33.00. Case contains 1,000.
Powder, 60 grains. Bullet, 260 grains.
Primed Shells, per 1,000, $18.00. Bullets, per 1,000, $9.00.
Adapted to the Lightning Magazine Rifles, .40–.60 Calibre.

Price per 1,000, $22.00. Case contains 2,000.

Powder, 40 grains. Bullet, 250 grains.

Primed Shells, per 1,000, $10.50. Bullets, per 1,000, $7.00.
Adapted to Army Double and Single Action Revolvers, .45 Calibre.

Price per 1,000, $33.00. Case contains 1,000.
Powder, 60 grains. Bullet, 300 grains.
Primed Shells, per 1,000, $20.00. Bullets, per 1,000, $9.00.
Adapted to the Lightning Magazine Rifles, .45–.60 Calibre.

Price per 1,000, $37.00. Case contains 1,000.
Powder, 85 grains. Bullets, 285 grains.
Primed Shells, per 1,000, $22.00. Bullets, per 1,000, $9.50.
Adapted to the Lightning Magazine Rifles, .45–.85 Calibre.

Price per 1,000, $40.00. Case contains 1,000.
Powder, 95 grains. Bullet, 300 grains.
Primed Shells, per 1,000, $23.00.
Bullets (hollow), per 1000, $11.00. Bullets (solid), per 1,000, $9.00.
Adapted to the New Lightning Magazine Express Rifles.

Savage Rifles

Savage Cartridges *Manufactured by the SAVAGE ARMS CO.*

For 303 Savage Magazine Rifle

No. 1

REGULAR SMOKELESS POWDER CARTRIDGE

Cartridges, per thousand, $38.00 | Primed shells, per thousand, $18.50
Bullets, per thousand . 10.50 | (Cases contain 1000 cartridges)

This load, with completely nickel-covered bullet, is for long range work. It is well adapted for wild geese shooting. The bullet does not expand on striking flesh or wood. It will give accurate results up to one thousand yards.

No. 2

EXPANDING SMOKELESS POWDER CARTRIDGE

Cartridges, per thousand, $38.00 | Primed shells, per thousand, $18.50
Bullets, per thousand . 10.50 | (Cases contain 1000 cartridges)

This cartridge, with its soft-nose bullet, is for use on big game. Because of its very high velocity (2000 feet a second), it mushrooms perfectly and expands upon striking flesh alone.

303 BULLET

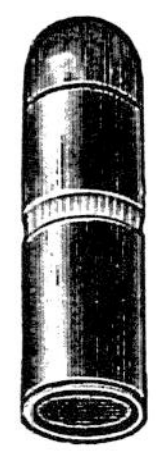

BEFORE EXPANDING

303 BULLET

AFTER EXPANDING

No. 4

MINIATURE SMOKELESS CARTRIDGE (Grooved Shell)

Cartridges, per thousand, $30.00 | Primed shells, per thousand, $18.50
Bullets, per thousand . 3.00 | (Cases contain 1000 cartridges)

This load has a small bullet with reduced powder charge and is accurate up to 60 yards. It is an auxiliary cartridge and can be successfully used for small game and short range shooting. Barrel should be kept clean when using this load.

No. 6

MINIATURE METAL-COVERED CARTRIDGE

Cartridges, per thousand, $35.00 | Primed shells, per thousand, $18.50
Bullets, per thousand . 5.00 | (Cases contain 1000 cartridges)

A suitable cartridge, with small metal-cased bullet, for ranges up to 125 yards. The jacket on the bullet prevents any possible leading of the barrel and permits of the use of a greater powder charge than that used in cartridge No. 4.

Savage Arms Co. Utica, N.Y. U.S.A.

SINGLE SPECIMENS

The cartridges following are singles that do not belong with the rifle makers' families, or are commercial loads of black powder military cartridges. Smith & Wesson created one rifle cartridge, Evans Rifle Manufacturing Company designed two, Dangerfield & Lefever just the one big .58 for their break-open single shot rifle they called the "Destroyer." Among them there will be dozens of primer, bullet and headstamp variations so only representative samples are shown to identify the case.

There are still unknowns out there, and it is not very likely that I have included every specimen that should be here, so I would be gratified to learn about them for future editions.

Headstamps (or no headstamps), primer types and bullet variations add hundreds of specimens to be sought after. The commercial headstamps on the military cartridges offer a collecting specialty for those who want to try something different. The Milbank primed cartridges produced by Winchester are very difficult to find; display board dummies produced by Winchester, U. M. C. and U. S. Cartridge Company form another area for collecting.

You will find no .236 Lee Navy, no .236 Remington, no .22 Savage High Power, no Newton cartridges in this book. There is enough material to fill another volume on the smokeless powder developments from the late 19th century to WW II. That will come next.

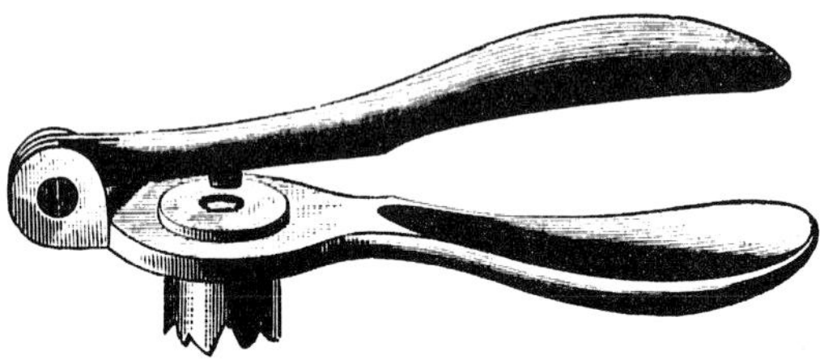

No. 1180.

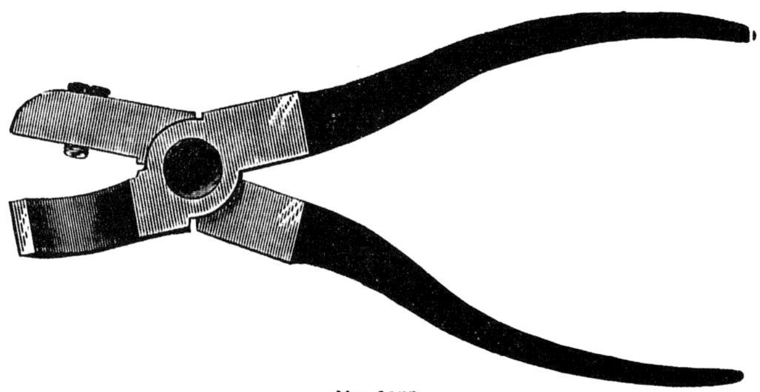

No. 1182.

No. 1183.

No. 1181.

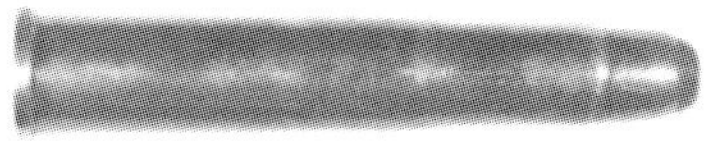

.25-20-85, 1.63". This was a privately developed cartridge adopted by most makers of single shot rifles. It has been included with the Winchester Model 1885 single shot cartridges as a chambering available with the announcement of that rifle.

.32 Long Rifle, 82" case. An early version of the similar .32 Long rimfire, this one is loaded with an inside lubricated bullet.

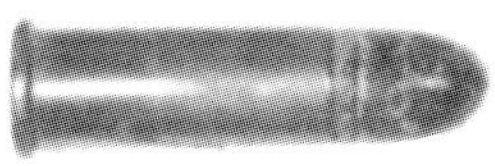

.32-13/16". Used in various small single shot rifles.

.32 Smith & Wesson Rifle. The nose of the bullet in this cartridge is flush with the case mouth. It first appears in the Winchester catalogue in January 1884 and was dropped in 1910.

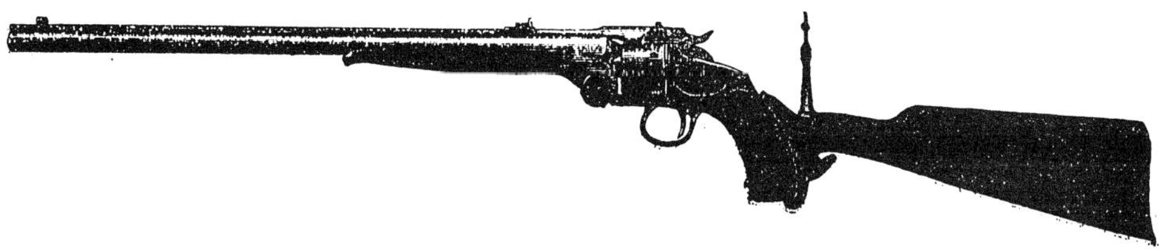

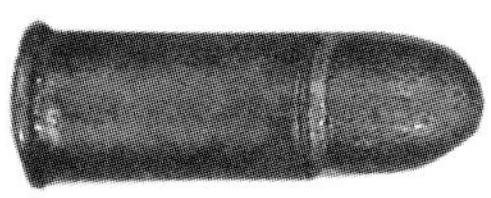

.38 Short or Long. Mentioned as options for small-frame single shot rifles by Ballard and other makers but soon dropped.

.44 Evans Old Model, .99" case. Used in the Warren Evans buttstock magazine rifle patented September 19, 1871.

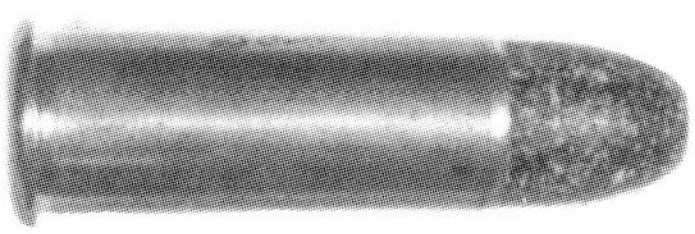

.44 Evans New Model, 1.54". Used in the same rifle.

.50 so-called Meigs. One of many experimentals by United States Cartridge Company. Plain flat head with slightly dished primer.

.42 Russian Berdan No. 2 (10.66x57R). Commercial cartridges were loaded for this single shot, bolt action military rifle—example W. R. A. Co. 42 RUSS. (See Vol. Two, p. 190-193).

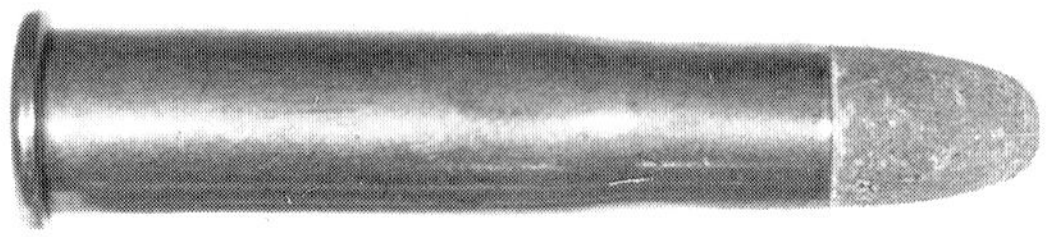

.43 Spanish Remington (11x57R). This cartridge was used as a military load by Spain and Argentina and also in United States militia Remington rolling block rifles. It was a popular commercial round and specimens will be found headstamped W. R. A. Co. 43 SPANISH, also .43-77. (See Vol. Two, pp 235-238).

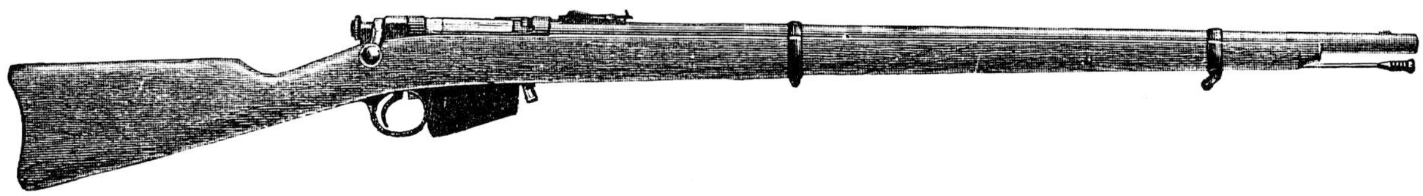

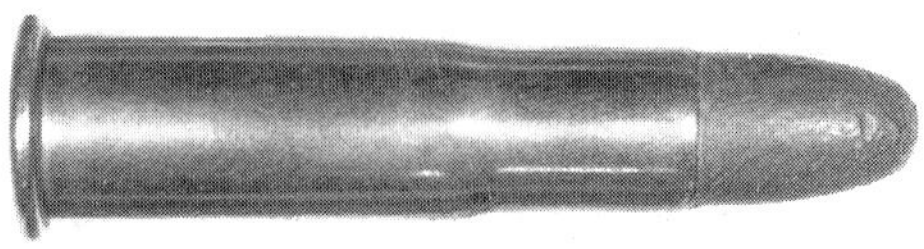

.43 Remington carbine. Used in the Remington rolling block and similar Whitney carbines.

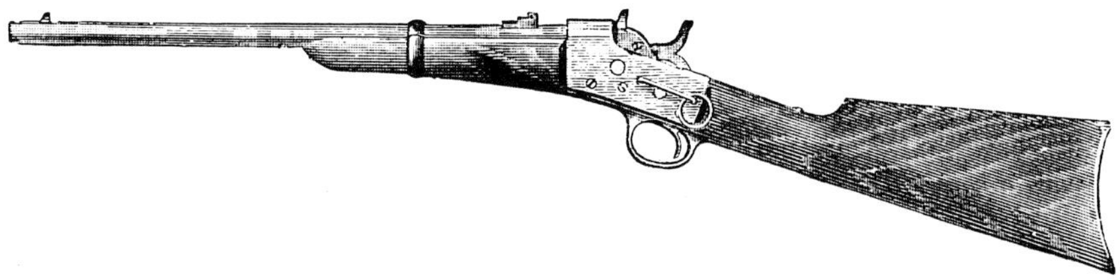

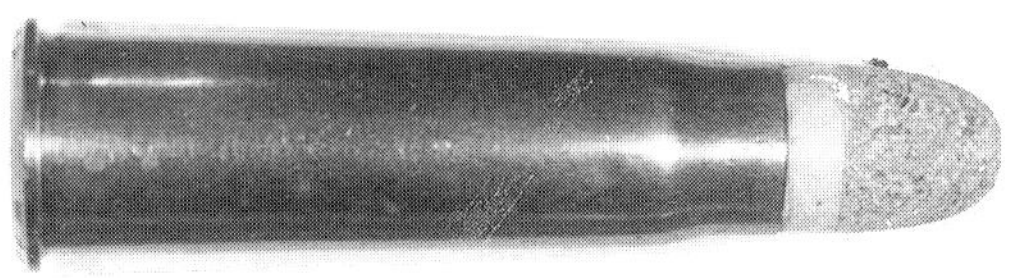

.446 Brazilian Comblain (11.4x53R). This cartridge has a raised headstamp—W. R. A. Co. COMBLAIN. This curious single shot, falling block rifle was used by Brazil and Chile, and with other cartridges in Europe and elsewhere.

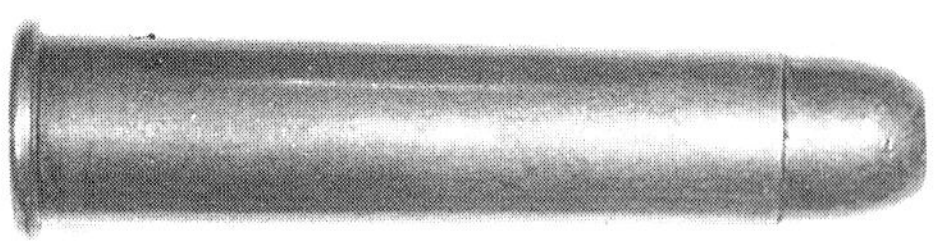

.45-70-405 or 500 Springfield. The old war horse of the United States Army from 1873 to introduction of the .30-40 Krag-Jorgensen rifle in the late 1880s. It has been used in single shot and lever action magazine, even in double rifles, and will no doubt survive into the 21st century.

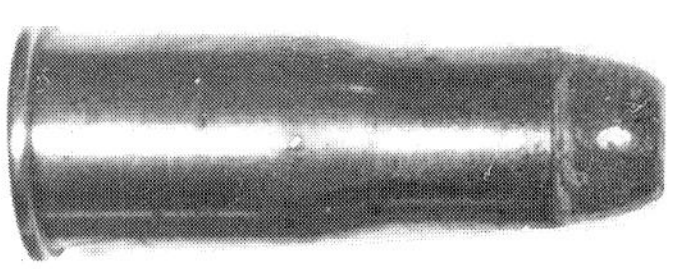

.45-1.452. Plain head, Berdan primer, flat-nosed lead bullet. Specimens exist with a paper-patched bullet and the same flat nose. This cartridge was made for some magazine rifle, and its similarity to the rimfire .56-46 Spencer with long neck is apparent but no sure identity has been established.

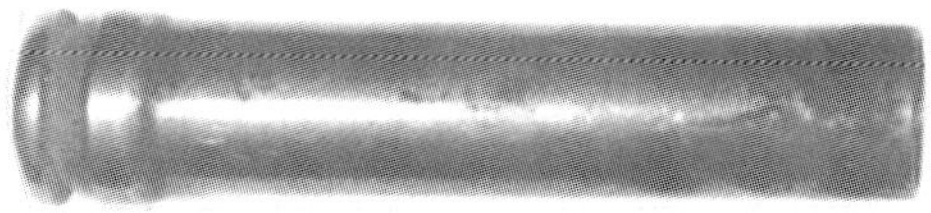

.40-2 7/16" Roper. This is the steel case for use in the rifle invented by Sylvester H. Roper and patented April 10, 1866 (U. S. no. 53,881). It had a finned revolving cartridge carrier housed in a metal casing and fired from the open-bolt position. When the trigger was pulled the hammer, bolt and cartridge were driven forward to chamber the cartridge, which was then fired. The bullet lay flush with or below the case mouth. (Source: **Spencer Repeating Firearms** by Roy M. Marcot.

.45-2 14" Van Choate. A bolt action military rifle offered in the U. S. 1872, it did enter production and was purchased by Panama. The case is slightly longer and larger in diameter than the .45-70, and the bullet is paper patched. The plain head with raised ring indicates manufacture by Union Metallic Cartridge Company. (See Vol. V,1, no. 121, **Collector's Ammunition** and John Witzel's index to same for other listings.

.45-78- 2 3/8" Wolcott. Also found with the same bullet in a longer 2 5/16" case. This cartridge was named for Mr. H. H. Wolcott, President of the Starr Arms Company; it evidently was intended for competitive shooting.

.45-2.6" Brown Standard. The bullet shape sets this cartridge apart from the .45-2.6" Sharps, and the bullet is patched down to the base, not over it. The base of the bullet is marked in raised letters—H. F. CLARK, POKEEPSIE, N. Y. The flat-based, Berdan-primed case has no headstamp.

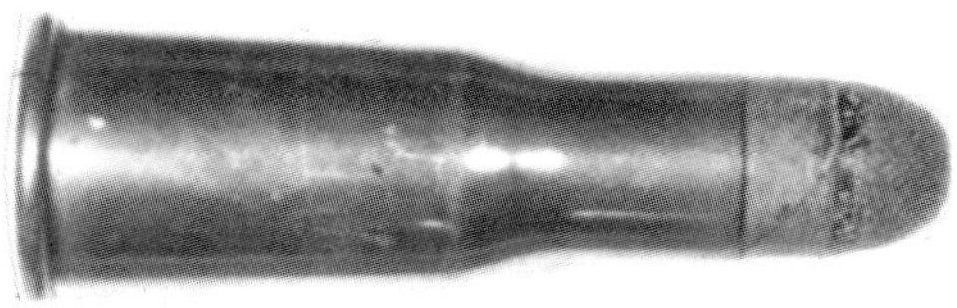

.45-85 Ward Burton. Probably never a commercial cartridge, the designers of the rifle necked the .58 Musket cartridge down to .45 and competed for U. S. military adoption in the 1870s without success (See Vol. Two, p. 43).

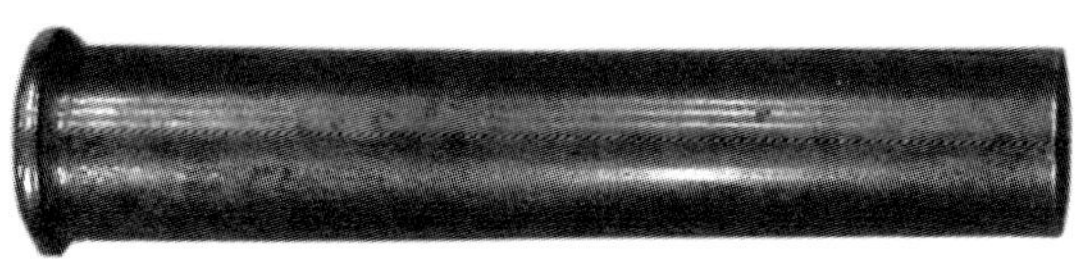

.45-2.81" "Holden." This case is smaller in rim dimensions than the .44 and .45 Sharps cases but is larger in head and body dimensions. The head on this fired Winchester-made case has the slightly rounded rim with raised ring around the Boxer primer pocket. Rob Buttweiler reports in Vol. VIII, No. 3 (234 and 235) that Cyrus B. Holden worked for Frank Wesson in the early 1870s, then began making his own single shot rifles. RTB also shows a nickeled specimen of this case. Tom Rowe reports a 48" barrel specimen of the rifle (see **Boys' Single Shot Rifles**, James Grant).

.50 Springfield carbine. The cartridges were loaded commercially by Winchester, U.M.C. and no doubt U. S. Cartridge Company for use in surplus carbines.

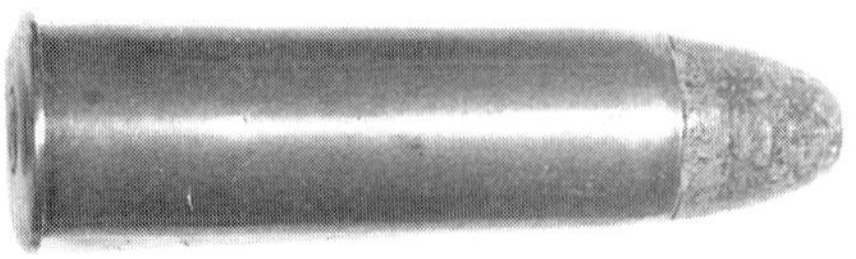

.50-70 Spingfield. Used not only in surplus rifles but chambered in Peabody, Remington and other single shot hunting and target rifles. (See the Peabody section).

.58 carbine. Also loaded commercially for use in surplus military firearms.

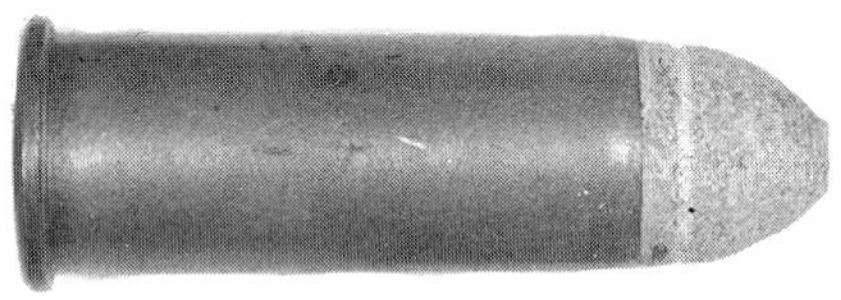

.58 Musket. The head at the left is the early Winchester Milbank, not a success, a later Winchester, and a Berdan, presumably by U. M. C.

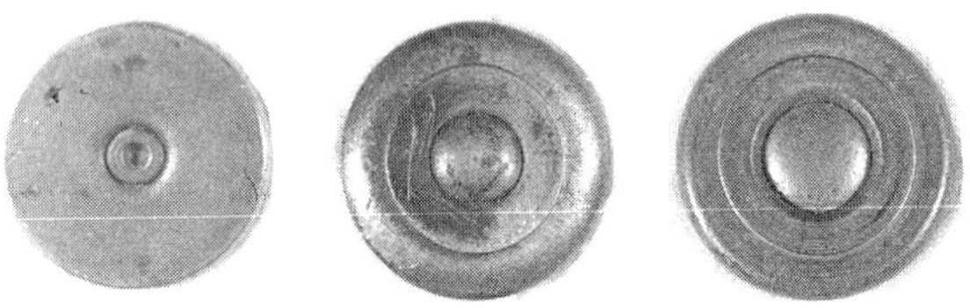

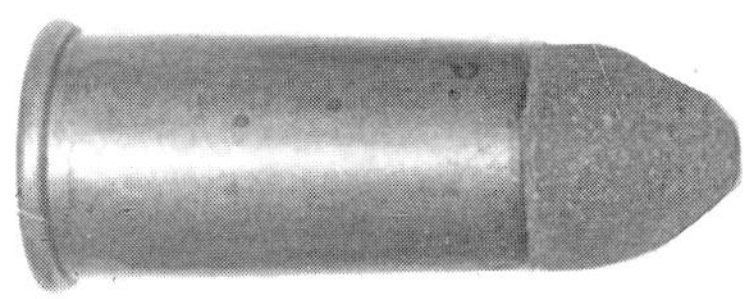

.58 Roberts. A case somewhat larger in head dimensions than the .58 Musket. It was made for the Roberts muzzle-loading musket conversion. This specimen has a large Berdan primer, ringed head, no marks. (See Vol. Two, p. 42).

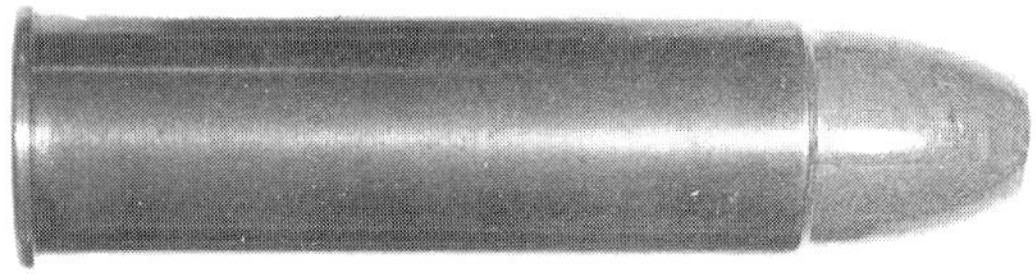

.58 Dangerfield & Lefever. This massive cartridge with its 2.1" case and flat-nosed bullet was produced by U. M. C. for a hinged, break-open single shot rifle produced at Syracuse, New York in the 1870s by Francis Dangerfield and Daniel Lefever. It was covered by U. S. patent no. 130,984 granted September 3, 1872. The makers called it the "Destroyer."

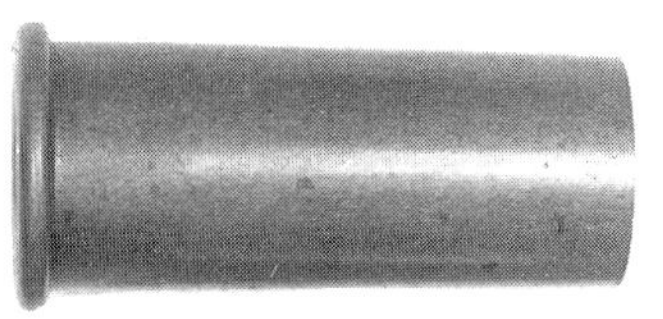

.60 or .61 ? Primed empty case with large brass Berdan primer,no headstamp, ringed head. We have all associated it with the Roberts and the Russian and Spanish Berdan cartridges but no rifle is yet known for it. (See "Collector's Ammunition," Vol. V-3, no. 156).

U.M.C.

TRADE U.M.C. MARK.

U.M.C. PERCUSSION CAPS.

	Per M in tenth tin bxs.	Cont. of Case.	Weight of Case.
	s. d.	M	Lbs.
F.C. TRIMMED EDGE Percussion Caps	1 8	100	71
F.C. TRIMMED EDGE Percussion Caps, CENTRAL-FIRE	1 8	100	70
Sizes: 10, 11, 12, 13, and 18.			
U.M.C. GROUND EDGE, foil-l'd, WATERPROOF	2 1	100	99
U.M.C. GROUND EDGE, foil-l'd, WATERPROOF, CENTRAL-FIRE	2 1	100	98
Sizes: 9, 10, 11, 12, 13, 14, and 18.			
U.M.C. DOUBLE WATERPROOF, CENTRAL-FIRE EXTRA HEAVY	4 7	100	119
Sizes: 9, 10, 11, 12, 13, 14, and 18.			

Above caps packed in quarter tin boxes, if required, at same prices.

When ordering, give assortment and state *if ground edge U.M.C. or trimmed edge F.C.* caps are required.

Nos. 9, 13, 14 and 18 are only packed when specially ordered.

COLT'S PISTOL CAPS.

In quarter boxes	2 11	100	67
In tenth ..	3 2	100	70

MUSKET CAPS.

In quarter boxes, tin	2 3	100	122
In tenth ,, tin	2 4	100	128
In .. .. paper	2 3	100	85

Numbers of U.M.C. caps correspond with the English.

PRIMERS.

		Per M.	Wgt. of Case.
		s. d.	Lbs.
A	Berdan Prim. No. 1, for Sp. & Milit. Ctgs.	7 1	85
	Berdan .. No. 1½, for Met. Shot Shells	7 1	80
B	Berdan .. No. 2	7 1	52
C	U.M.C. Prim. No. 00	8 4	50
D	U.M.C. .. No. 0	8 4	57
E	U.M.C. .. No. 1	8 4	65
F	U.M.C. .. No. 1½, brass	8 4	66
G	U.M.C. .. No. 2 cop. for SHOT SHELLS	8 4	86
H	U.M.C. .. No. 2½ brs. for S. and M. Ctg.	8 4	89

"U.M.C." Primers will interchange with same sizes of other makers.

Exact size of No. 2 U.M.C. Primer. Guaranteed sure fire. Enlarged view showing Improved Anvil.

RIM-FIRE CARTRIDGES.

BULLET BREECH (B. B.) CAPS.

	Per M.	Weight of Case.
	s. d.	lbs.
In quarter boxes (250)	16 8	37
In tenth .. (100)	16 8	38

CONICAL BULLET (C. B.) CAPS.

In quarter boxes (250), swaged bullets	18 9	55
In tenth .. (100) ..	18 9	56

B. B. and C. B. caps are packed 10 M. in a Case.

When ordering, state if wanted tenth or quarter boxes.

Blank cartridges list same as ball cartridges except 22 and 23.	Per M.	Cont. of Case.	Weig't of Case.	Primed Shells. Per M.
	£ s. d.	M.	Lbs.	£ s. d.
22 Short, 50 in box	1 0 10	10	67	0 8 4
22 Short, 100 ..	1 0 10	10	66	0 8 4
22 Long, 50 ..	1 5 0	10	74	0 10 5
22 Long, 100 ..	1 5 0	10	73	0 10 5
22 Colt's M. R. Short, 50 in box	1 0 10	10	67	0 8 4
22 Colt's M. R. Short, 100 ..	1 0 10	10	66	0 8 4
22 Colt's M. R. Long, 50 ..	1 5 0	10	74	0 10 5
22 Colt's M. R. Long, 100 ..	1 5 0	10	73	0 10 5
22 Long, Rifle, 50 in box	1 5 0	10	90	0 10 5
22 Long, Rifle, 100 ..	1 5 0	10	89	0 10 5
22 Extra Long	1 17 6	10	93	0 12 6
25	1 15 5	5	47	0 12 6
30 Short	1 17 6	5	64	0 12 6
30 Long	2 3 9	5	83	0 14 9
32 Extra Short	2 1 8	5	65	0 12 6
32 Short	2 1 8	5	86	0 14 9
32 Long	2 7 11	5	100	0 16 8
32 Extra Long	3 8 9	2	45	1 7 1
38 Short	3 6 8	3	82	1 0 10
38 Long	3 15 0	3	96	1 5 0
38 Extra Long	5 4 2	2	77	1 19 7
41 Short	3 2 6	3	78	0 18 9
41 Long	3 12 11	3	96	1 2 11
42 Cal. No. 64, Forehand and W.	6 5 0	2	88	1 11 3
44 Short	4 15 10	2	81	1 5 0
44 Long	5 4 2	2	92	1 11 3
44 Henry, Flat	5 0 0	2	87	1 11 3
44 Henry, Pointed	5 4 2	2	86	1 11 3
44 C. and W.	5 4 2	2	90	1 11 3
44 Howard, Long	5 12 6	1	51	1 17 6
44 Extra Long	7 10 0	1	57	2 5 10
46 Pistol, Short	5 4 2	2	94	1 11 3
46 Rifle, Long	6 17 6	2	123	1 15 5
46 Extra Long	9 11 8	1	71	2 5 10
56-46 Spencer Sporting	8 6 8	1	71	2 10 0
56-50 Spencer Carbine	8 6 8	1	73	2 10 0
56-52 Spencer Rifle	8 6 8	1	78	2 10 0
56-56 Spencer Carbine	8 6 8	1	77	2 10 0

TRADE U.M.C. MARK.

C.F. PISTOL AND RIFLE CARTRIDGES.

Primer	Blank cartridges list same as ball cartridges.	Per M.	Quan. Powd.	Wgt. Bullet	Cont. Case.	Wgt. Cage.
		£ s. d.	Grs	Grs.	M	Lbs.
E	22 Extra Long	2 14 2	9	45	...	...
E	22 Winchester	3 2 6	14	45	2	38
D	32 Short Revolver	2 5 10	9	80	5	89
D	32 Long Revolver or Rifle	2 10 0	13	90	5	100
D	32 Smith and Wesson	2 5 10	10	88	5	99
D	300 Rook Rifle	3 2 6	10	80	5	90
E	32 Winchester	3 6 8	20	100	2	57
E	32 Colt's M.R. for Colt's Lightn'g Rifle	3 6 8	20	100	2	58
F	32 Extra Long Ballard	3 15 0	18	105	3	80
F	32 ,, ,, Smith & Wesson Rifle	4 3 4	17	98	3	86
E	32-44 S. & W. Gallery, round ball	3 15 0	6	50	3	48
E	32-44 ,, Target, grooved	3 19 2	11	83	3	...
D	38 Smith & Wesson	2 16 3	15	146	2	67
E	38-44 S. & W. Target, grooved	4 15 10	23	146	2	...
D	38 Merwin, Hulbert & Co.	2 16 3	15	142	2	63
D	38 Short Revolver	2 16 3	15	125	4	112
D	38 Long, for Colt's D.A. Pistols, etc.	3 0 5	18	152	3	98
E	38 Winchester	3 19 2	40	180	2	96
E	38 Colt's M.R. for Colt's Lightn'g Rifle	3 19 2	40	180	2	96
E	38 Extra Long	4 15 10	31	146	2	81
D	41 Short, Colt's Single Action	3 2 6	15	163	2	67
D	41 ,, ,, Double ..	3 2 6	15	163	2	66
D	41 Long ,, ,, ..	3 12 11	22	196	2	83
D	44 S. & W. No. 3, American	3 19 2	23	218	2	93
G	44 ,, ,, 3, Russian	4 3 4	23	256	2	106
G	44 ,, Russ. Gallery, round ball	4 11 8	7	120	2	...
G	44 ,, ,, ,, grooved	5 4 2	7	110	2	...
E	44 Merwin, Hulbert & Co.	4 3 4	28	222	2	97
G	44 Colt's Old Model Revolver	4 3 4	23	210	2	94
E	44 Long, Ballard	4 11 8	35	227	2	99
E	44 Extra Long, Wesson	6 5 0	50	257	2	123
E	44 ,, ,, Ballard	6 5 0	50	265	2	126
E	44 Winchester	3 19 2	40	217	2	107
E	44 Colt's M.R. for Colt's Lightn'g Rifle	3 19 2	40	217	2	107
H	44 Evans' New Model	5 8 4	43	276	2	124
E	44 ,, Old ,,	4 11 8	33	220	2	95
E	44 Webley	3 8 9	19	200	2	83
E	44 Bull Dog	3 0 5	15	170	2	69
E	45 Webley	4 3 4	20	230	2	97
E	450 Revolver	3 2 6	13	225	2	92
E	45 Smith & Wesson	5 4 2	30	250	2	110
G	45 Colt's U. S. A.	4 11 8	40	250	2	118
G	45 .. round ball	4 11 8	...	138	2	...
A	50 Pistol	5 16 8	20	300	1	69

C.F. CARTRIDGES, FOR TARGET AND SPORTING RIFLES.

Primer	Calibre.	All packed 1000 in a case. Blank cartridges list same as ball cartridges. Express bullets, 14s.7d. extra per M.	Per M.	Lgth. Shell.	Quan. Powd.	Wgt. Bullet	Style of Bullet.	Wgt. Case.
			£ s. d.	Inch.	Grs.	Grs.		Lbs.
F	32-35	Stevens	5 12 6	1 7/8	35	165	Grvd.	...
F	32-40	Marlin & Ballard	5 12 6	2 3/32	40	165	,,	59
F	32-13	M. & B. short range	5 12 6	2 3/32	13	98	,,	...
F	32-40	Ballard	5 12 6	2 3/32	40	185	Patch	58
F	32-40	Bullard Rifle	6 11 3	1 25/32	40	150	Grvd.	55
F	38-45	.. ..	6 11 3	1 25/32	45	190	,,	64
F	38-55	Marlin & Ballard	6 17 6	2 3/32	55	255	,,	74
F	38-20	M. & B. short range	6 17 6	2 3/32	20	155	,,	...
F	38-55	Ballard, straight shell	6 17 6	2 3/32	55	255	Patch	75
F	38-40	Rem. Hep. No. 3 Rifle	6 11 3	1 3/4	40	245	Grvd.	69
F	38-40	.. .. ,, ..	6 11 3	1 3/4	40	255	Patch	70
F	38-50	,, ...	7 5 10	2 1/4	50	255	,,	...
F	38-50	.. ..	7 5 10	2 1/4	50	245	Grvd.	...
H	38-56	Winchester	6 17 6	2 1/16	56	255	,,	...
H	38-56	Colt's M. R.	6 17 6	2 1/16	56	255	,,	...
H	38-90	Winchester Express	10 8 4	3 1/4	90	217	,,	...
A	40-50	Sharps & Rem. straight	7 5 10	1 7/8	50	265	Patch	74
H	40-65	.. .. ..	8 6 8	2 1/2	65	330	,,	92
H	40-70	.. .. ..	8 6 8	2 1/2	70	330	,,	93
H	40-65	.. .. ..	8 2 6	2 1/2	65	330	Grvd.	...
H	40-70	.. .. ..	8 2 6	2 1/2	70	330	,,	...
A	40-50	.. only, necked	7 5 10	1 11/16	50	265	Patch	72
A	40-70	.. .. ..	8 6 8	2 1/4	70	330	,,	91
A	40-90	.. .. ,,	9 7 6	2 5/8	90	370	,,	107
H	40-90	.. straight	10 8 4	3 1/4	90	370	,,	127
A	40-50	Reg. & Rem. necked	7 5 10	1 11/16	50	295	,,	82
A	40-70	.. ,, ...	8 6 8	2 1/4	70	365	,,	102
A	40-90	.. ..	10 8 4	2 5/8	90	365	..	108

TRADE U.M.C. MARK

CENTRAL-FIRE CARTRIDGES

FOR TARGET AND SPORTING RIFLES.

Primer p. 62.	Calibre.	All packed 1000 in a case. Blank cartridges list same as ball cartridges. Exp. bullets 14/7 extra per M.	Per M. £	s.	d.	Length Shell.	Quan. Powd.	Wgt. Bullet.	Style of Bullet.	Wgt. Case.
			£	s.	d.	Inch.	Grs.	Grs.		Lbs.
F	40-70	Ballard, straight	8	6	8	2 3/8	70	330	Patch	100
F	40-85	Ballard	10	8	4	2 15/16	85	370	,,	107
F	40-90	,, ...	10	8	4	2 15/16	90	370	,,	107
F	40-60	Winchester Rifle	6	13	4	1 7/8	60	210	Grvd.	77
H	40-65	.. ..	6	17	6	2 1/16	65	260	,,	...
H	40-75	.. Express ...	8	2	6	2 4/16	75	260	,,	...
H	40-82	.. Mod. 1886	7	10	0	2 4/16	82	260	..	90
H	40-110	.. Express ...	16	13	4	3 1/4	110	260	,,	...
F	40-75	Bullard Rifle...........	6	17	6	2 1/16	75	258	,,	81
H	40-90	,, ...	8	6	8	2 1/16	90	300	,,	96
F	40-60	Marlin Magazine Rifle	6	17	6	2 1/16	60	260	,,	83
F	40-60	Colt's .. ,,	6	17	6	2 1/16	60	260	,,	83
A	40-70	What Cheer, necked...	8	6	8	1 3/4	70	380	Patch	108
A	40-90	.. ..	10	8	4	2 1/16	90	500	,,	133
A	44-60	Sporting	8	2	6	1 9/16	60	395	Grvd.	102
A	44-77	Sharps only	9	3	4	2 1/4	77	405	Patch	106
A	44-90	.. ..	11	13	4	2 5/8	90	500	,,	126
A	44-105	.. ..	12	1	8	2 5/8	105	520	,,	...
A	44-77	Reg. and Rem.	9	3	4	2 1/4	77	470	,,	119
A	44-90	.. ..	9	7	6	2 5/8	90	470	,,	123
A	44-90	Rem. Special, ...	11	13	4	2 7/16	90	550	,,	...
A	44-95	What Cheer, ...	10	8	4	2 5/16	95	550	,,	139
A	45	Peabody Sport'g.. ...	7	1	8	1 17/24	50	290	,,	80
A	45-70	Sharps, straight	7	10	0	2 1/16	70	420	,,	109
A	45-70	.. Express st'ght.	8	4	7	2 1/16	70	357	,,	109
A	45-75	.. straight	7	10	0	2 1/16	75	420	,,	106
A	45-90	.. ..	12	1	8	2 6/16	90	500	,,	130
A	45-105	.. ..	12	10	0	2 7/8	105	550	..	130
H	45-125	.. & Winch., st'ght	16	13	4	3 1/4	125	500	..	...
F	45-70	Marlin Magazine	7	5	10	2 1/16	70	405	Grvd.	106
F	45-85	.. ..	7	14	2	2 1/16	85	285	..	93
A	45-40	.. Shot, l'dwithshot	10	8	4	2 1/2	40	...	...	...
H	45-70	Gov. Pchd. Bul. Ex.st.	8	4	7	2 1/16	70	345	Patch	...
H	45-70	.. Grvd.	8	0	5	2 1/16	70	345	Grvd.	...
H	45-55	.. Carbine...........	7	5	10	2 1/16	55	405	..	98
H	45-70	.. Solid Head	7	5	10	2 1/16	70	405	..	106
H	45-70	.. S. H. New Mod.	7	18	4	2 1/16	70	500	..	121
H	45- 5	.. Armoury Practice	6	5	0	2 1/16	5	140	Rnd.	55
H	45-60	Winchester	6	17	6	1 7/8	60	300	Grvd.	88
H	45-75	.. ..	7	5	10	1 7/8	75	350	..	96
H	45-82	.. ..	7	10	0	2 4/16	82	405	..	...
H	45-85	.. ..	7	10	0	2 4/16	85	350	,,	...
H	45-85	.. Express ...	7	18	4	2 4/16	85	300	..	...
H	45-90	.. Mod. 1886	7	10	0	2 4/16	90	300	..	95
H	45-125	.. Express ...	16	13	4	3 1/4	125	300	..	...
H	45-60	Colt's M. R.	6	17	6	1 7/8	60	300	..	88
F	45-85	.. ..	7	14	2	2 1/16	85	290	..	92
F	45-85	Bullard Rifle...........	7	14	2	2 1/16	85	290	..	92
H	50-115	.. Express	18	15	0	2 3/16	115	300	..	120
H	50-115	.. Solid Ball ...	18	15	0	1 3/16	115	346	..	133
A	50-90	Sharps, straight	12	1	8	2 1/2	90	473	Patch	130
H	50-95	Winchester Express...	8	6	8	1 15/16	95	300	Grvd.	95
H	50-95	.. Solid Ball	8	6	8	1 15/16	95	312	..	101
H	50-95	Colt's M. R. Express	8	6	8	1 15/16	95	300	..	95
H	50-95	.. Solid Ball	8	6	8	1 15/16	95	312	..	101
H	50-70	Sporting	9	3	4	1 3/4	70	425	Patch	...

PRIMED SHELLS AND BULLETS

FOR CENTRAL-FIRE PISTOLS AND RIFLES.

Primer p. 62.		Primed Shells. £	s.	d.	Bullets. £	s.	d.	Wgt. of Bullet.
		£	s.	d.	£	s.	d.	Grains.
E	22 Extra Long	1	17	6	0	10	5	45
E	22 Winchester........	1	17	6	0	10	5	45
D	32 Short Revolver	1	5	0	0	12	6	80
D	32 Long Revolver or Rifle........	1	7	1	0	12	6	90
D	32 Smith & Wesson	1	5	0	0	12	6	88
E	32 Winchester........	1	17	6	0	16	8	100
E	32 Colt's M. R. for Colt's Lightning Rifle	1	17	6	0	16	8	100
F	32 Extra Long Ballard	1	13	4	0	14	7	105
F	32 .. .. Smith & Wesson Rifle	2	1	8	1	0	10	98

U.M.C.

TRADE U.M.C. MARK

PRIMED SHELLS AND BULLETS

FOR CENTRAL-FIRE PISTOLS AND RIFLES.

Primer p. 62.		Primed Shells. £	s.	d.	Bullets. £	s.	d.	Wgt. of Bullet.
		£	s.	d.	£	s.	d.	Grains.
E	32-44 S. & W. Gallery, round ball ...	2	1	8	0	16	8	50
E	32-44 .. Target, grooved	2	1	8	1	0	10	83
D	38 Smith & Wesson	1	11	3	0	18	9	146
E	38-44 S. & W. Target, grooved	2	10	0	1	0	10	146
D	38 Merwin, Hulbert & Co.	1	13	4	0	18	9	142
D	38 Short Revolver	1	9	2	0	16	8	125
D	38 Long, for Colt's D.A. Pistols, &c.	1	11	3	0	18	9	152
E	38 Winchester........	2	1	8	1	0	10	180
E	38 Colt's M. R. for Colt's Lightning Rifle	2	1	8	1	0	10	180
E	38 Extra Long........	2	5	10	1	0	10	146
D	41 Short, Colt's Single Action	1	11	3	0	16	8	163
D	41 .. .. Double	1	11	3	0	16	8	163
D	41 Long	1	13	4	1	5	0	196
D	44 S. & W. No. 3, American	1	15	5	1	5	0	218
G	44 ,, .. 3, Russian	1	19	7	1	7	1	256
G	44 .. Russ. Gallery, round ball	1	19	7	1	5	0	120
G	44 .. .. .. grooved...	1	19	7	1	11	3	110
E	44 Merwin, Hulbert & Co.	2	1	8	1	7	1	222
G	44 Colt's Old Model Revolver	2	1	8	1	5	0	210
E	44 Long, Ballard........	1	19	7	1	5	0	227
E	44 Extra Long, Wesson........	2	18	4	1	17	6	257
E	44 .. .. Ballard........	2	18	4	1	17	6	265
E	44 Winchester........	2	1	8	1	5	0	217
E	44 Colt's M. R. for Colt's Lightning Rifle	2	1	8	1	5	0	217
H	44 Evans' New Model........	2	5	10	1	13	4	276
E	44 .. Old	1	19	7	1	7	1	220
E	44 Webley	1	13	4	1	5	0	200
E	44 Bull Dog........	1	11	3	1	0	10	170
E	45 Webley	1	19	7	1	5	0	230
E	45 Smith & Wesson	2	3	9	1	9	2	250
G	45 Colt's U. S. A.	2	3	9	1	9	2	250
G	45 .. round	2	3	9	0	18	9	138
A	50 Pistol	2	14	2	1	13	4	300

TRADE U.M.C. MARK

PRIMED SHELLS AND BULLETS

FOR CENTRAL-FIRE TARGET AND SPORTING RIFLES.

Primer p. 62.	Calibre.	Express bullets, 14/7 extra per M.	Shells, Per M. £	s.	d.	Bullets, Per M. £	s.	d.	Wgt. of Bullet.	Style of Bullet.
			£	s.	d.	£	s.	d.	Grains.	
F	32-35	Stevens	3	15	0	1	10	2	165	Grvd.
F	32-40	Marlin & Ballard	3	2	6	1	10	2	165	..
F	32-13	M. & B. short range ...	3	2	6	1	5	0	98	..
F	32-40	Ballard	3	2	6	2	0	7	185	Patch.
F	32-40	Bullard Rifle	4	3	4	1	10	2	150	Grvd.
F	38-45	.. ..	4	3	4	1	13	4	190	..
F	38-55	Marlin & Ballard	3	15	0	1	19	7	255	..
F	38-20	M. & B. short range ...	3	15	0	1	9	2	155	..
F	38-55	Ballard, straight shell	3	15	0	2	5	10	255	Patch.
F	38-40	Rem. Hep. No. 3 Rifle	4	3	4	1	17	6	245	Grvd.
F	38-40	.. ..	4	3	4	2	5	10	255	Patch.
F	38-50	.. ..	4	7	6	2	5	10	255	..
F	38-50	.. ..	4	7	6	1	17	6	245	Grvd.
H	38-56	Winchester........	3	15	0	1	19	7	255	..
H	38-56	Colt's M. R........	3	15	0	1	19	7	255	..
H	38-90	Winchester Express ..	5	8	4	1	19	7	217	..

TRADE U.M.C. MARK

PRIMED SHELLS AND BULLETS

FOR CENTRAL-FIRE TARGET AND SPORTING RIFLES.

Primer p. 62.	Calibre.	Express bullets, 14/7 extra per M.	Shells, Per M. £	s.	d.	Bullets, Per M. £	s.	d.	Wgt. of Bullet.	Style of Bullet.
			£	s.	d.	£	s.	d.	Grains.	
A	40-50	Sharps & Rem. straight	4	3	4	2	5	10	265	Patch.
H	40-65	,, ...	4	11	8	2	7	11	330	,,
H	40-70	,, ...	4	11	8	2	7	11	330	..
H	40-65	.. .. ..	4	11	8	2	5	10	330	Grvd.
H	40-70	.. .. ..	4	11	8	2	5	10	330	..
A	40-50	.. only, necked ...	4	3	4	2	5	10	265	Patch.
A	40-70	.. ..	4	7	6	2	7	11	330	,,
A	40-90	.. ..	5	4	2	2	12	1	370	..
H	40-90	.. straight	5	8	4	2	12	1	370	..
A	40-50	Reg. & Rem. necked ...	4	3	4	2	7	11	295	..
A	40-70	.. ..	4	7	6	2	10	0	365	..
A	40-90	.. ..	5	4	2	2	10	0	365	..
F	40-70	Ballard, straight	3	19	2	2	7	11	330	..
F	40-85	Ballard	5	8	4	2	12	1	370	..
F	40-90	.. ..	5	8	4	2	12	1	370	..
F	40-60	Winchester Rifle	4	3	4	1	13	4	210	Grvd.
H	40-65	.. ..	3	15	0	1	17	6	260	..
H	40-75	.. Express ...	5	0	0	2	5	10	260	..
H	40-82	.. Mod. 1886	5	0	0	1	17	6	260	..
H	40-110	.. Express ...	8	15	0	2	5	10	260	..
F	40-75	Bullard Rifle	3	15	0	1	17	6	258	..
H	40-90	.. ..	4	11	8	2	3	9	300	..
F	40-60	Marlin Magazine Rifle	3	15	0	1	17	6	260	..
F	40-60	Colt's	3	15	0	1	17	6	260	..
A	40-70	What Cheer, necked ...	3	19	2	3	2	6	380	Patch.
A	40-90	.. ..	4	7	6	4	3	4	500	..
A	44-60	Sporting	4	7	6	2	10	0	395	Grvd.
A	44-77	Sharps only, necked ...	4	13	9	3	2	6	405	Patch.
A	44-90	.. ..	5	18	9	3	15	0	500	..
A	44-105	.. ..	5	18	9	3	15	0	520	..
A	44-77	Reg. and Rem.	4	13	9	3	6	8	470	..
A	44-90	.. ..	5	18	9	3	6	8	470	..
A	44-90	Rem. Special, ...	5	18	9	3	19	2	550	..
A	44-95	What Cheer, ...	4	11	8	4	15	10	550	..
A	45	Peabody Sprt'g	3	19	2	2	10	0	290	..
A	45-70	Sharps, straight........	4	11	8	3	2	6	420	..
A	45-70	.. Express, st'ght	4	11	8	3	4	7	357	..
A	45-75	.. straight........	4	11	8	3	2	6	420	..
A	45-90	.. ..	5	18	9	3	15	0	500	..
A	45-105	.. ..	6	5	0	4	3	4	550	..
H	45-125	.. & Winch. st'ght	8	15	0	3	15	0	500	..
F	45-70	Marlin Magazine	4	11	8	2	10	0	405	Grvd.
F	45-85	.. ..	4	11	8	1	19	7	285	..
A	45-40	,, Shot........	5	18	9	...	...	...	...	...
H	45-70	Gov. Patched Ex. st....	4	11	8	3	17	1	345	Patch.
H	45-70	.. Grooved	4	11	8	3	2	6	345	Grvd.
H	45-55	.. Carbine	4	11	8	2	10	0	405	..
H	45-70	.. Solid Head........	4	11	8	2	10	0	405	..
H	45-70	.. S. H. New Model	4	11	8	3	2	6	500	..
H	45- 5	.. Armoury Practice	4	11	8	0	18	9	140	Round
H	45-60	Winchester........	4	3	4	1	17	6	300	Grvd.
H	45-75	.. ..	4	3	4	1	19	7	350	..
H	45-82	.. ..	5	0	0	2	10	0	405	..
H	45-85	.. ..	5	0	0	1	19	7	350	..
H	45-85	.. Express ...	5	0	0	2	5	10	300	..
H	45-90	.. Mod. 1886	5	0	0	1	17	6	300	..
H	45-125	.. Express ...	8	15	0	2	5	10	300	..
H	45-60	Colt's M. R.	4	3	4	1	17	6	300	..
F	45-85	.. ..	4	11	8	1	19	7	290	..
F	45-85	Bullard Rifle	4	11	8	1	19	7	290	..
H	50-115	.. Express........	10	8	4	2	5	10	309	..
H	50-115	.. Solid Ball......	10	8	4	1	17	6	346	..
A	50-90	Sharps, straight........	6	13	4	3	6	8	473	Patch.
H	50-95	Winchester Express ...	4	15	10	2	5	10	303	Grvd.
H	50-95	.. Solid Ball	4	15	10	1	17	6	312	..
H	50-95	Colt's Express	4	15	10	2	5	10	300	..
H	50-95	.. Solid Ball	4	15	10	1	17	6	312	..
H	50-70	Sporting	4	13	9	3	2	6	425	Patch.

APPENDIX A—PRIMITIVES

During the past 15 years many primitive cartridges have been discovered that were not available in the early 1980s for inclusion in Volumes One and Two. From our late friend, Jim Tillinghast, came over 50 photographs of rare paper cartridges from the Alan Kelley collection, from which these following are just a sample. Ultimately they will appear in some future work in their full color. Mick McLaughlin provided the specimen with Japanese characters on the case. Randy Long took the photographs of the groups. Some of the specimens appear in Volumes One and Two but the Johnston & Dow, Doremus & Budd, and Bartholow patent cartridges have never been brought together before for a comparative study.

The four groups following the combustibles illustrate the tremendous variety of ideas that were tried out, gradually leading to the self-contained center fire cartridge. For the needle fire collection we are indebted again to Jim Tillinghast, Paul Klatt, Peyton Hall and Guenther Wohl.

Below is the first primer--a length of matchlock cord from the 16th century made of twisted tow, courtesy of Charles R. Suydam, with illustrations of how it all started, taken from **Historia Arma De Fogo Portatil** by Captain Luiz Mardel, Portuguese Army, 1887, courtesy of Roy M. Marcot.

18mm (.709) for Model 1830 Danish musket. The round ball is tied off from the powder charge, buff white paper case.

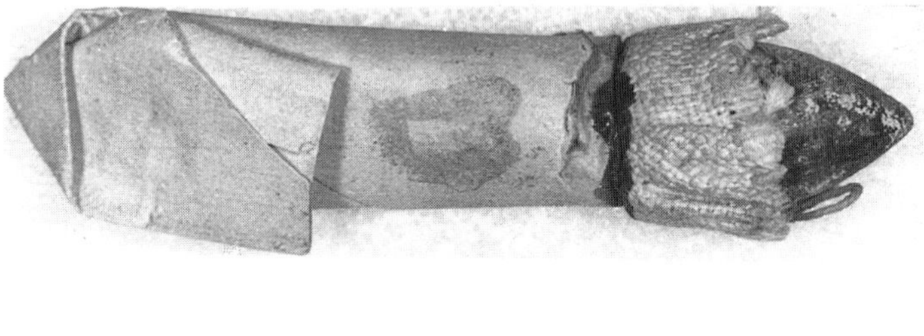

19mm (.748) for Wall Rifle, attributed to Norway but not confirmed. The conical bullet contained in what appears to be heavy muslin is tied off from the powder charge. When the powder had been poured down the muzzle the bullet in its muslin patch was rammed down on the charge.

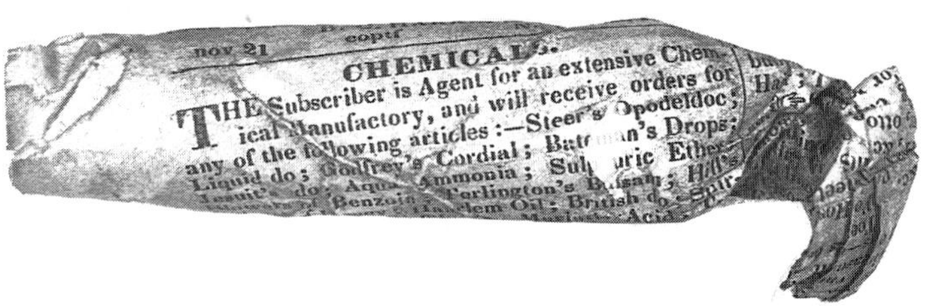

A home-made .69 flintlock musket cartridge in newsprint advertising the products of a Chemist (pharmacist). The paper case is glued over the round ball and the tail is twisted closed.

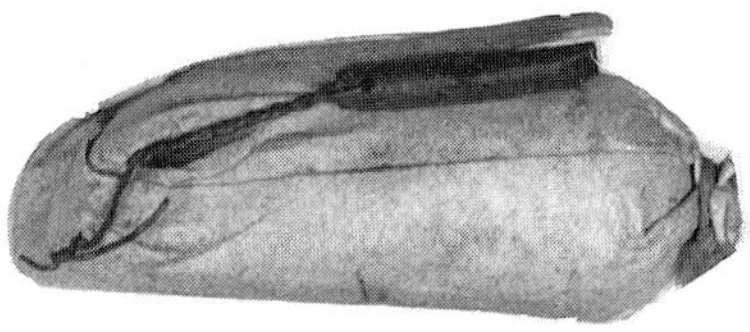

16.3mm (.642) pistol, round ball in buff white paper. There is a Console-system tube ignitor made of brass foil rolled into a tube just visible under the folded tail of the case. The tube was inserted in the touch hole and crushed by the falling hammer to set off the charge. See Volume One, pp. 17 and 24.

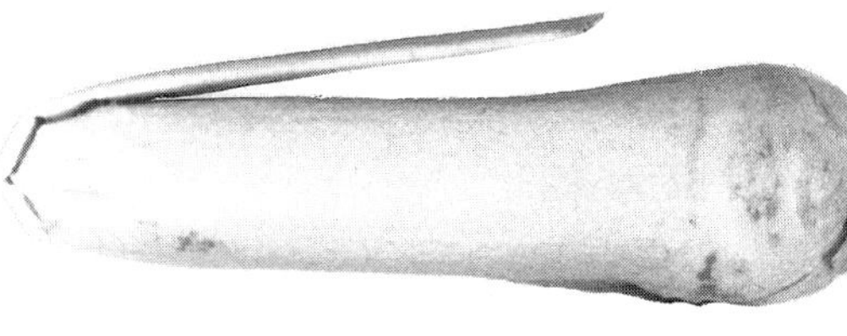

17.5mm (.689) Swiss musket cartridge, Model 1842, with lead ball in white paper case glued closed over the ball and tail twisted and folded along the case. The packet label reads—"Canton Du Geneve, Cartouches & Infanteri Du Cont ngent-Poudre de Berne No. 7-50 CART es a la Livre, Balles de 18 ½ La Livre."

Here is a .545" (13.84mm) cartridge of Japanese origin in white paper. Translation of the writing on the case may tell something about the rifled musket for which it was made.

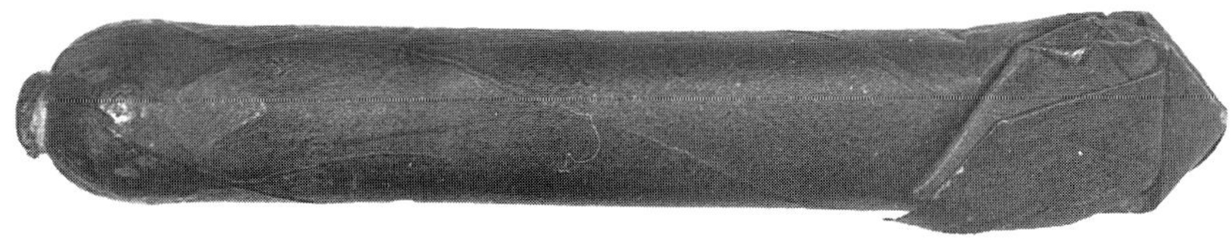

17.11mm (.669) Danish flintlock musket cartridge, tied closed over the round ball, very dark brown paper. Large charge of powder, part of which was used to prime the pan.

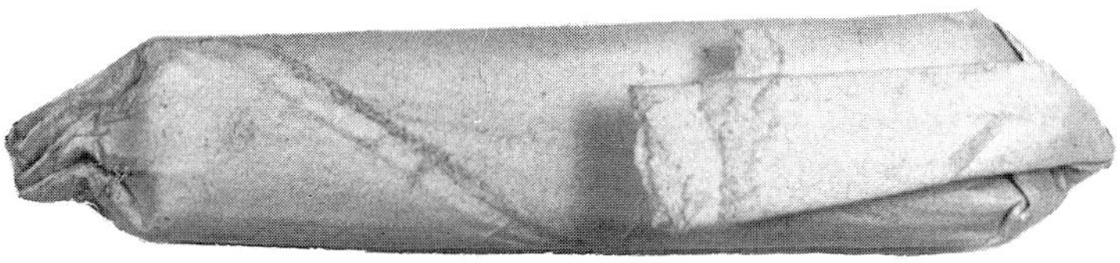

16.5mm Danish (?), round ball, buff white paper. This one is twisted closed over the ball.

.69 for the English Brunswick rifle, 1831. The case is tied over the ball and folded back, and also tied behind the ball.

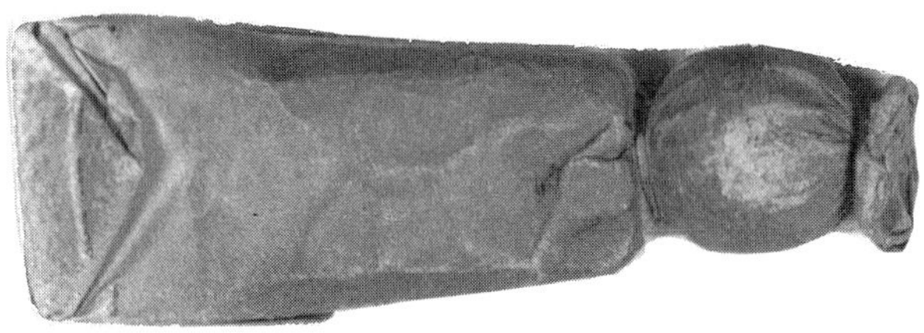

17.5mm (.689) musket, round lead ball is tied off from the powder charge. The case was torn open at the other end, the powder poured down the muzzle, and the bullet, paper and all, rammed home on the powder.

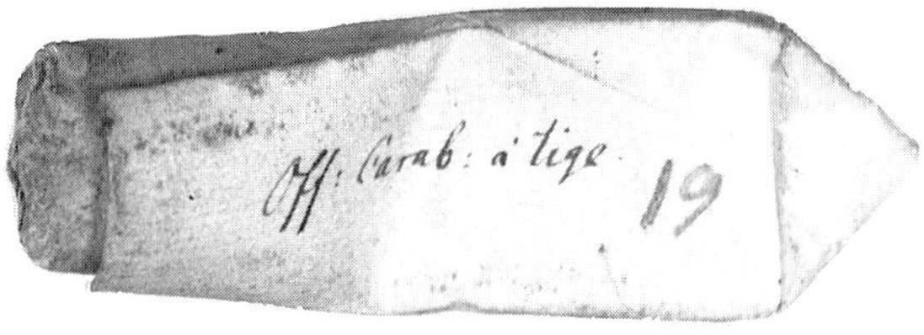

17.3mm (.681) Danish musket, lead bullet in the base, white paper. The inscription, "Off Carab: a ' tige" suggests that this was the cartridge for an officer's carbine with the post, or tige, projecting up from the base of the powder chamber against which the bullet was rammed to spread it into the rifling.

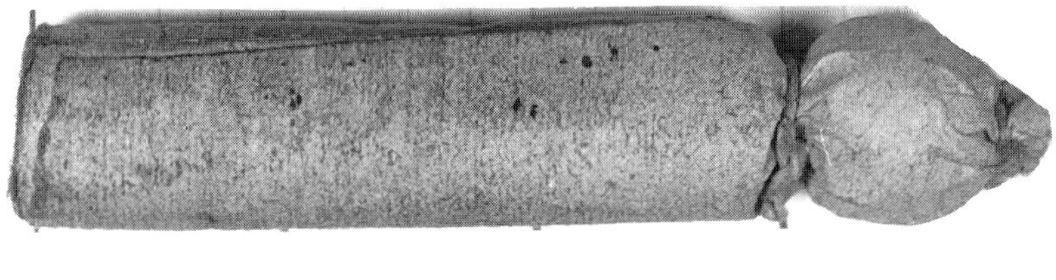

19.2mm (.756) Norwegian wall gun musket. Lead round ball tied off from the powder charge in tan paper. There is a felt wad exposed in the base.

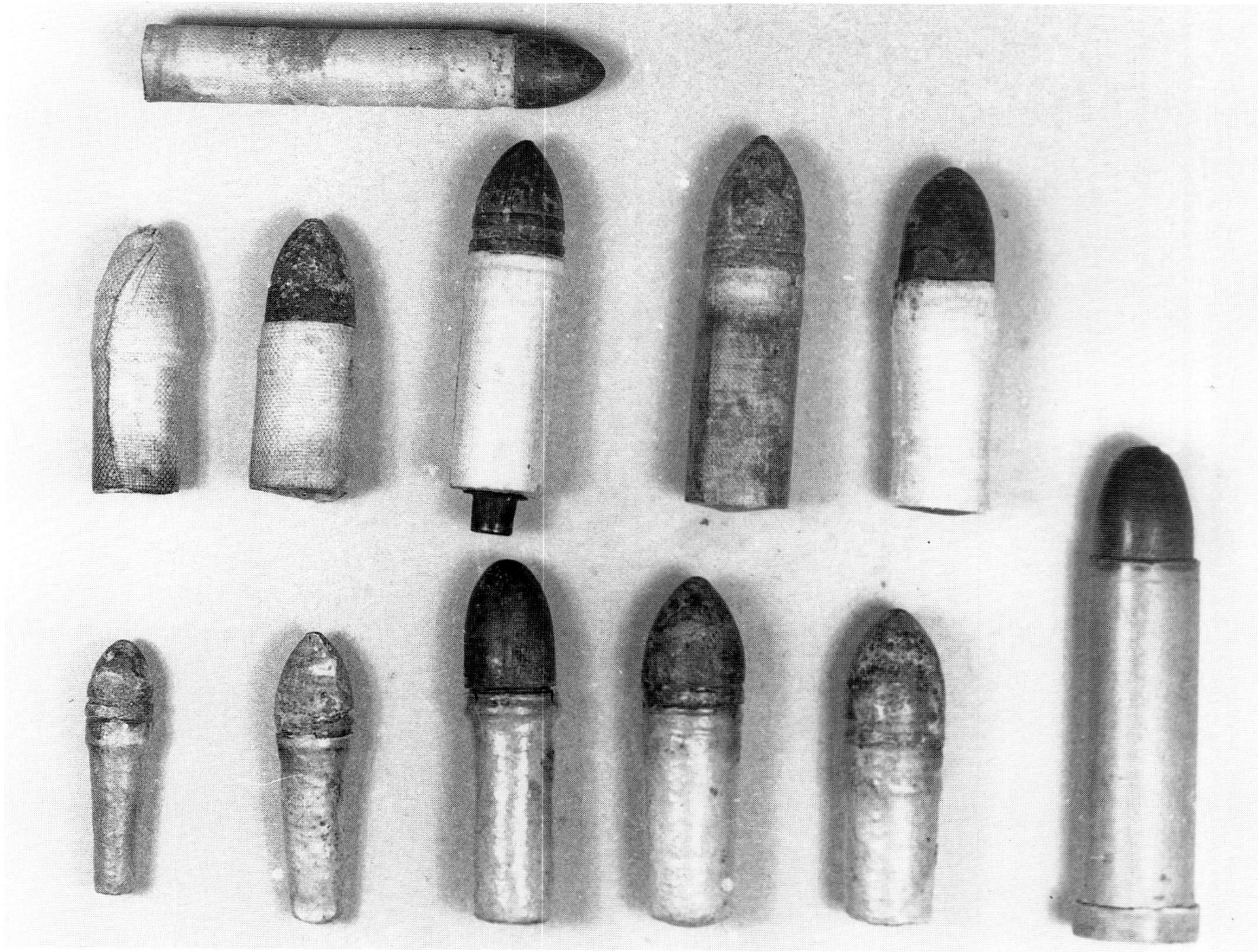

Top: .50 cal. Perry carbine. Second row, l to r: .50 cal. Jenks carbine, linen case; .50 Cosmopolitan carbine (Gwyn & Campbell), linen; .52 Sharps carbine, percussion cap attached to base, linen; same cartridge, no cap; .54 Starr carbine, linen. Third row, l to r: .36 cal. Remington Navy, Johnston & Dow patent collodion-covered case; .44 Colt Army, same patent; .50 Johnston & Dow; .574 for Springfield or Enfield rifled musket, same patent; .58 cal. rifled musket, same. .54 cal. Rochaz-Lindner, 1859, an early centerfire (see Vol. 1, pp 66-68 and Vol. 2, p. 200).

Top row, l to r: .705 cal. (17.9mm) Austrian musket, Johnston & Dow patent (U. S. 33,393, October 1, 1861) using an imflammable coating of the case so the cartridge could be loaded intact. The following six all made under the Roberts Bartholow patent (U. S. 32,345, May 21, 1861) for a compressed powder cemented to the bullet and water proofed with collodion tincture of shellac; .58 rifled musket, .54 rifled musket, .52 Sharps, .44 Colt Army, .36 Colt Navy, .32 Colt pocket revolver.

Second row: seven cartridges made under the Doremus & Budd patent (U. S. 34,725 and 34,744, March 18 and 26, 1862 covering a compressed powder charge affixed to the bullet in the form shown, and brushed with collodion. L to r: .31 for revolver, .44 Colt Army; .52 Hall carbine, .54 rifled-musket, .52 Sharps, .58 rifled musket, .69 musket.

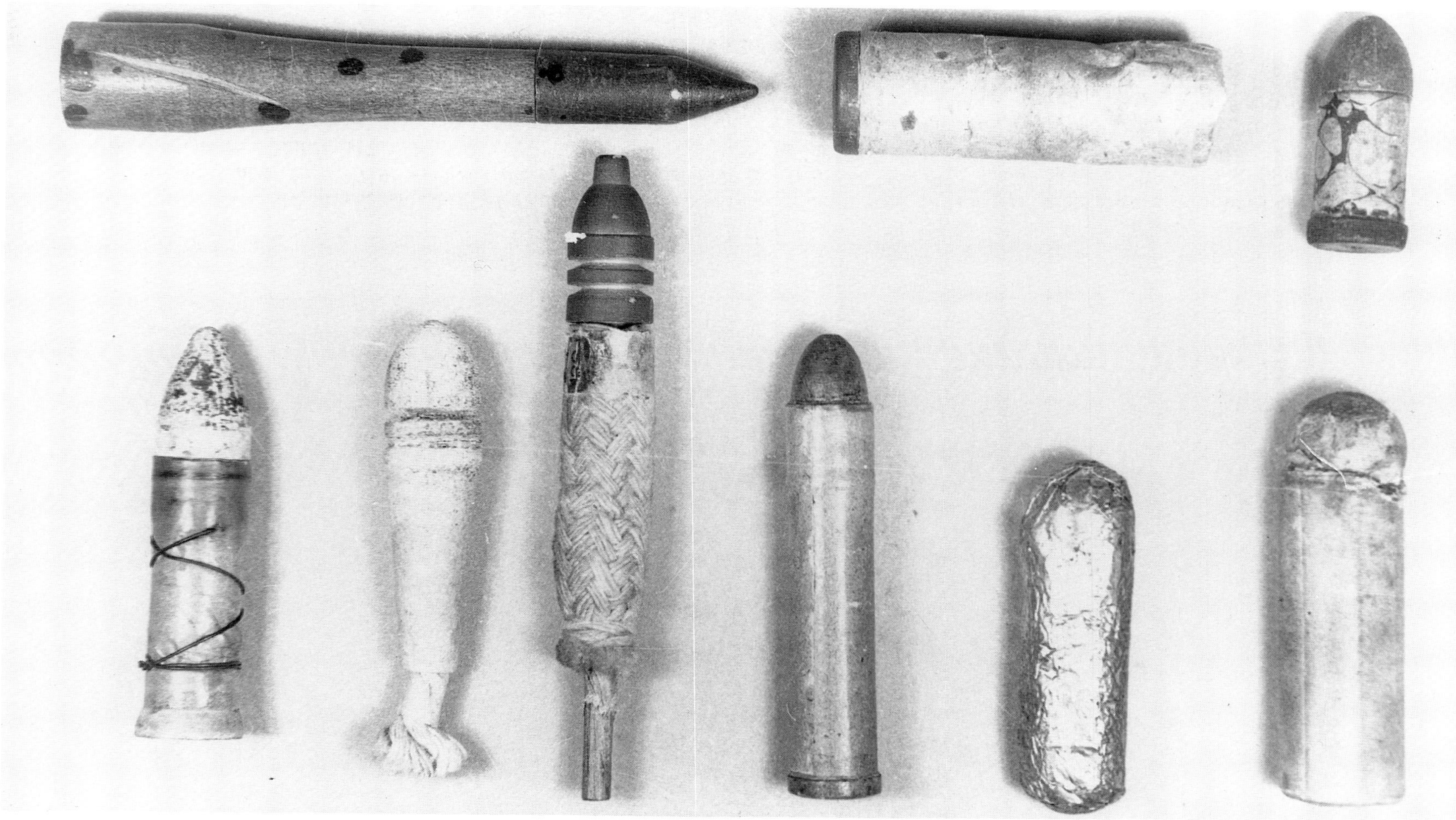

Top row: U. S. Mefford sub caliber projectile (Civil War era, see Vol. 1, p. 47); 12 gauge Needham needlefire shot cartridge; .54 Marston;

Second row: .56 English for Sharps rifle, 1860 (Vol. One, p. 96). The wire wrapped around the case is attached to the bullet, and was intended to pull unburned paper out the bore; .58 Lindner yarn stopper cartridge, 1861 (U. S. patent 32949). (V. 1, 62); 13.7mm von Lenk gun cotton cartridge (Austria), 1846, developed by Schoenbein, a Swiss inventor, V.1, 47; .47 12mm Lenoir (France) horizontal pinfire, paper case, fiber base. C. l. 2.422; .64 Colt tinfoil cartridge (V. 1, 26); 16.26 Galy-Cazalat, 1826, a primitive centerfire with percussion primer in base of parchment case (V. 2, 6).

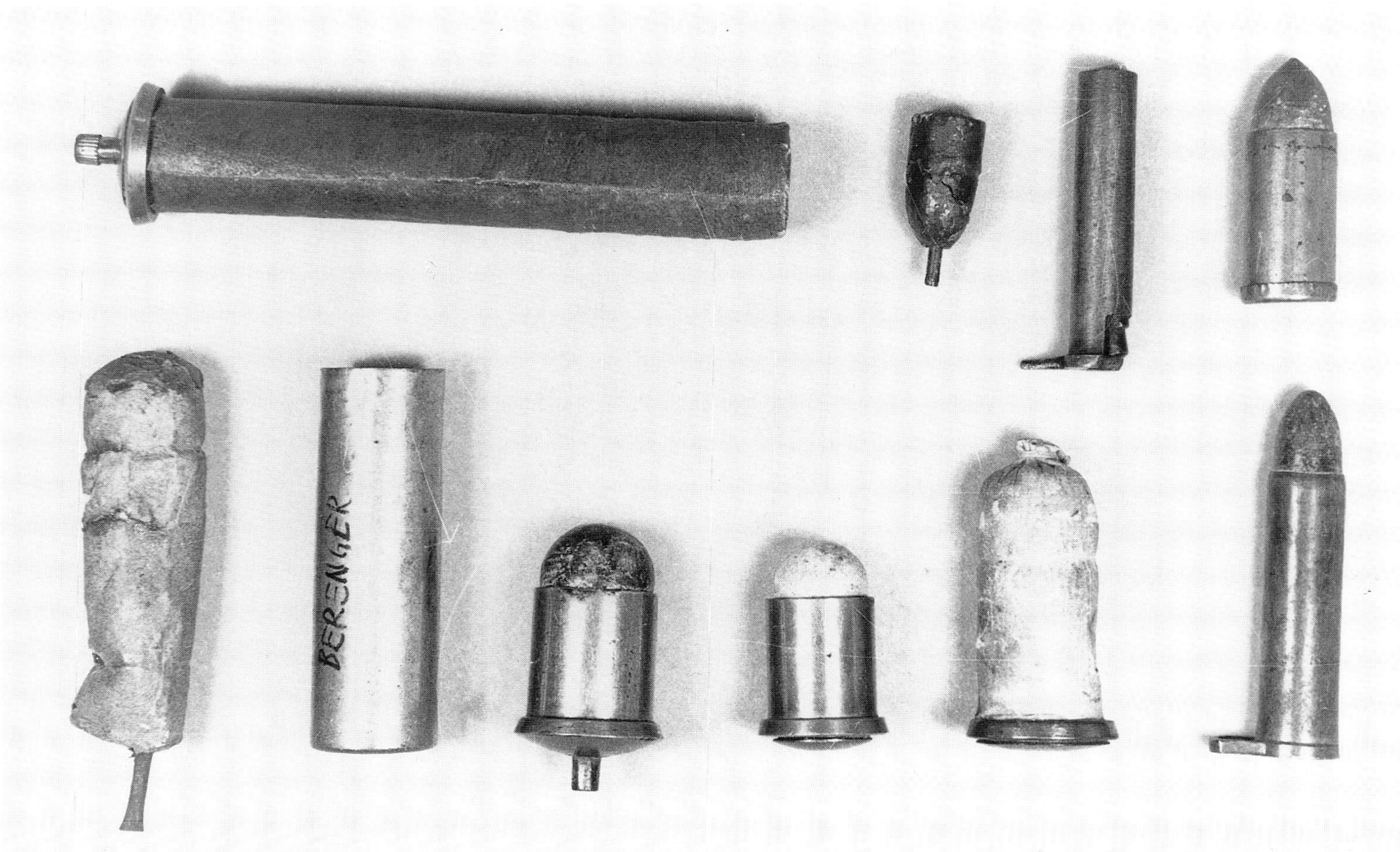

Top row: .64 Pauly, France, shotshell (J. Pauly English patent 4026, 1816), the second type with brass head, percussion nipple, green paper case. C. l. 3.719. Genhart tube ignition for pistol, U. S. patent 16477, 1857. The detonating tube is affixed to the powder charge. Bullet .417, c. l. .7968 (V. 1, 101). Third specimen, no information. .46 Maynard 1856, first type, U. S. patent 15141. C. l. .968 with cap.

Second row: Demondion type tube ignition, 1836 (V.1, 102); 16 gauge Berenger, 1850, rimless brass case centerfire, French paent 5545. Headstamp is "Berenger Bt Paris." Head .711, c.l. 2.172. 16.26 Pauly, 1816 (V. 2, 5); 14.5mm Pauly, 1812 (V. 2, 3); 1 16.26mm Pauly, 1812 (V. 2, 5). 9mm Cent Garde, 1858, enclosed pinfire (V. 1, 118-119).

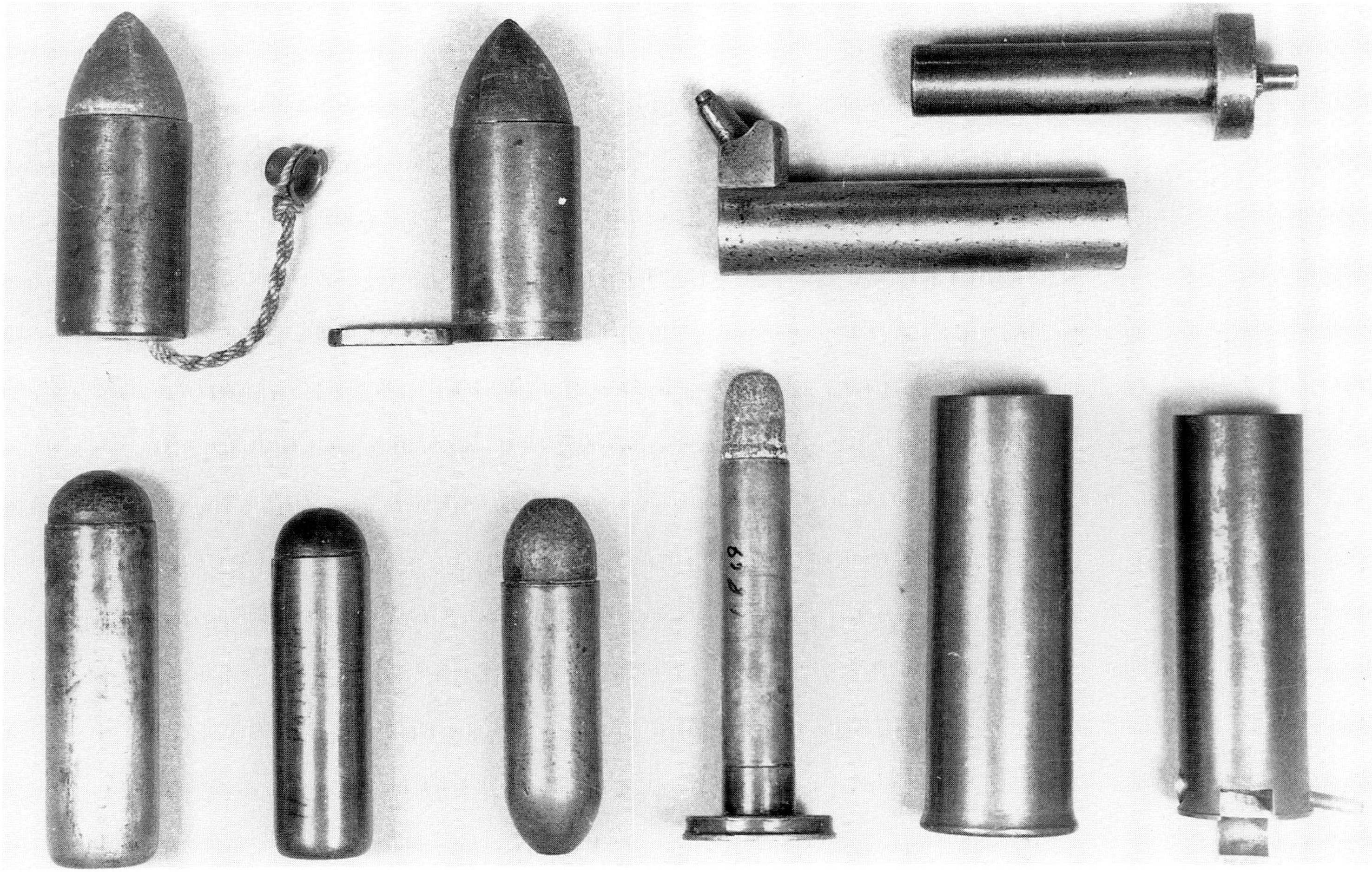

Top row, left to right: .69 Maynard, cord extractor, 1863, case length, 1.25"; .69 Maynard flop ear, 1863, c.l. 1.2343" (made for 1858 trials, patented in 1863); .44 (?) Samuel Adams 1838 patent (U. S. #960), steel case with percussion nipple, head .540", c.l. 2.234; Moses patent .38 or .40, M. Moses, U. S. patent 36571. Head .423, rim, .727, c.l. 1.859.

Second row: .58 Gallagher (?). Deep dish base with percussion orifice. Case looks like steel or iron but is not. Case length 1.937", mouth .613"; .50 Gallagher, 1860, deep dish base. Head .549", mouth .543", c.l. 1.687; .50 Gallagher, head .551", mouth .550", c.l. 1.547"; a .35-40 Maynard case with Parker-type head and rim modification; 12 gauge Lancaster shotshell, 1852, copper case, inside-primed centerfire, head .795", c.l. 2.45"; 16 ga. Roberts, 1847, French patent 3601 by Chaudin. Headstamp, "Brevete Liege R." This is a trapdoor pinfire for quick replacement of primer. Head .709, c.l. .2.25".

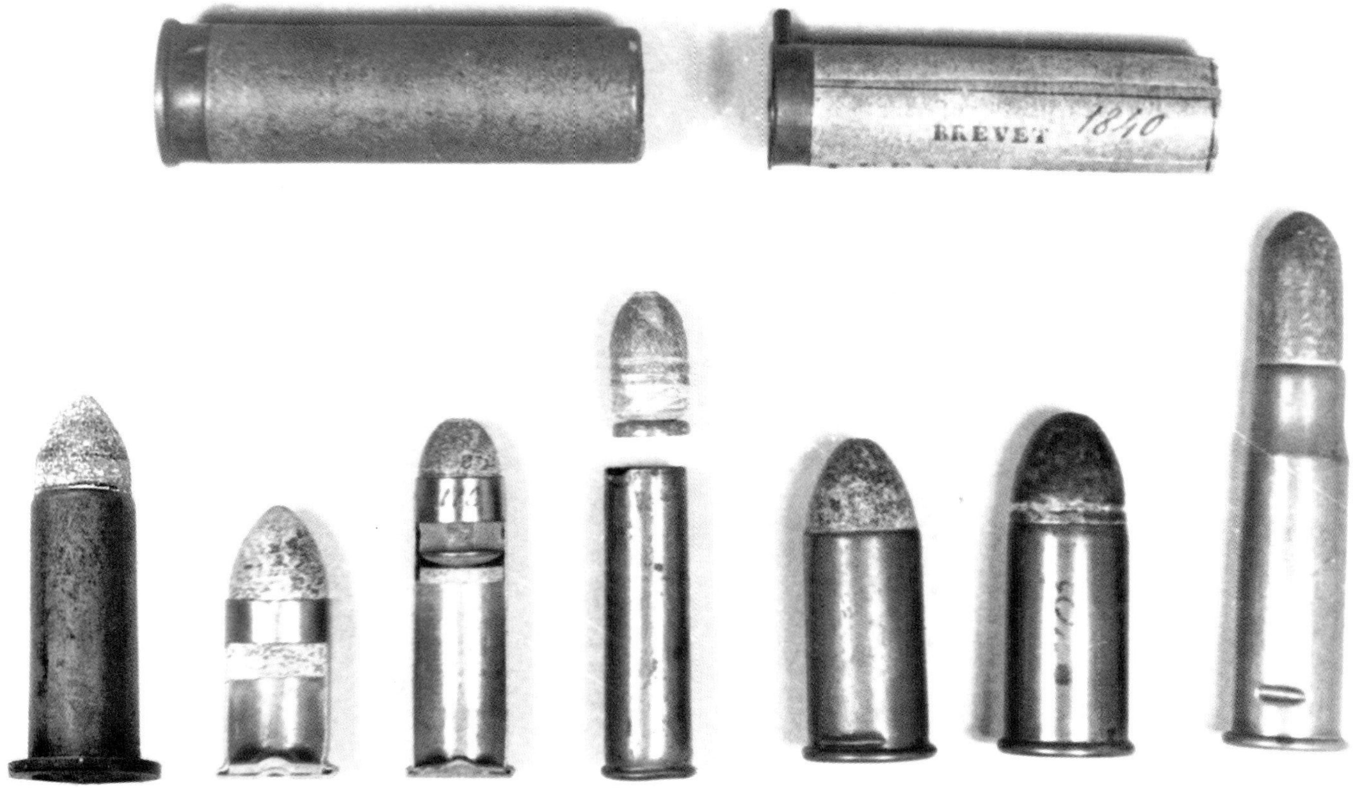

Top row, l to r: 10 gauge Leet, 1869 (U. S. Leet-Hotchkiss patent 98278). Headstamp is "C. D. Leet No. 10, Springfield, Mass." Red paper case. 16 ga (?) Le Faucheux (French) pinfire shotshell, 1835. Head .653, c.l. 2.297".

Second row from left: (1) .54 Morse. (2) .50-9531", copper-case Frankford Arsenal sectioned specimen. Full case crimp around base. Head indented for priming, the priming mixture smeared into this indent; (3) .44 (?), no information but appears to be similar construction the .50 at left; (4) .42 disc anvil, 1862, patented by Moffat in 1866, U. S. 53168. Head .453", rim .506", c.l. 1.656. (5) Copper case with Williams cleanout bullet, patent 37145 of Dec. 9, 1862. (5) .58 Musket copper case, bar primed, two crimp. A Frankford Arsenal experimental headstamped ".58 CAL." (6) .58 Musket, no information given but looks like later Winchester commercial. (7) .45 steel cup anvil, Frankford Arsenal experimental with two short crimps .298" from rim. C. l. 2.016", overal 2.844".

Here is a group of the drawn case and primer experiments of Captain Josiah V. Meigs to add to those reported in Volume Two (pp 22-27). The top two rows show the rimless case with cutaways of the case and primer ideas. The lower section shows various attempts to produce workable primers. United States Cartridge Company produced these cartridges and cases but abandoned the Meigs ideas when better primers were developed.

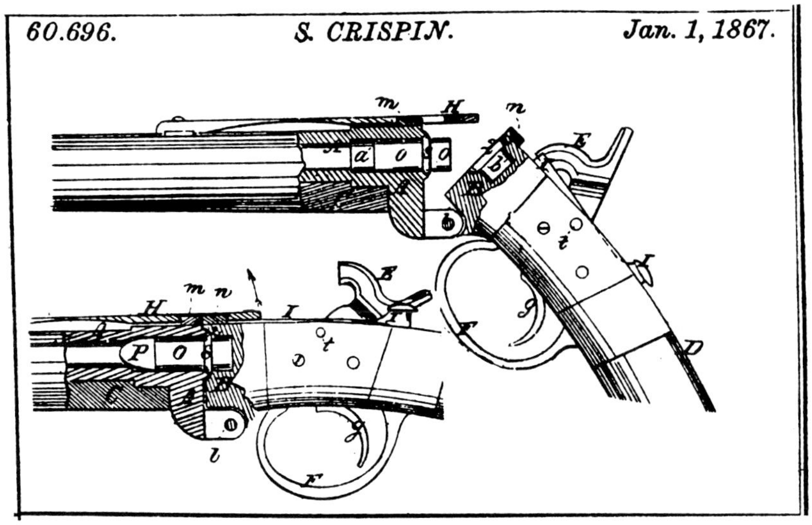

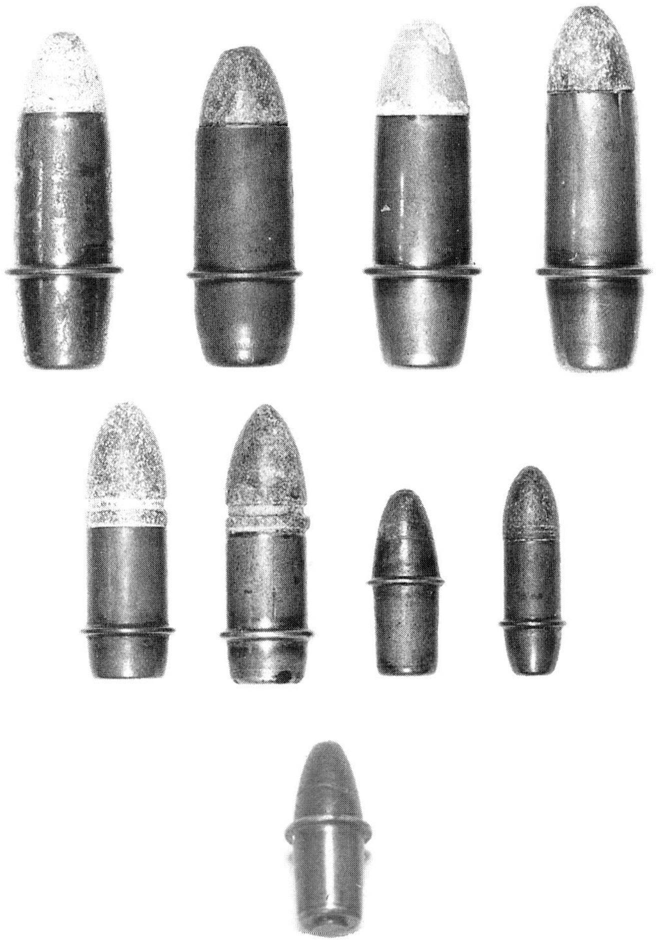

This is the family of annular rimfires for Silas Crispin's modification of the Smith percussion breech-loading carbine. The base of the cartridge fit in the hollowed face of the breech with the rim exposed at the juncture of breech and barrel. At top from left are three variations of the intermediate case; at right the .50-1 ½" Long. At the center are the pistol sizes: from left, two .44 cal. with slightly varying bases and bullets, and the .36 and .31 calibers. At bottom is an attempt to convert the .36 Crispin to centerfire by drilling the base for a small percussion cap. The annular rim would have no fulminate.

THE NEEDLE FIRE CARTRIDGE

The Prussian needle fire rifle (zuendnadelgewehr) and its cartridges, and those of other European countries, were described in Volume One of this series published in 1981. Little did we know then the extent to which this system, which ultimately led to self-contained centerfire ammunition, was applied to sporting rifles, shotguns and handguns.

The specimens illustrated here were produced between 1827 and about 1875. Records of these cartridges, and the firearms they were made for, are almost non existent. What we know about them is found in the few illustrated studies produced in German during the 19th century, and by x-raying or sectioning them for study.

The drawings below are from "Historia da Arma De Fogo Portatil" by Captain Luiz Mardel, Lisbon, Portugal, 1887. Above is the Prussian von Dreyse needle fire rifle action; center the French Chassepot, and the Italian Carabine De Bersaglieri converted to needle fire from flintlock musket, courtesy of Roy M. Marcot.

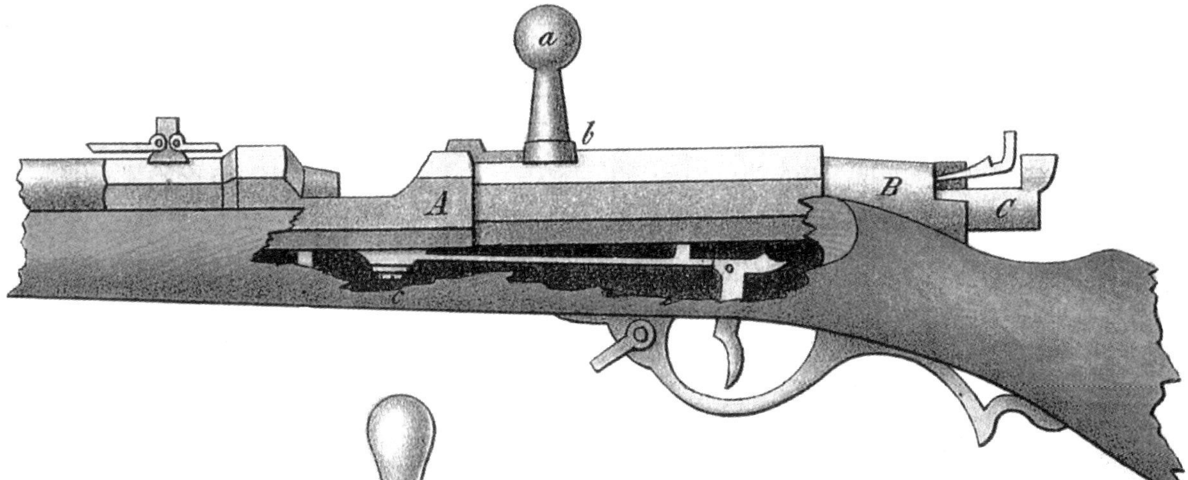

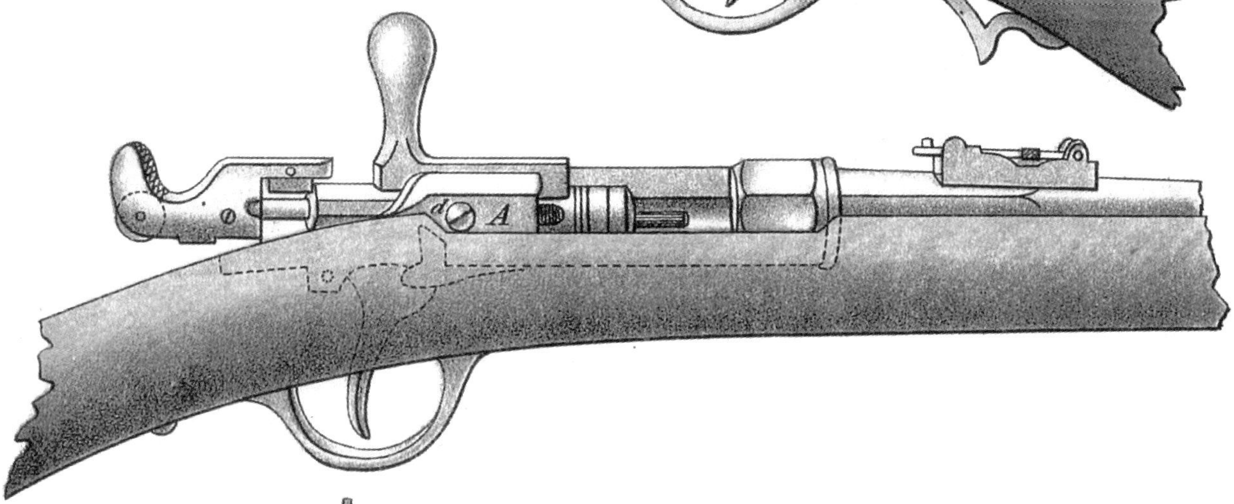

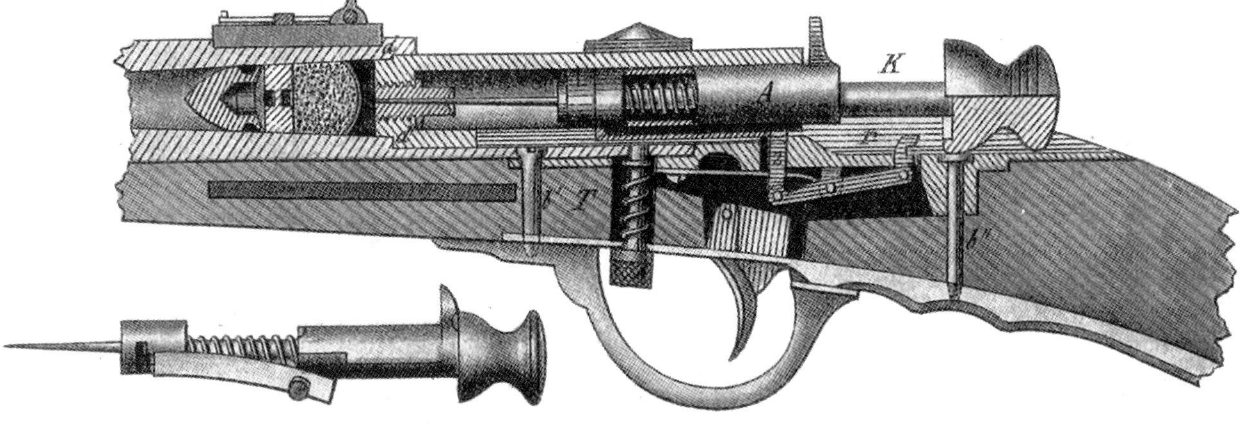

A 15.28x35 mm round ball load in off white paper with separate base disc of the same paper. The x-ray shows the papier mache sabot holding the projectile and the primer capsule ahead of the powder charge. The sectioned projectile and sabot below show a different arrangement, with primer or incendiary compound extending through the sabot and into the projectile. This is most likely the cartridge Nikolaus von Dreyse devised for use in his first muzzle-loading needle fire rifle of 1827, which he soon abandoned.

A 15.42x55 specimen with the penned inscription “zundnadelpatr.” (needle fire cartridge). The x-ray shows an early form of bullet seated in the papier mache base with primer at the rear ahead of the powder charge. The bullet in this specimen would directly contact the rifling of the weapon. Date—ca 1840-1847.

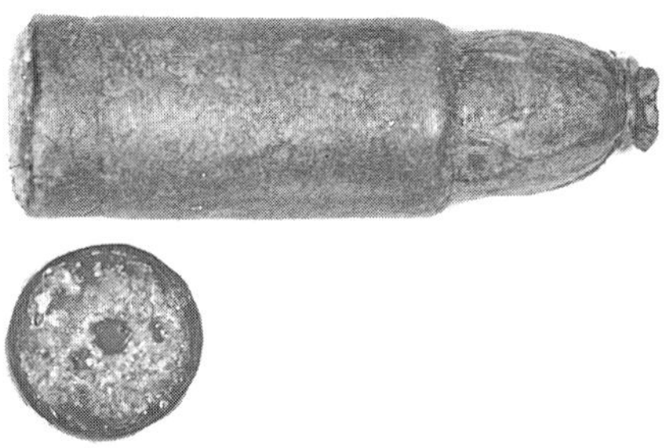

Another similar specimen with a soft lead base. This metal base was no doubt intended to obturate the breech and prevent the escape of gas around the bolt face. (John Belton collection).

Still another, similar, with normal paper base closure but with an odd grooved bullet. Since only the sabot rode on the rifling, it is not clear what purpose the grooves served. Others experimented with grooves to help stabilize the flight of the projectile.

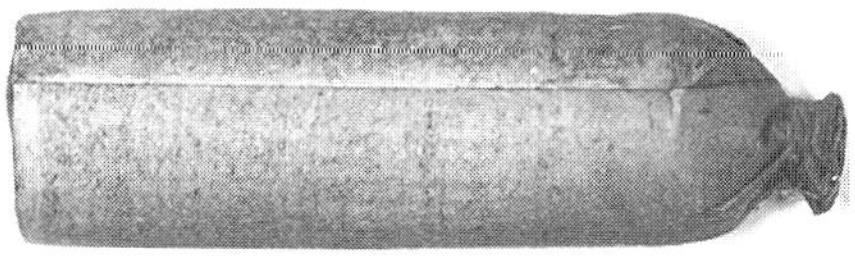

A 15.42x61mm round ball load in gray paper, twisted closed and tied over the ball. There is a heavy coating of grease or beeswax over the projectile. Origin and firearm unknown.

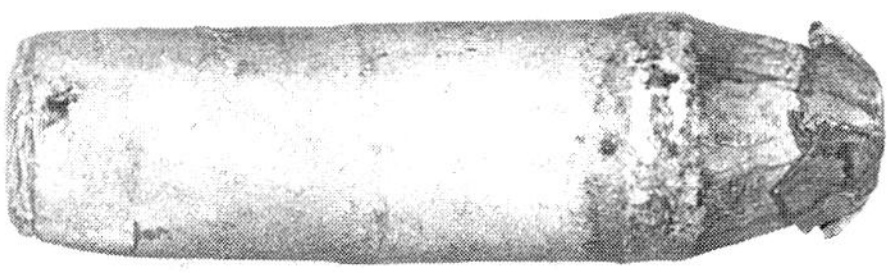

Another measuring 15.42x61mm with the egg-shaped bullet, paper tied at the nose and folded back.

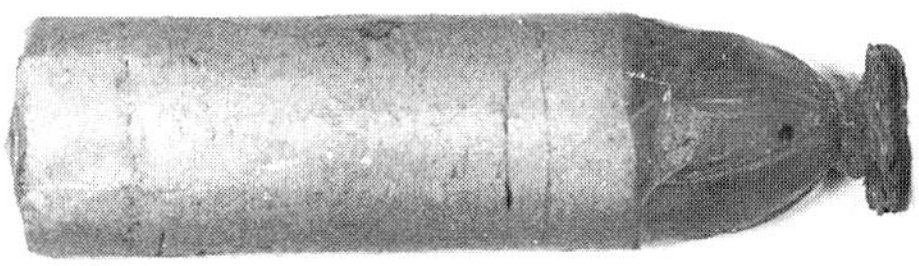

This is a standard Model 1855 load but with a leather washer attached to the heavily waxed paper tied over the bullet. This washer could have been intended to make the cartridge easier to grasp and pull out of a covered ammunition pouch.

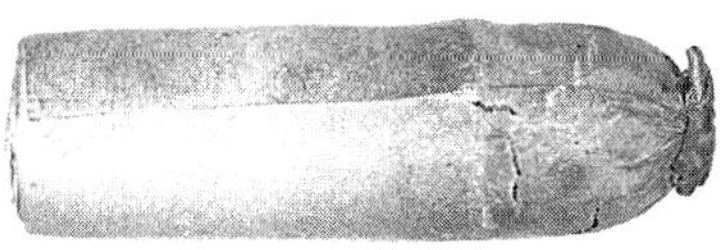

A 14 x 48mm von Dreyse-type cartridge in white paper presumably for pistol.

A 15.42mm Dreyse blank cartridge in off-white paper.

A dark blue 14.5x50mm specimen with grooved bullet and inverted percussion primer inside the base. This is most likely a late commercial cartridge for a sporting rifle or carbine.

Cutaway specimen of a cavalry pistol cartridge.

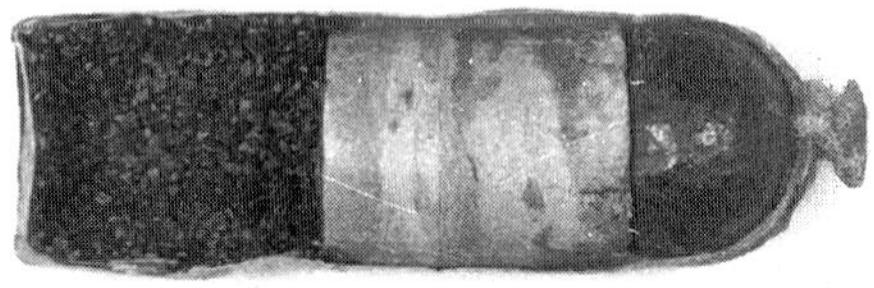

Another, probably for pistol or carbine loaded with the round ball which is the same diameter as the sabot, and thus contacted the rifling directly.

Another specimen, unidentified by the list with wide groove in the bore-diameter bullet. The primer, missing, sat in the shallow cavity in the base of the sabot.

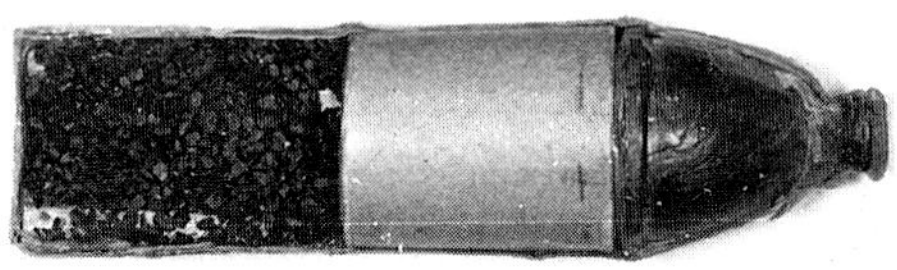

Yet another cutaway specimen, unidentified, with bore-size projectile seated in a heavy, cardboard sabot with inverted percussion primer cap seated in its base.

The case of this specimen is light cream, darkened over the bullet with wax. The ligature over the case is in the rear groove of the projectile. The primer is seated nearly against the base of the bullet.

This 15.43x61 cartridge looks like an ordinary specimen, but the x-ray reveals another type of bore-diameter bullet with narrow belt around its center to take the rifling. The inverted musket cap in the sabot suggests that this was a later experiment.

This is the flare load for the 15.43mm Dreyse needle fire rifle. The extended case contains several pyrotechnic elements. The front two thirds of the cartridge is choked on to the powder and sabot section with string, and the nose is waxed.

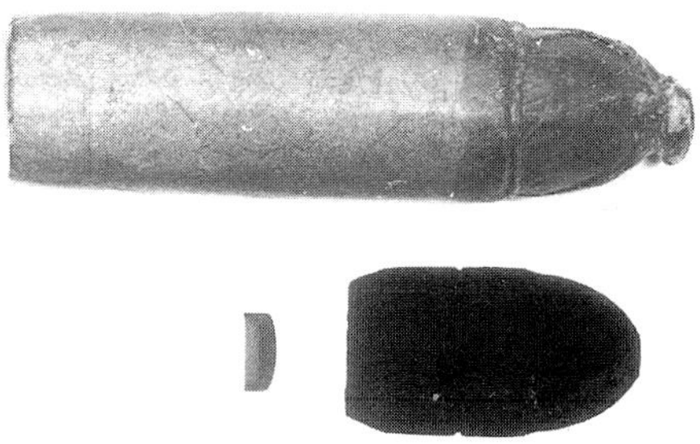

This blue paper specimen is described on the list as "raketengeschoss, M 1855/62"—rocket cartridge, possibly another flare. The x-ray shows an unusual shape of bullet and the primer behind it. The rocket or flare material would be in the hollow base of the bullet.

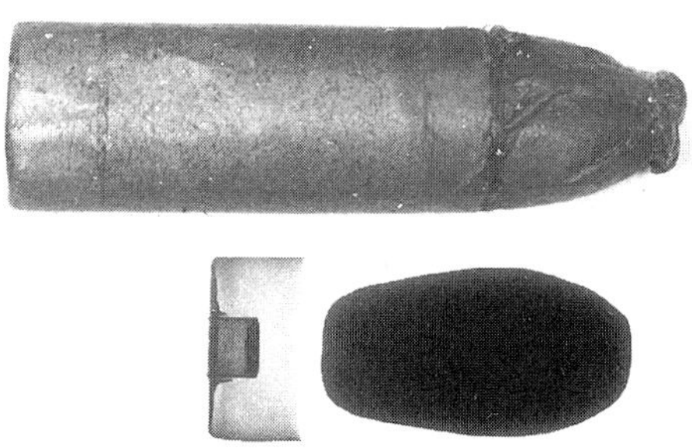

This specimen has a deep rose-colored case and is described as "explosiogeschoss, M1855/60." If there is an explosive element it would be buried in the hollow base of the projectile.

This is a deep blue specimen tied at the nose, loaded with a two-groove round nosed, flat base bullet. There is no visible primer compound in this x-ray. Could it be embedded in the base of the bullet?

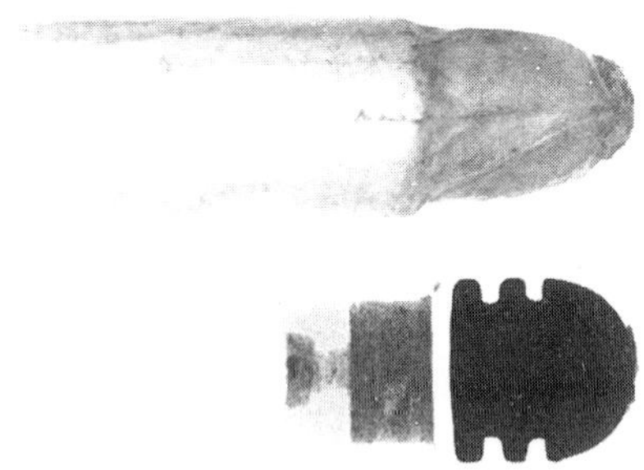

A 15.43x58mm white paper case specimen with odd grooved projectile. Since the bullet is less than bore size, the purpose of the grooves is not evident. Also, there is a metal element in the base ahead of the primer whose purpose is not apparent.

Another specimen with two-groove bullet that is bore size.

A grooved bullet similar to the one above seated in its papier mache sabot. The primer is an inverted metal percussion cap, indicating that this was a late development. Identity or use unknown.

Another bullet which may be the same as the one above showing the heel on which the sabot was fitted.

An egg-shaped projectile with hole drilled from end to end. The purpose of the hole is not apparent but may have been an attempt to improve flight stability.

The normal shape of the 13.6mm diameter bullet used in the 15.43mm Dreyse needle fire cartridge—heavier in the fore end. The bullet got its stabilizing spin from the sabot holiding it on its flight down the rifled barrel. The sabot dropped away when it left the barrel, leaving the bullet to go on its way.

An elongated egg-shaped specimen with paper patch around its middle.

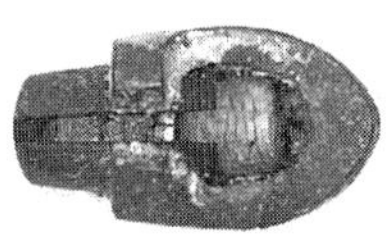

An all-iron specimen with what looks like a floating striker in the cavity that must have contained an explosive charge.

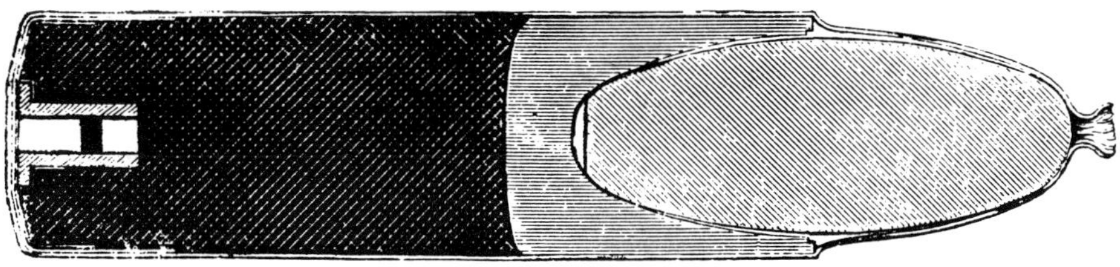

The 25mm Wall Gun cartridge is loaded with an iron, egg-shaped projectile. The papier mache sabot, being larger and bore fitting, the iron did not contact the rifling to any degree but was spun by the rifling biting into the sabot which fell away when the projectile left the bore. The sectioned specimen is substantially longer than the loaded one above it, probably because all of these cartridges were hand assembled. The rifle had a hinged, screw breech and a dog leg stock fitted with a spring-loaded recoil-absorbing butt. For an illustration see **Firearms Past and Present** by Jaroslav Lugs.

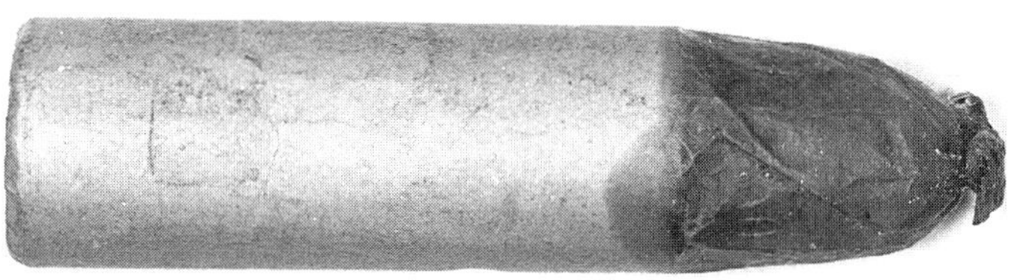

This is the 20mm Granatenpatrone (grenade cartridge) fired from a similar parapet weapon. The drawing shows an igniting device in the base of what appears to be an explosive-filled projectile. However, the x-ray of the specimen does not show this type of projectile, which may be solid lead. It does not take a magnet like the 25mm specimen.

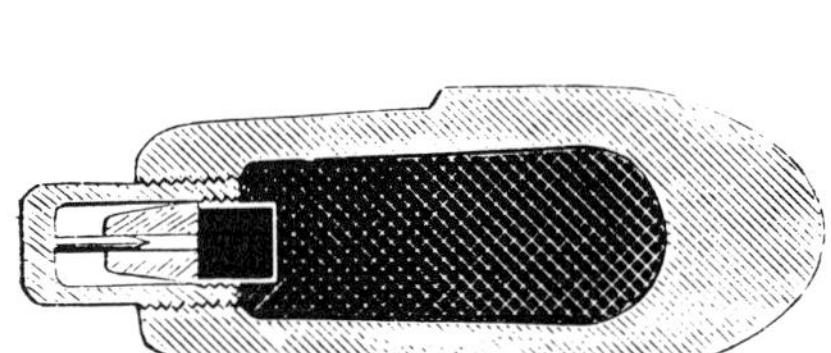

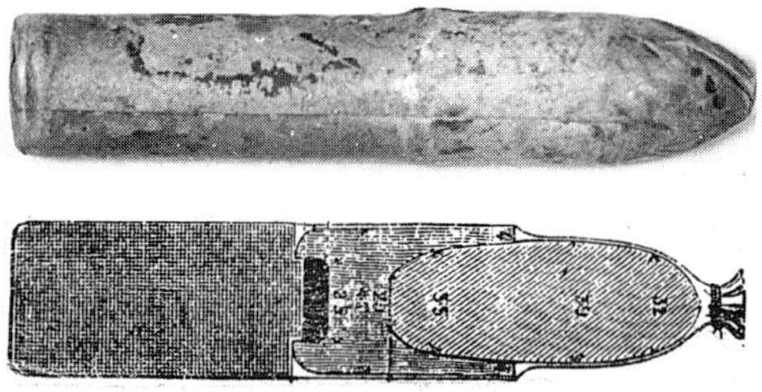

In his book "Das Zundnadel-gewehr", W. von Plonnies described this specimen as "Patrone kleinen Kalibers zum Zundnadelgewehr von Schilling." Small caliber cartridge for a needle fire rifle by Schilling. Schilling was a master German sporting gun maker during the mid to late 19th century.

This is a 13.2x50R specimen with very long, heavy bullet and brown paper powder and primer capsule apparently glued to the base of the bullet. The American Schroeder needle fire carbine produced during the Civil War, 1861-1865 was .52 caliber (13.2mm), and conceivably this may have been made for trials of that weapon. It does not look European.

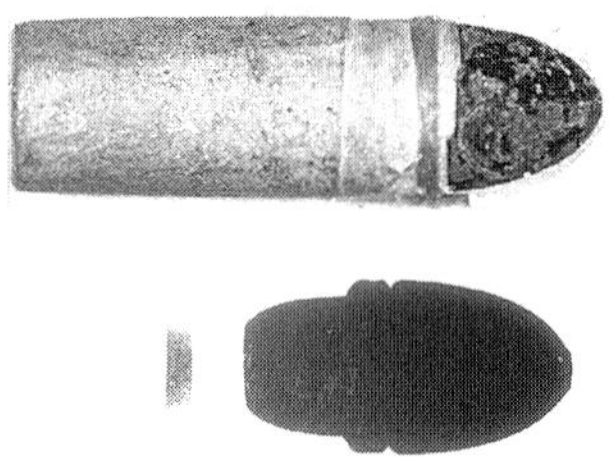

12mm and 41mm overall for a von Dreyse needlefire revolver with bore-diameter bullet.

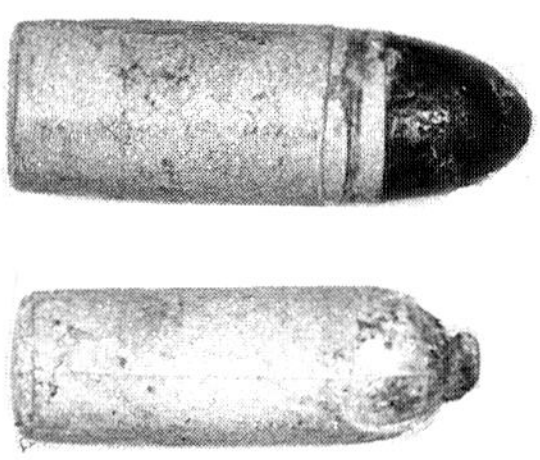

Another, 12mm x 36mm overall, and a 9 x 32mm for similar revolvers.

Here are two 9mm cartridges and one approximately 7.7mm for "damen pistolen" (ladies', or muff, pistols) made on the same von Dreyse fashion, the first two with round ball in a sabot, the last with cylindrical bullet.

This smallest of all needle fire cartridges was used in salon pistols and is also reported to have been used in a sub-caliber device with short length of barrel that chambered in the French 11mm Chassepot rifle.

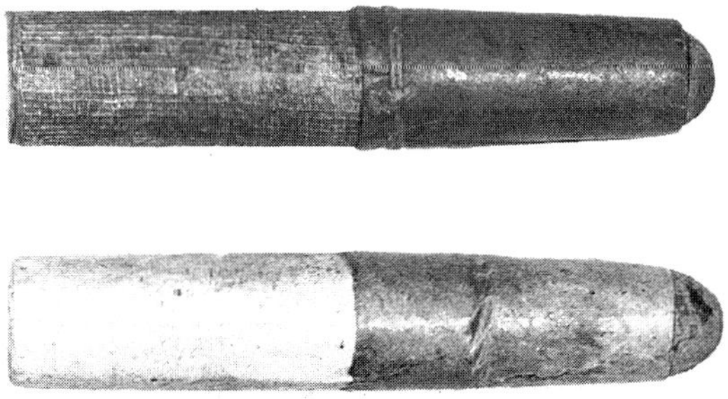

Two examples of th French Model 1866 Chassepot needle fire cartridge, the upper with linen-covered powder chamber and separate, tied-on bullet in full paper patch. The second has a white paper powder chamber and different paper covering over the bullet. These are just two examples of these hand-made cartridges for which many different fabrics were used.

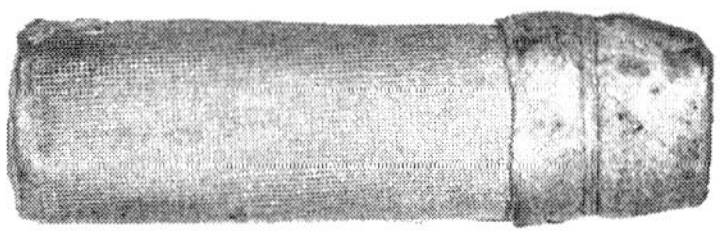

This is the common military blank with linen-covered powder chamber and paper tied over the case mouth.

This is a collodian-covered Chassepot cartridge, the clear covering intended to water proof the cartridge and increase its combustibility. The imprint reads—"F. Beaumont Inventeur 42 Rue ? 42, Paris."

Another collodian-covered specimen slightly different in dimensions but for the same rifle. There is no case wall inscription.

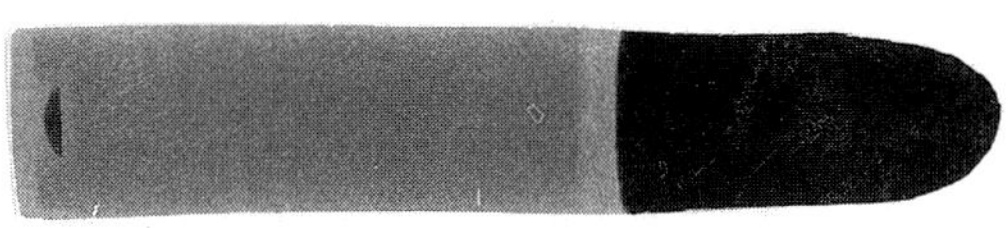

X-ray of the collodian-covered Chassept cartridge showing the priming capsule in the base of the case, not at the base of the bullet as in the von Dreyse Prussian type.

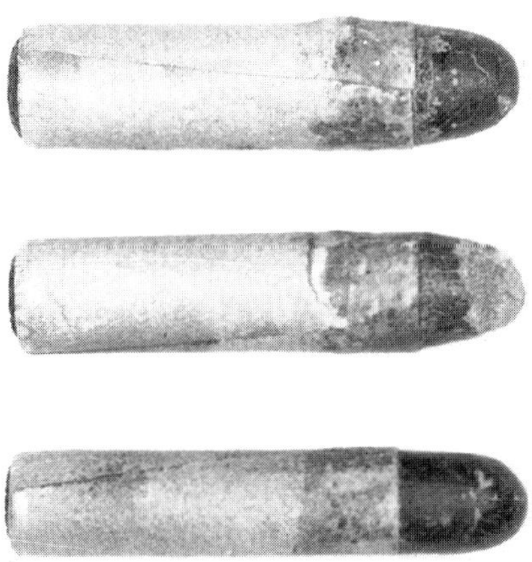

These three cartridges are believed to be of French origin for a so-far unidentified type of sporting rifle. They are from above, 13.2mm, 11.7mm and 10.2mm, all with 40-42mm light tan paper cases. They all have identical bases with a black gutta percha ring surrounding an exposed red covering over the primer.

A 17.5x62mm unidentified specimen with light brown to gray case material with cord tied into a groove in the bullet.

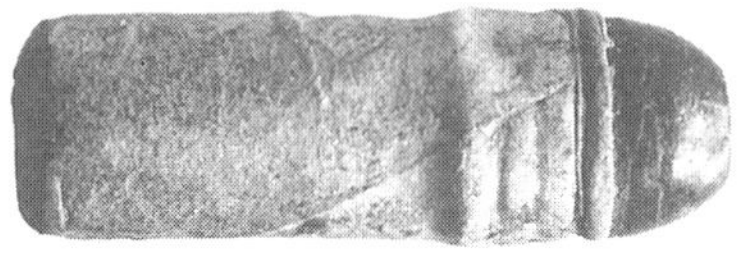

Another, 16x47.5mm in white case with ligature in upper of three grooves in the bullet. There is a small hole in the base. This and the specimen above are bore diameter bullets.

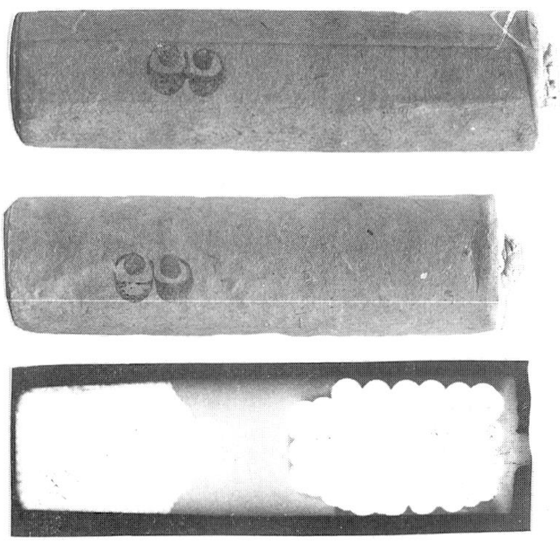

Two shot loads, 15.8x62mm and 15.8x59mm, white paper cases and a dark "3" stamped on the case wall. These two specimens have the old-style base closure where the disc is inserted into the case and glued without a fold over the base.

Another shot load, 16.4x52mm in white paper with the shot load in a sabot. This specimen has the Model 1872 base closure where the bottom of the case wall is folded over the base disc and glued.

All specimens and photographs on this page courtesy Guenther M. Wohl.

Two shot loads for a von Dreyse needlefire gun with the shot in a papier mache sabot carrying the primer at its base. The paper of the upper is a pale green, the lower is tan, soaked with tallow over the shot charge. The lower specimen has a rectangular paper patch on the base marked "6."

This is a 12 gauge shot cartridge in light tan paper case with a zinc base closure which has a central perforation to allow the needle to penetrate and ignite the primer just inside the head. This is attributed to Joseph Needham, the English gun maker.

A needle fire shot load for an unidentified French firearm with the marking on the case wall-- "LEROY MONTIGNY brevete' Fontaine-L Eveque CAL. 14."

Stages in the development of shot loads for needlefire shotguns:

This 15.4mm white-case specimen is tied closed at the nose and is made with a heavy cardboard base over which is fitted a brass cap, the 8mm center of which is slightly depressed with a central hole to admit the needle. The specimen has a raised headstamped— DREYSE 0 - 71. There is no rim on the brass base.

On this green-case specimen, which is a bit smaller in diameter, the brass base over cardboard has a slight rim and what looks like an ordinary centerfire primer. This is a thin brass covering over the central hole leading to the internal needlefire primer. There is no headstamp.

This is the final version of this type of cartridge. The tan case is crimped at the mouth over a white over-shot wad marked "7 2/1/2/M/M." There is a strip of black metal foil wound over the base and a one-piece brass base fitted over this, with central perforation. Within the hole, nearly flush with the base, is visible the head of an internal firing needle. The shotgun for this cartridge had a striker which struck the internal needle in the cartridge itself to ignite the primer. The headstamp on this specimen is "W. COLLATH Sohne 6 6 Frankfurt A/o." Collath made shotguns with their own chamber sizes from O through 8. The rim and case diameters of the Collath 6 is the same as 20 gauge.

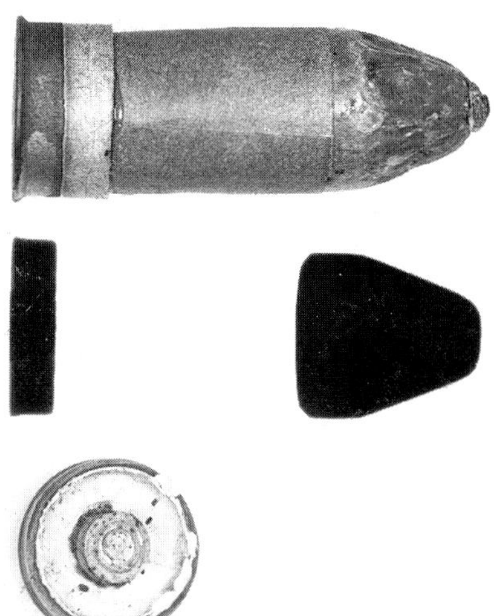

A green-case bulleted cartridge made in the same manner as the green shot specimen above. The conical bullet is completely covered with tan, tallow-covered paper tied at the nose. The cartridge looks like an ordinary centerfire, but what appears to be a primer is a thin cover over the needlefire primer within. The needle penetrated this thin cover to reach and ignite the primer. Below the specimen is the internal construction of the head of this and the top three shot cartridges showing the papier mache capsule containing the primer. The primer capsule projects about 6mm into the powder charge.

A 10mm Tesching needlefire cartridge loaded with a round ball in papier mache sabot, primer in its base.

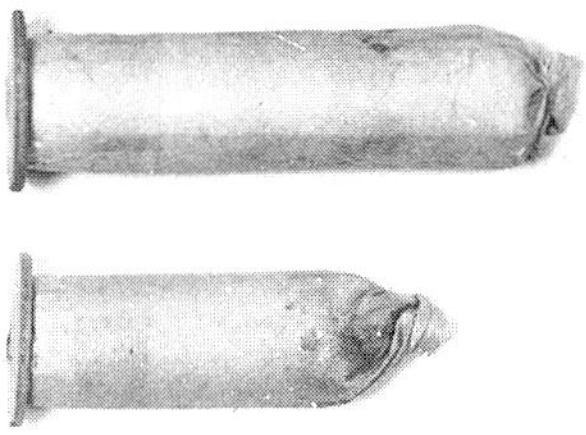

Two others 9.95mm, the upper one from the 25-round box below. It has a cardboard base rim with center hole to admit the needle. These were loaded with bullet or shot.

This is a group of needlefire cartridges for single shot small game, or "garden" guns, produced in England from mid 19th century into the 1890s, perhaps later. They are found in 75, 90 and 110 bore, equivalent to .41, .36, and 31 calibers. All were loaded by Eley Bros. in blue, tan and orange cardboard cases, usually stamped on the base, ELEY'S NEEDLE GUN CARTRIDGE. These 75 bore specimens are round ball (two), a conical bullet load and a blank. They will also be encountered with paper-patched and hollow-nosed lead projectiles. See Volume Three (p. 4) for others and a box label.

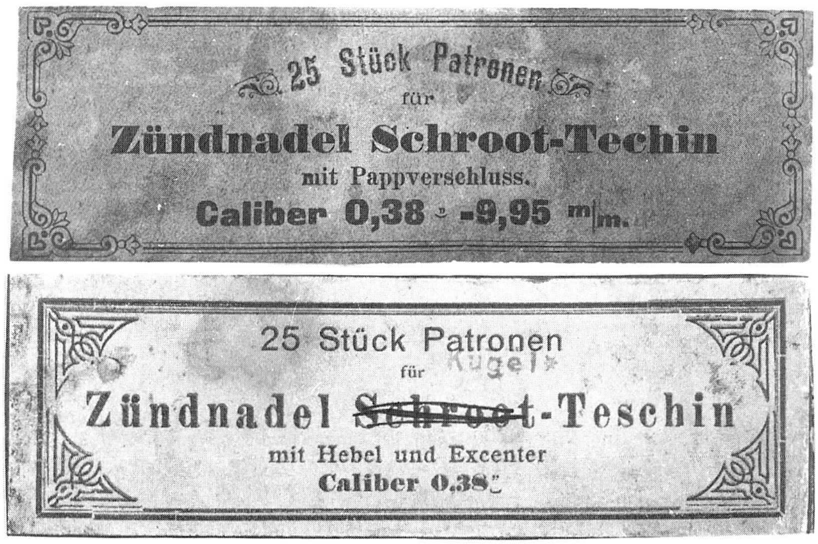

"25 Each Cartridges for Needlefire Small Caliber Shotgun With Paper Case (upper specimen), with rim (lower specimen). The lower label has "Schroot" (shot) lined out and "Kugel" (bullet) stamped above it. The word for small caliber rifle has evolved from Techin to Teschin. The modern spelling is Tesching.

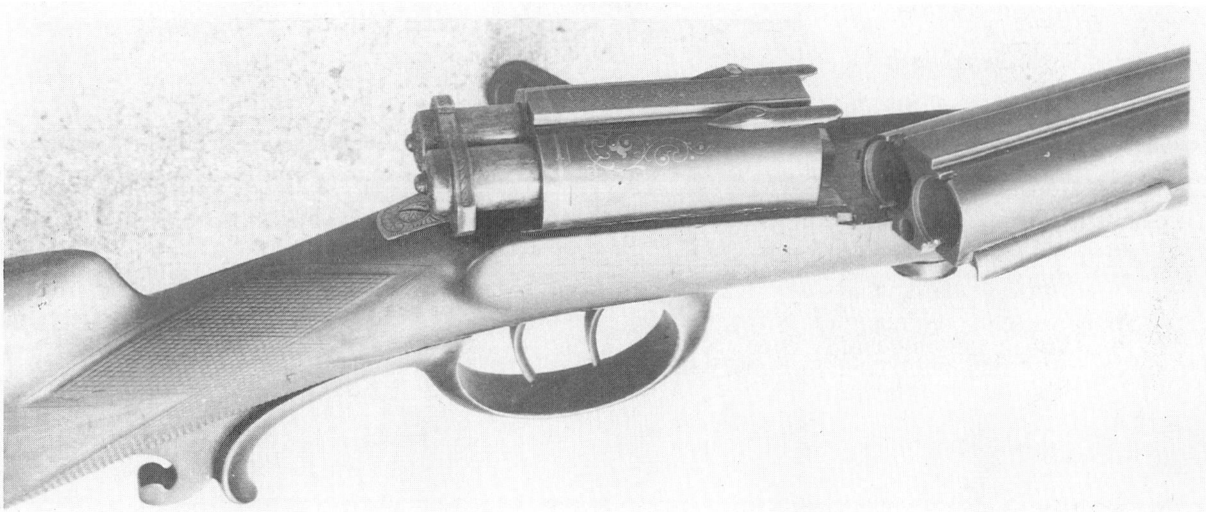

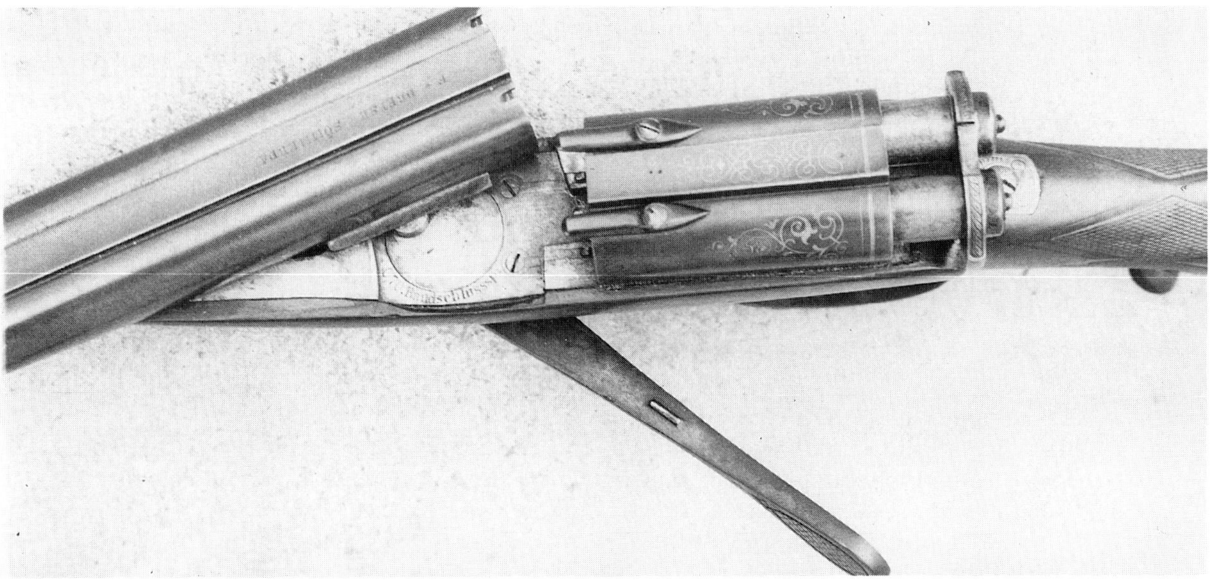

This is a von Dreyse double barrel needle fire shotgun made for the type of cartridge with the slight flang, or rim. The opening lever is swung counter clockwise from under the forend to swivel the chambers to the right for loading and extraction of fired empties. Opening the gun draws back and cocks the needles. Note the spring latches on top of each chamber that fit over the barrels when the gun is closed. DREYSE SOMMERDA is stamped on the rib between the barrels. Courtesy of Elizabeth Woodin.

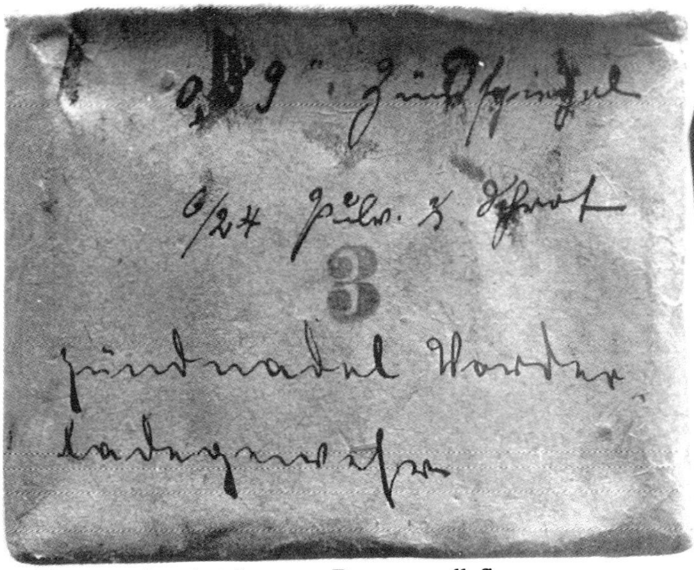

Ten-round packet for a von Dreyse needlefire shotgun, the "3" evidently being shot size.

Military ten-round packet for the von Dreyse needlefire carbine.

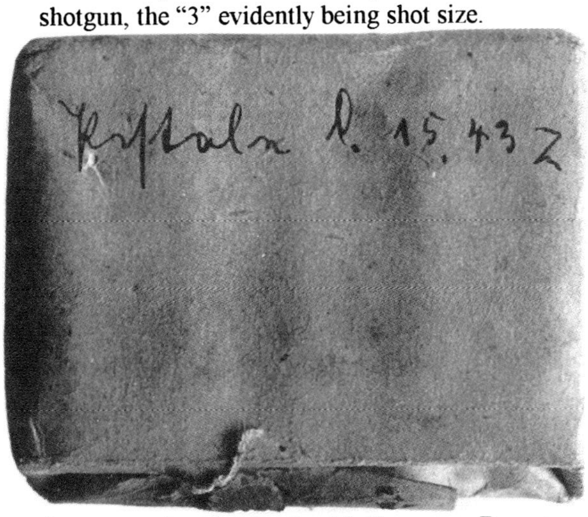

A ten-round packet for the 15.43mm von Dreyse needlefire pistol. All three packets are medium gray paper. Guenther Wohl collection and photos.

A 9-round brick of French Chassepot cartridges.

Orange label for the rifle cartridge shown below the progression of shotshells, reading approximately: "50 powder case and bullet cartridges with rimmed base for needlefire rotating double rifle (same action as that of the shotgun). Manfred Beutter collection and photo.

APPENDIX B—PATENTS

The following United States patents have been selected from the 248 dating from 1848 to 1877 found in **Cartridges For Small Arms, Patents,** published by the United States Patent Office in 1878. Another edition carried the work up to 1885. Those included here cover some of the most important ideas leading to perfection of the primer and center fire cartridge case.

This is the only work of its kind produced by the United States Patent Office. In 1985 Armory Publications reprinted the book in an edition of 825 copies. Out of print.

In 1874 V. D. Stockbridge, Examiner, U. S. Patent Office in charge of Class of Fire Arms &c., privately published his **Digest of Patents Relating To Breech-Loading and Magazine Small Arms (Except Revolvers).** This book was re-issued by Norm Flayderman in 1963.

Other patents relating to small arms and ammunition are: **Patents For Inventions, Class 9, Ammunition, Torpedoes, Explosives, and Pyrotechnics**, 1855-1900. This study was issued by the British Patent Office and was reprinted by S. James Gooding, Museum Restoration Service, P.O. Box 390, Bloomfield, Ontario, Canada KOK 1GO in seven hardcover volumes. Includes American and continental European as well as British patents. Still in print.

In 1993 **Patents for Inventions, Small Arms, Class 119**, 1855-1930, also issued by the British Patent office, was re-published by Armory Publications in seven 8 ½" x 11" hardcover volumes together with an alphabetical index. Includes American and continental European patents as well as the British. Still in print.

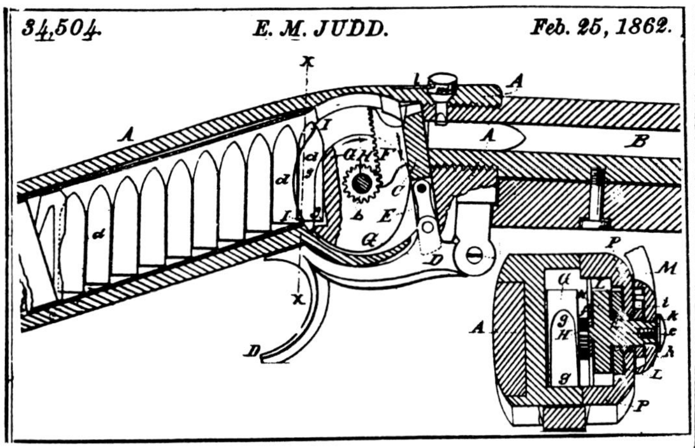

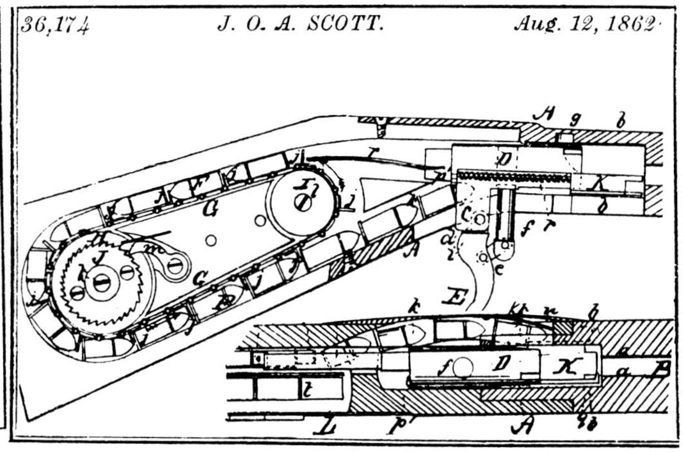

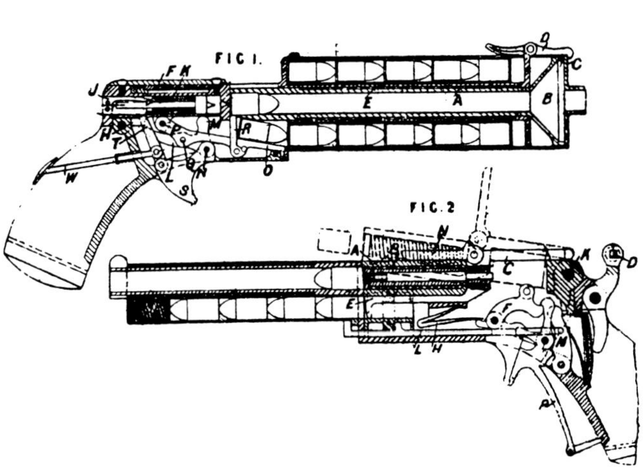

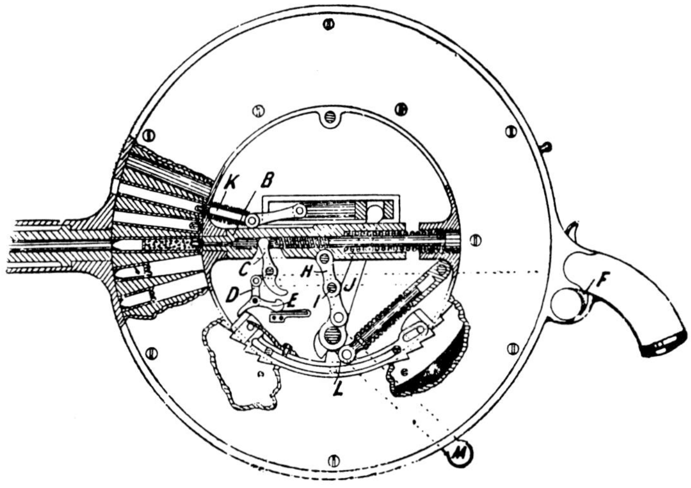

UNITED STATES PATENT OFFICE.

EDWARD MAYNARD, OF WASHINGTON, DISTRICT OF COLUMBIA.

IMPROVEMENT IN CARTRIDGES.

Specification forming part of Letters Patent No. 15,141, dated June 17, 1856.

To all whom it may concern:

Be it known that I, EDWARD MAYNARD, of the city and county of Washington, in the District of Columbia, have invented and constructed a new manufacture in the shape of an Improved Article of Fixed Ammunition for Breech-Loading Fire-Arms; and I do hereby declare that the following is a full and exact description thereof, reference being had to the accompanying drawings, making a part of this specification—

Figure 1 being a longitudinal section in a line passing through the center of a piece of my improved fixed ammunition, and Fig. 2 a rear-end view of a piece of said ammunition.

Each piece or cartridge of my said improved fixed ammunition consists of a cylindrical-shaped shell, *a*, of brass, or some other tough and stiff metal, having a centrally-perforated bottom, rendered impervious to air and water by internal packing, then charged with powder, and then combined with a projectile, *b*, of the shape and in the manner substantially as hereinafter set forth.

The exposed portion of the projectile *b* is of a pointed or semi-oval shape, and the portion of said projectile that enters the shell is of a cylindrical shape, of such a size as to closely fit within the shell, and having a sufficient length of bearing-surface to insure the point of the projectile being retained in a line with the axis of the shell. To insure a perfectly tight joint between the outer periphery of the projectile and the inner periphery of the shell, annular grooves are formed in the cylindrical portion of the projectile, and filled with any suitable greasy composition. The peripheries of the ledges between the grooves in the cylindrical portion of the projectile, fitting closely against the inner periphery of the shell, will prevent the grease working inwardly to injure the powder, or outwardly to soil anything that the ammunition may be brought in contact with. The said greasy composition serves the purpose of rendering the joint between the projectile and the shell perfectly tight while the ammunition is in a fixed state; and when the ammunition is discharged, the said greasy composition serves to lubricate the bore of the gun.

The perforated back end of the shell may be closed and made perfectly impervious to air or water by placing one or more disks of waxed or gummed paper against its inner surface before placing the powder within the shell.

The projectile should be pressed into the shell with a sufficient degree of force to insure close contact between the inner end of the projectile and the powder, care being taken in so doing not to turn the point of the projectile out of line with the axis of the shell.

After a cartridge has been discharged, the shell *a* should be removed from the chamber of the gun.

In using my improved fixed ammunition, the projectile is set more accurately within the chamber of the fire-arm than it is possible to set it in the best muzzle-loading target-rifles. The cartridges may be exposed with impunity in any weather, and may be handled roughly with less liability to injury or accident than would be likely to occur from such usage of any other fixed ammunition for small-arms known to me.

The shells may be charged a great number of times, with less apparatus and at less cost than that of the ordinary cartridges, and the expense of the shells is so small as to be quite insignificant when taking into account the great advantages possessed by this ammunition.

As a matter of course, the shells of my improved cartridges must be of such a size as to fit accurately within the chamber of the gun the said cartridges are prepared for.

I am aware that cartridge-cases of a tapering shape have been made of sheet-copper or other hard metal, combined with soft-metal rings, as described and represented in a patent granted to A. E. Burnside, March 25, 1856; and I am also aware that the said cartridge can only be used in a movable breech-piece, and that it does not possess in other particulars the peculiar advantages which distinguish my improved cartridge for breech-loading fire-arms; therefore,

What I claim as my invention, and desire to secure by Letters Patent, as a new manufacture, is—

My improved cartridge for breech-loading fire-arms, composed of a hard-metal cylindrical case, charged with powder and combined with a projectile of such a shape that, whether the case receive a large or a small charge of powder, the said projectile is self-retained in contact with the powder, in such a position that its point must be coincident with the axis of said case, and a perfectly tight joint formed between said projectile and case, by filling the grooves in the former with greasy matter, substantially as herein set forth.

The above specification of my improved fixed ammunition for breech-loading fire-arms signed this 16th day of April, 1856.

EDWARD MAYNARD.

Witnesses:
Z. C. ROBBINS,
M. H. MANSFIELD.

E. Maynard,
Cartridge.

No 15141

Patented June 17 1856.

Fig. 1

Fig 2

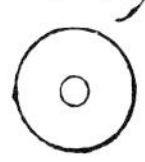

725. Maynard, E. March 22.

FIG. 5.

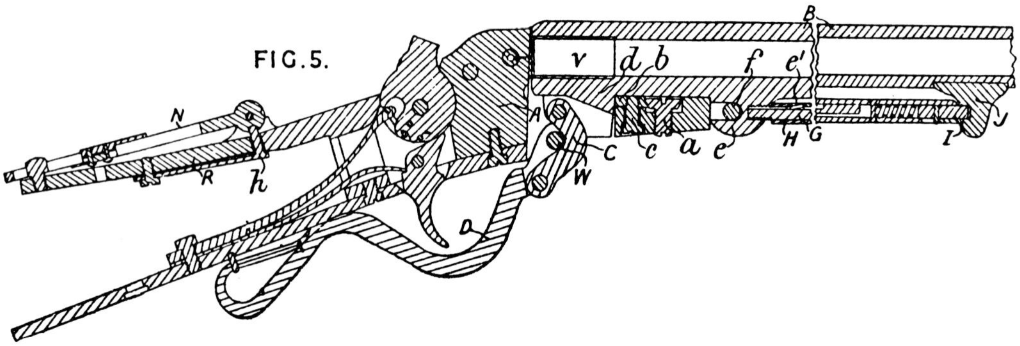

Breech actions, drop-down barrel; ramrods; cleaning; barrels.—The barrel B, Fig. 5, has a hook *e* engaging a pin *f* and fitting between lugs on the breech-piece A. The lever D, forming a trigger guard and working on a slotted pin W, is connected to a link C, which, when the gun is to be opened, acts on a shoulder *d* on the barrel. The link may be slightly elastic. A tight joint between the barrel and breech-piece is effected by means of a wedge *b*, adjusted by screws *a*, *c*, and a loose flanged metal cup *v*, attached to the pin *f* by a chain or strap and used with loose ammunition. The inner end of the barrel is chamfered to facilitate the removal of the cup *v*. An arm b^1, Fig. 9, on the end of the pin W, bears on the cover of the cap magazine in the closed position. The barrel may be detached by partially withdrawing the pin W; this pin is secured by an external set-screw. The nipple *s*, Fig. 8, is held by a side screw *t*. The hammer is connected by a link *r*, Fig. 6, to the mainspring *n*, and actuates a ratchet-wheel *h* and cap-feeding wheel by a pawl *j*. The barrel is supported by a tube H, Fig. 5, applicable as a ramrod or cleaning-rod, and provided with a spring bolt I engaging a lug J; the tube encloses a rod G, and is automatically locked thereon when drawn out, by means of a projection on the rod engaging a slot in the tube. A feather e^1 also engages a groove in the rod. A short projecting tube is inserted in the mouth of the chamber prior to cleaning.

Sights.—The back-sight leaf N, Fig. 5, is mounted on the tang R, and is held elevated or prone by a spring catch *h*. The bar consists of front and back plates j^1, l^1, Figs. 11 and 12, having inwardly-projecting guides, and held at any desired elevation by a set-screw n^1. The sight plate k^1 is adjusted laterally by means of a slot and set-screw m^1.

FIG. 9.

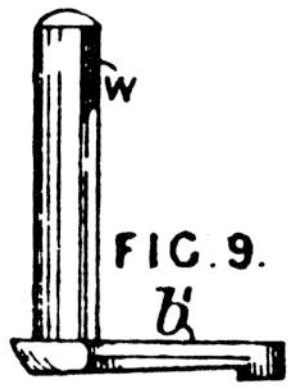

FIG. 6.

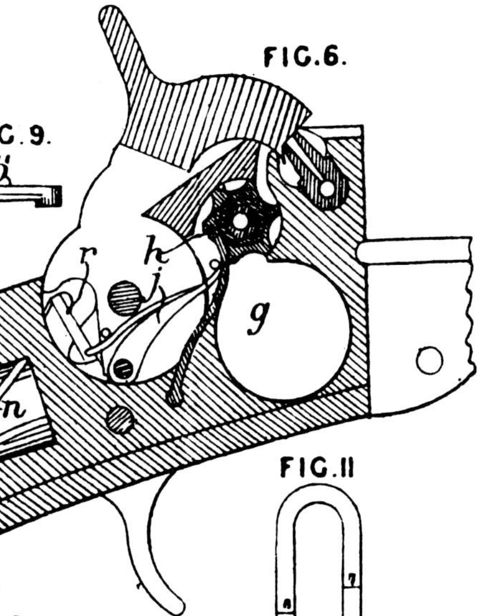

FIG. 8.

FIG. 12.

FIG. 11

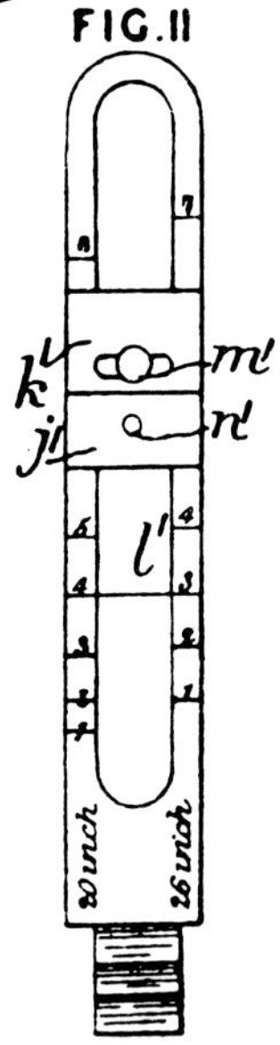

UNITED STATES PATENT OFFICE.

EDWARD MAYNARD, OF WASHINGTON, DISTRICT OF COLUMBIA.

IMPROVEMENT IN METALLIC CARTRIDGE-CASES.

Specification forming part of Letters Patent No. 22,565, *dated January* 11, 1859.

To all whom it may concern:

Be it known that I, EDWARD MAYNARD, of the city and county of Washington and District of Columbia, have invented an Improved Metallic Cartridge for Breech-Loading Fire-Arms; and I do hereby declare that the following is a full and exact description thereof, reference being had to the accompanying drawings, making a part of this specification—

Figure 1 being a bottom view of said cartridge; Fig. 2, a top view of the same; and Fig. 3, a section in the line *a a* of Fig. 2.

My improved metallic cartridge is produced by soldering a steel disk to the exterior surface of the bottom of a brass cup, which is of less diameter than the said disk, and then drilling a small central aperture through the combined brass and steel bottom of said cartridge.

Brass is of such a nature that it is not injured by contact with powder, and it is also of such a nature that when a cartridge made of said metal is loaded and placed in the chamber of a gun it will expand, at the instant of discharging the same, to such a degree as to fill the said chamber and prevent the rearward escape of the gases, and it will almost as instantly resume its normal size again; consequently, the said cartridge can be loaded and discharged a very great number of times without injury to its shape or strength; and the said cartridge may also remain loaded for any length of time without being chemically or mechanically injured by the action of the powder.

The steel disk *b* I shall generally bring to a pretty high temper, and then combine it with the bottom of the brass cup *c* by the process of soldering, which method of procedure will impart to the said disk a spring temper of the proper degree of elasticity. I do not, however, intend to limit myself to the particular temper of the steel disk nor to the manner of combining the same with the brass cup. The projecting periphery of the disk *b* forms a flange which enables the cartridge to be readily taken hold of by the thumb and finger for the purpose of withdrawing it from the chamber of a gun after it has been discharged, and the said flange also serves the purpose of guarding the bottom of the cartridge against being bruised or indented by rough handling when a large number of the cartridges are transported in boxes or barrels. Another useful quality possessed by the steel portion of my improved cartridge is its power of preserving the small size of the vent-hole after a great number of discharges, which enables said vent-hole to be closed perfectly water-tight by simply pressing a little wax into the same with the finger, either before or after charging the cartridge. And again, when one of my improved cartridges has been placed in the opened chamber of the Maynard breech-loading fire-arm, the movement which brings the barrel thereof to its proper position for firing produces a severe friction between the bottom of the cartridge and the solid breech-piece of said fire-arm, which friction and pressure is not uniform over the entire surface of the bottom of the cartridge, and, therefore, it will readily be perceived that the said strain would distort and destroy either the barrel or the bottom of a soft metal cartridge thus situated, while it produces no injury upon any portion of my improved cartridge.

The composition of the brass and the temper of the steel portion of my improved cartridge, and also the method of combining the one with the other, I shall vary to suit the requirements of practical experience.

What I claim as my invention, and desire to secure by Letters Patent as a new manufacture, is—

An improved metallic cartridge composed of a brass cup combined with an exterior steel disk, substantially as herein set forth.

EDWARD MAYNARD.

Witnesses:

J. H. H. DOLY,

SAML. DRURY.

E. Maynard,

Cartridge.

No. 22,565. Patented. Jan. 11. 1859.

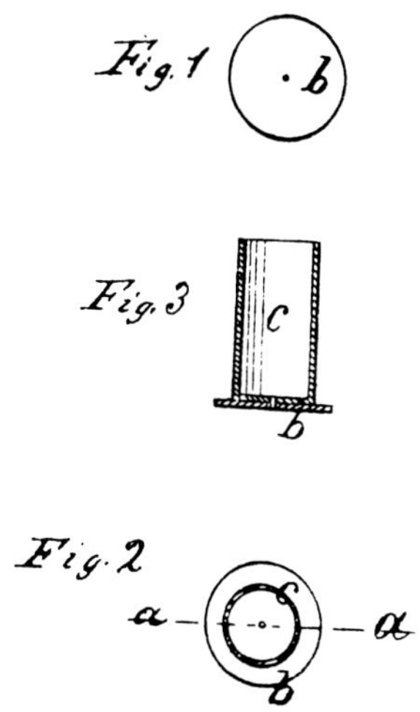

UNITED STATES PATENT OFFICE.

EDWARD MAYNARD, OF WASHINGTON, DISTRICT OF COLUMBIA.

IMPROVEMENT IN METALLIC CARTRIDGES.

Specification forming part of Letters Patent No. 39,823, dated September 8, 1863.

To all whom it may concern:

Be it known that I, EDWARD MAYNARD, of the city of Washington and District of Columbia, have invented certain new and useful improvements in the method of priming or combining fulminating-powder with loaded metallic or otherwise durable and impermeable cartridges; and I hereby declare that the following is a full, clear, and exact description thereof, reference being had to the accompanying drawings, which form a part of this specification, in which—

Figure 1 is a side elevation and Fig. 2 a vertical section of a cartridge arranged in accordance with my invention; Fig. 3 a side elevation, and Fig. 4 a top view of another form of my invention; Fig. 5 a side elevation of still another arrangement thereof; Fig. 6 a top or interior view and Fig. 7 a vertical section through the line *x x* of Fig. 6, representing my improved metallic cartridge with the percussive powder arranged in the retracting-arm thereof; Fig. 8 a top or interior view, and Fig. 9 a vertical section in the line *y y* of Fig. 8, showing another combination of the fulminate with the retracting-arm of a metallic cartridge.

Similar letters indicate like parts in each of the drawings.

The object of my invention is to combine a fulminating compound with a loaded metallic cartridge in such a m anner as to diminish or prevent the danger of an accidental explosion of the cartridge before it is placed in the gun, and thereby remove a serious objection to the general use of primed and loaded metallic cartridges.

To accomplish this end I propose to place the fulminate without the cartridge by combining it with an exterior arm or cord, which will subserve the additional purpose of a retractor, whereby the cartridge may be withdrawn from the gun after its discharge.

The simplest form of my invention is represented in Figs. 1, 3, and 5 of the accompanying drawings. I attach to a thin metallic cartridge, A, a suitable cord or thong, B, of such a length as to answer fully the purpose of a retracting-arm therefor, and to the end of this flexible arm or cord I attach securely an ordinary percussion-cap, C, either by slipping it into a noose formed at the end of the cord and folding over it a portion of the edge of the cap, as shown in the drawings, or in other suitable manner. The cord may be readily and cheaply secured to the cartridge by passing it through a simple aperture in the side of the cartridge near its base, knotting the end within the cartridge to prevent it from slipping back, and then imbedding this knot in such gum, wax, or cement, *k*, as may be found sufficiently cheap, incombustible, and durable for the purpose, or by attaching it to an eye or loop of wire the ends of which either pass through and are soldered down within the cartridge, as illustrated in Figs. 3 and 4, or are soldered within a recess formed upon its outer circumference, as shown in Fig. 5.

In this manner many of the advantages found in cartridges which contain in their base the necessary fulminating compound are obtained without the danger of premature explosion, which must prevent their general use, for the cap can be readily and quickly placed upon the gun without detaching it from the cord; and the embarrassment, impediment to rapid firing, and difficulty experienced from dropping percussion-caps in attempting to place them upon the nipple under the excitement of an action, or in very cold weather, are thereby in a measure obviated.

This feature of my invention is peculiarly adapted for use in connection with my improvements in fire-arms patented October 30, 1860; but I contemplate also the extension of the invention to that class of fire-arms in which the cartridge is fired without the intervention of a cone or nipple, and to this end I place the fulminating compound in a rigid projection or retracting-arm, as shown in Figs. 6, 7, 8, and 9.

In the cartridge shown by a vertical section at Fig. 9, a round arm, D, of suitable length is fitted and soldered into the side of the cartridge, close to its base, with its inner head beveled off, as shown in Fig. 9, to occupy the least possible space. A small aperture, *a*, is pierced centrally from the inner end of the arm outwardly a short distance, and there communicates with a recess, *b*, formed in the side of the arm in a position to be struck either mediately or directly by the hammer of the gun.

It is evident that by merely placing the fulminate in a very small compass, and at a single point, a great advantage is obtained over the usual mode of arranging the detonating compound around the entire base of the cartridge; but in order to obviate entirely the danger of an accidental discharge of the cartridge from premature explosion of this fulminate when occupying even the smallest possible space, I prefer the expression of my invention illustrated in Figs. 6 and 7. Within the arm E of the cartridge, (made thick enough for the purpose,) I form a small recess, *c*, (Fig. 7,) at a point which will be struck by the hammer of the gun when the cartridge is placed in its proper position relative thereto. This recess terminates in an aperture formed through the arm as illustrated in Fig. 6, and is filled with a suitable fulminating compound. Upon the side of the cartridge A, at a point opposite the mass of powder contained therein when loaded, I pierce a simple aperture, *d*, (to be filled with wax or grease,) and in order to connect this aperture with the recess in the arm containing the detonating compound, I pierce an aperture in the gun-barrel, (as illustrated by the dotted line in Fig. 7,) in such a position as that when the cartridge is placed in the chamber of the barrel with the priming-recess in its retracting-arm in place to be struck by the hammer of the gun, (or any substitute therefor,) the two openings in the guiding aperture will correspond and coincide with the priming-recess *c*, at one end, and with the opening *d* pierced in the side of the cartridge at the other, so that the spark will be conducted from the one to the other, and into the cartridge to fire its charge.

It is clear, however, that if by any means the detonating compound be exploded before the cartridge is in its proper position in relation to the gun-barrel, the spark will not reach the powder contained in the cartridge, but will glance harmlessly off.

I am aware that cartridges have heretofore been constructed in which a sliding pin, subserving the purpose of a retracting-arm, has been so combined therewith as that when struck by the hammer of the gun it will be driven in upon and explode a charge of fulminate placed within the cartridge. Hence I do not claim broadly the combination of fulminating-powder with the retracting-arm of a cartridge, however they may be arranged.

The important and novel feature of my invention is placing the fulminate outside of the cartridge, which permits it to be loaded with as great a degree of safety as an ordinary paper cartridge, and with far greater rapidity. The facility with which the fulminate may be subsequently added to my improved cartridges renders it unnecessary to prime them until they are required for immediate use, and they can, therefore, be transported without a priming of fulminate, completely avoiding the danger from this source, while, as has already been stated, the form of cartridge which I prefer (Figs. 6 and 7) will, when primed and ready for use, take fire from the fulminate only when placed properly in the gun.

Having thus fully described my invention, what I claim therein as new, and desire to secure by Letters Patent, is—

Combining any suitable detonating compound with a metallic or otherwise solid and durable cartridge, by means of an arm, cord, or thong, substantially in the manner and for the purpose herein set forth.

This specification of my new and useful improvement in the mode of priming metallic cartridges signed by me this 28th day of May, A. D. 1863.

EDWARD MAYNARD.

Witnesses:
RANDOLPH COYLE, Jr.,
DAVID A. BURR.

E. Maynard.
Cartridge.

No. 39823. Patented Sep. 8. 1863.

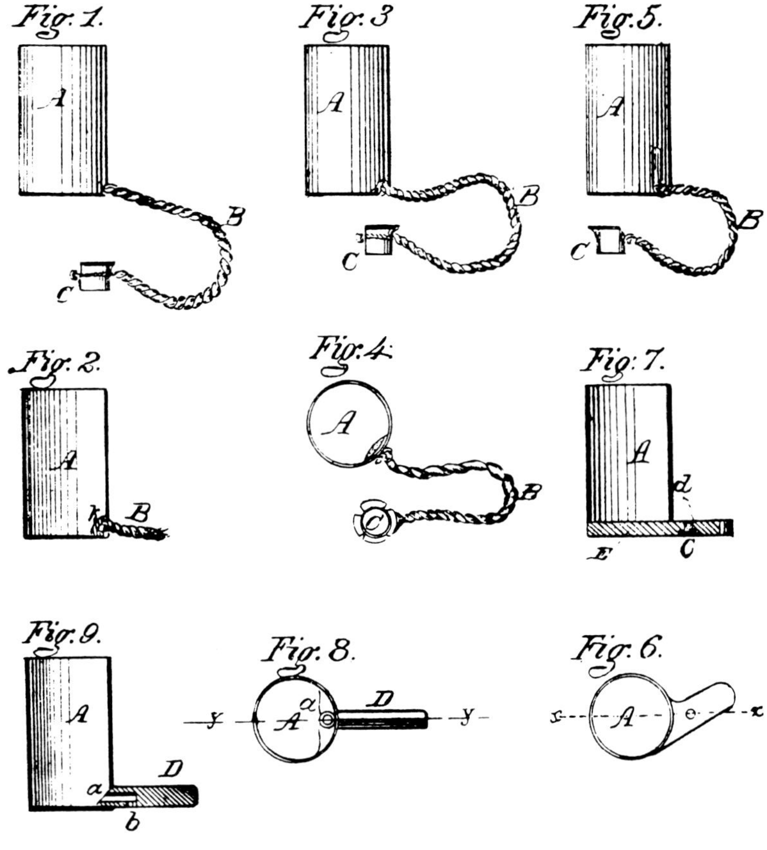

Witnesses: Randolph Coyle Jr, Thomas Adams

Edward Maynard By Robbins & Burr Atty's.

UNITED STATES PATENT OFFICE.

EDWARD MAYNARD, OF WASHINGTON, DISTRICT OF COLUMBIA.

IMPROVEMENT IN CARTRIDGES.

Specification forming part of Letters Patent No. 40,111, *dated September* 29, 1863.

To all whom it may concern:

Be it known that I, EDWARD MAYNARD, of the city of Washington and District of Columbia, have invented a new and useful Improvement in the Manufacture of Fixed Ammunition for Shot-Guns; and I do hereby declare that the following is a full, clear, and exact description of the same, reference being had to the accompanying drawings, making a part of this specification, in which—

Figure 1 is a top view of a metallic cartridge in which the charge is secured by means of my invention, and Fig. 2 is a vertical section of the same in the line *x x* of Fig. 1.

Similar letters indicate like parts in each of the drawings.

My invention relates to metallic cartridges for shot-guns, and has for its object fastening and retaining the charge in such cartridges cheaply, permanently, and securely, without in the least affecting the facility of their discharge in the gun.

It consists of a series of tongues, *a a a*, which are formed in the upper end of a metallic cartridge, B, by simply cutting slits of any desirable length on either side of the tongue to be formed.

When the cartridge is properly loaded, as illustrated in Fig. 2, and a suitable wad, *c*, placed upon the charge of shot, the tongues *a a a* may be bent over upon it, and will thus securely retain the load within the cartridge, and enable it to be carried or jolted about in any position without danger of its becoming loose or unfixed. When, however, the cartridge is fired, the explosion of the charge will straighten out the tongues *a a a* into their original position without breaking them, so that the cartridge may be again reloaded and secured with facility.

Having thus fully described my improvement, what I claim therein as new, and desire to secure by Letters Patent, is—

The formation of one or more clasping or retaining tongues in the upper rim of a metallic cartridge by slitting the edge of the same, substantially in the manner and for the purpose herein set forth.

This specification of my new and useful improvement in fixed ammunition for shot-guns, signed by me this 28th day of May, A. D. 1863.

EDWARD MAYNARD.

In presence of—

RANDOLPH COYLE, Jr.,
DAVID A. BURR.

E. Maynard,
Cartridge.

No 40.111. Patented Sep. 29. 1863.

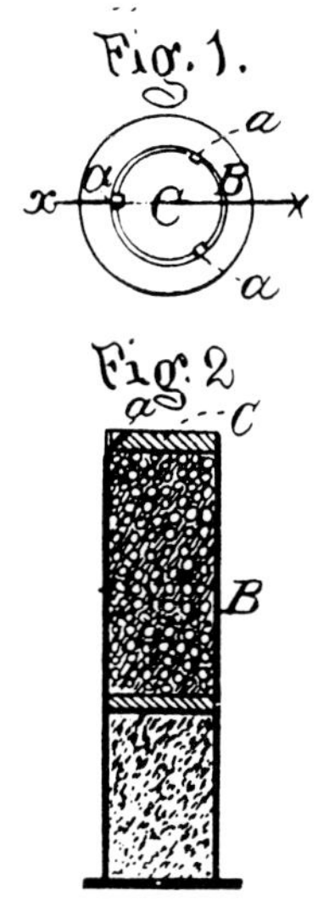

Witnesses: Thomas Adams

Edward Maynard
By Robbins & Burr
Atty's.

UNITED STATES PATENT OFFICE.

EDWARD MAYNARD, OF WASHINGTON, DISTRICT OF COLUMBIA.

IMPROVEMENT IN METALLIC CARTRIDGES.

Specification forming part of Letters Patent No. 40,112, dated September 29, 1863.

To all whom it may concern:

Be it known that I, EDWARD MAYNARD, of the city of Washington and District of Columbia, have invented certain new and useful Improvements in Metallic or Semi-Metallic Cartridges for Breech-Loading Fire-Arms; and I do hereby declare that the following is a full, clear, and exact description thereof, reference being had to the accompanying drawings, which form a part of this specification, and of which—

Figures 1, 8, 9, 10, and 11 illustrate different modes of applying a cord or thong to a metallic cartridge to subserve the purpose of a retracting-arm thereto, Fig. 9 being a vertical section of the cartridge represented by Fig. 8, and Fig. 11 a top or interior view of Fig. 10. Figs. 2, 3, 4, 5, 6, 7, 12, and 13 are different views of my improved cartridges, having rigid retracting-arms, Figs. 3, 5, 7, and 13 being, respectively, top or interior views of the cartridges represented in Figs. 2, 4, 6, and 12. Figs. 14, 15, 16, and 17 show my improved cartridges as combined with inflexible vibrating arms, Fig. 15 being a top view of one of said cartridges, Fig. 14 a vertical section of the same in the line *y y*, of Fig. 15, and Fig. 17 a top or exterior view of the cartridge shown in Fig. 16.

The object of my invention is to lessen the cost and weight of that form of metallic or similarly durable and impermeable cartridges which, from the nature of the fire-arm in which they are used, require a retracting-arm.

Heretofore it has been customary to make the bottom or base of these cartridges of a thickness greatly exceeding that of their cylindrical sides, involving the necessity of making the two separately, to be afterward united by soldering. The retracting-arm of such a cartridge is usually made in one piece with the heavy bottom thereof, and the whole is far more expensive and costly than a cartridge which requires no arm.

I lessen this expense and render the cost of retracting-arm cartridges but little greater than of those which are made without retracting-arms, by certain novel and peculiar modes of combining (either rigidly or with a certain play or vibrating movement, as may be found most convenient) a suitable flexible or inflexible retracting-arm, with a simple thin metallic (or otherwise durable) cartridge.

Figs. 1, 8, and 10 of the accompanying drawings represent my new and useful combinations of a flexible arm, cord, chain, or thong with a simple metallic cartridge. In Fig. 1 the wire or cord B is attached to an eye or loop of wire *a*, the ends of which are soldered down within a recess, *b*, formed for the purpose upon the exterior surface of the cartridge-case A, near its base.

In Fig. 8 the retracting cord or thong B is merely passed through a simple aperture pierced in the side of the cartridge near its base, and knotted within the same to prevent its slipping back. The knot is then to be embedded in such wax, gum, or cement, *c*, as may be found sufficiently cheap, incombustible, and durable, as shown in the vertical section represented at Fig. 9. Fig. 10 represents the retracting wire or cord B secured to an eye of wire, *d*, the ends of which pass through and are soldered upon the inner surface of the side of the cartridge-case, as seen in the interior view, Fig. 11. Instead of the cord or chain B, (shown in Fig. 11,) an inflexible arm, C, Fig. 16, of wire may be attached to the cartridge by passing it through an eye, *d*, secured as just described. (See Fig. 17.)

I also propose to form a rigid retracting-arm for cartridges in a cheap and effective manner, by simply passing or looping a suitable wire, D, through wo small apertures pierced near the base of the cartridge, and then uniting the two ends outwardly to form the arm D, Fig. 2, by which it is to be removed and handled. The small apertures *e e'*, through which the wire passes, are closed completely, and the wire fixed by soldering upon the inside of the cartridge. And where it is advisable that the arm should have moderate play or vibration, any cheap, incombustible, and durable wax or cement may be substituted for the solder in closing the apertures *e e'*, as is represented in Figs. 14 and 15.

Again, a single wire of sufficient strength, E, Fig. 4, or a piece of curved metal, F, Figs. 6 and 7, may subserve the purpose of a retractor, by passing one end thereof into the cartridge through a suitable aperture pierced for the purpose near its base, to be then bent up and soldered against the side thereof, as seen in Fig. 4, or so shaped as to fit and be secured in the angle formed by the inner periphery of the cartridge with its bottom, as is clearly illustrated in the top or interior view, Fig. 7.

Figs. 12 and 13 represent my mode of uniting and combining a rigid arm of wire with the cartridge by means of a groove or recess formed about its circumference near its base. The wire G is placed in this groove and drawn and fastened tightly by twisting its ends together, the twist forming a rigid arm by means of which the cartridge is withdrawn from the gun.

I contemplate, in the application of my invention, cartridges made either wholly or in part of metal, or of papier-maché, or other material which can be produced and fashioned so cheaply, and be made so durable and impermeable, as to become a fit substitute for metal in this connection.

Having thus fully described my invention, what I claim therein as new, and desire to secure by Letters Patent, is—

Combining, with a metallic or otherwise durable cartridge, a suitable retracting arm, chain, thong, or cord, in such a manner as to avoid the necessity of a distinct or unusually thick bottom thereto, substantially in the manner hereinbefore described.

This specification of my new and useful improvement in metallic cartridges signed by me this 28th day of May, A. D. 1863.

EDWARD MAYNARD.

Witnesses:

RANDOLPH COYLE, Jr.,
DAVID A. BURR.

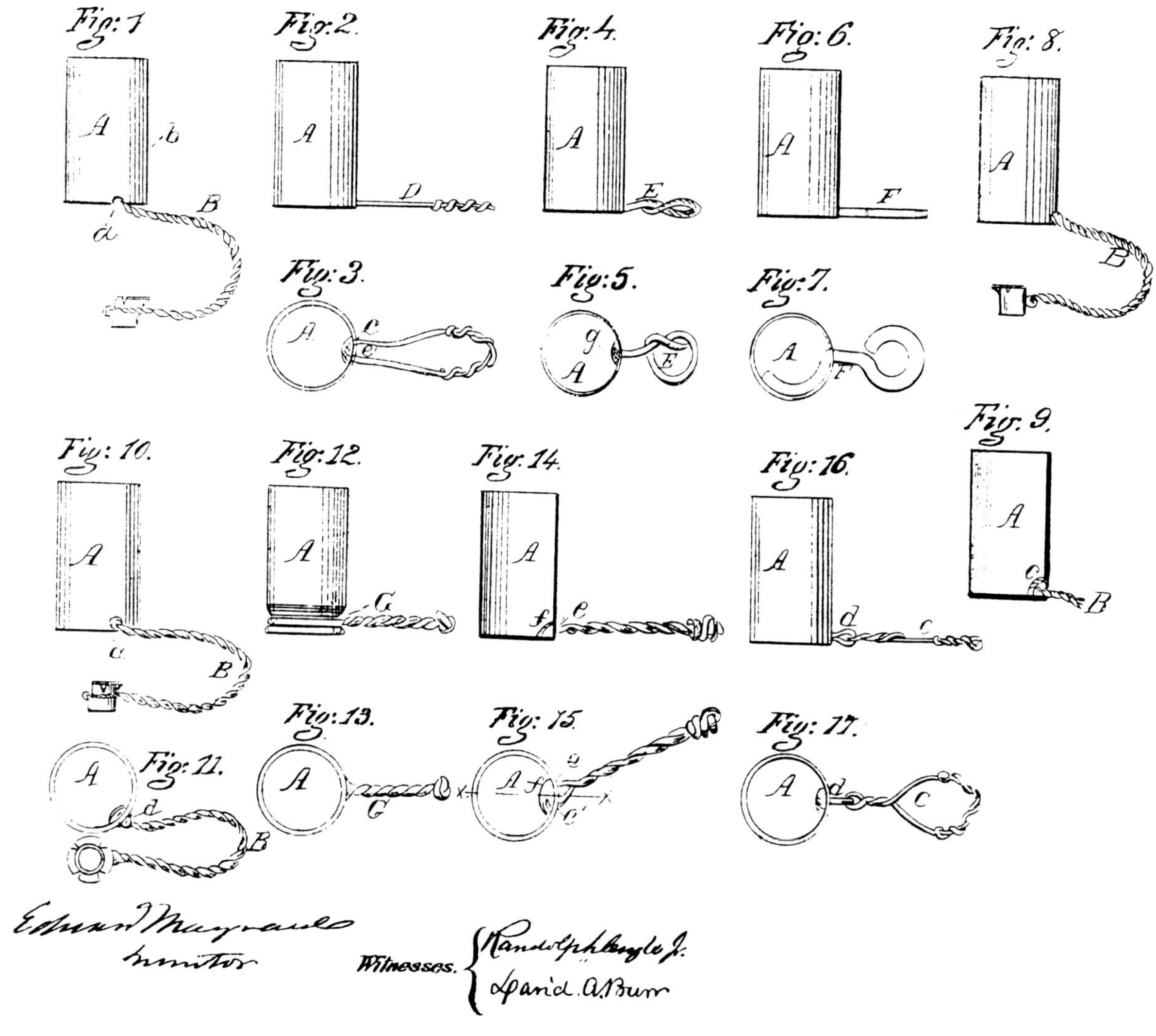

UNITED STATES PATENT OFFICE.

EDWARD MAYNARD, OF WASHINGTON, DISTRICT OF COLUMBIA.

IMPROVEMENT IN METALLIC WAD FOR CARTRIDGES.

Specification forming part of Letters Patent No. 42,388, dated April 19, 1864.

To all whom it may concern:

Be it known that I, EDWARD MAYNARD, of the city of Washington and District of Columbia, have invented a new and useful Improvement in Metallic Wads for Shot-Guns or Cartridges; and I do hereby declare that the following is a full, clear, and exact description thereof, reference being had to the accompanying drawings which form a part of this specification, and in which—

Figure 3 is a side elevation, and Fig. 5 a bottom view, of my improved metallic wad for shot-guns; Fig. 1, a side elevation, and Fig. 2 a vertical section, of the same, in combination with a metallic shot-cartridge, *c*, for a breech-loading gun, represented as partially pushed down to its proper position. Fig. 4 represents one of my improved wads with slitted sides; showing also another mode of combination with a cartridge, viz., with the open end thereof forward. Fig. 6 is a detached view of my improved wad when slitted.

Similar letters indicate like parts in each of the drawings.

The object of my invention is to provide a wad for shot-guns which shall economize space, confine the explosive and impelling gas produced by firing the charge in such a manner as to prevent its escape in advance of the shot, and, in combination with loaded metallic cartridges, afford them protection from dampness.

To attain this end, I have invented a cup-shaped wad, A, made of thin metal, by simply "striking up" cylindrical sides from a disk in the customary manner. The length of the sides of the wad should be so proportioned as that, when placed in the barrel of a gun, it shall occupy space enough longitudinally to prevent its turning therein when shot forward, as is usual with ordinary wads, and may properly be equal to at least half its diameter.

The cylindrical sides or rim of the wad not only serve as guides and stays to keep it exactly in a plane at right angles to the bore of the gun, but will, by their expansion when the gun is fired, prevent completely the escape of gas in advance of the shot, and thus insure their perfect and regular expulsion.

This expansion of the sides of the wad, which causes them to fit with the utmost nicety the bore of the gun, is attained, if the open end of the wad be placed outward when dropped in the gun (as shown in Fig. 4) by the resistance or inertia of the air confined in the barrel, as it is acted upon by the explosive force of the charge, a force which will itself cause their expansion when the wad is placed in the gun or cartridge with its open end or mouth inward, as in Fig. 1.

In order to accommodate a given wad in shot-guns of varying calibers, the rim thereof may be slitted (as is shown in the wad B of Figs. 4 and 6 of the accompanying drawings) to allow its greater expansion.

My improved wad, from its thin metallic structure, is peculiarly useful for combination with metallic or other shot-cartridges. Its novel form permits it to be securely fitted to the cartridge in such a way as to render it completely water-proof and secure from dampness, rendering it at the same time more neat and compact, but it is also most valuable in all cases where great accuracy and projectile force are required.

Having thus fully described my invention, what I claim therein as new, and desire to secure by Letters Patent, is—

My improved gun-wad, formed of a circular disk, with a projecting cylindrical edge or rim, substantially in the manner and for the purpose herein set forth.

This specification of my new and useful improvement in metallic wads for shot-guns, signed by me this 28th day of May, A. D. 1863.

EDWARD MAYNARD.

In the presence of—
RANDOLPH COYLE, Jr.,
DAVID A. BURR.

E. Maynard,

Cartridge.

No. 42,388. Patented. Apl. 19. 1864.

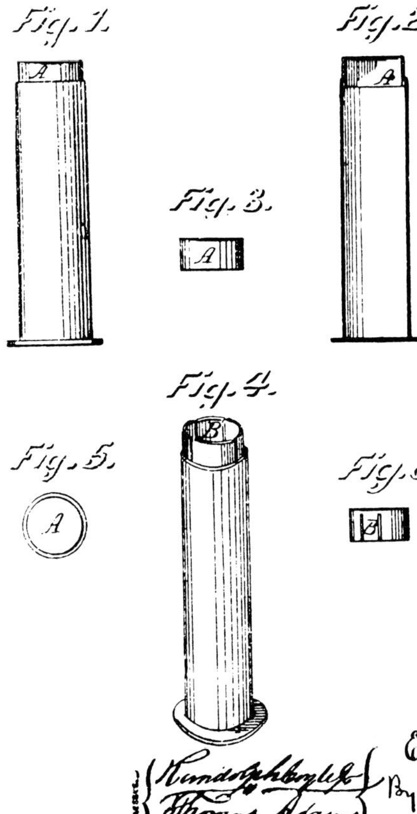

Witnesses { Randolph Coyle Jr. / Thomas Adams }

Edward Maynard
By Robbins & Burr
atty's.

UNITED STATES PATENT OFFICE.

EDWARD MAYNARD, OF WASHINGTON, DISTRICT OF COLUMBIA.

IMPROVEMENT IN PRIMING METALLIC CARTRIDGES.

Specification forming part of Letters Patent No. **45,420**, dated December 13, 1864.

To all whom it may concern:

Be it known that I, EDWARD MAYNARD, of the city of Washington, in the District of Columbia, have invented a new and useful Improvement in Primed Cartridges; and I do hereby declare that the following is a full, clear, and exact description thereof, reference being had to the accompanying drawings, making a part of this specification, and in which—

Figure 1 is an elevation of my improved cartridge; Fig. 2, a vertical central section of the same; and Fig. 3, a cross-section in the line *x x* of Fig. 1. (See thin drawing.)

Similar letters indicate corresponding parts in each of the figures.

The object of my invention is to produce cheaply a reliable cartridge primed at one point only, and which shall be peculiarly adapted for use in the United States muskets so altered as to become breech-loading guns, in the manner described in my patent of October 30, 1860, and in the specification of an improvement thereon, herewith filed, in a separate application for Letters Patent.

To accomplish my object, I combine a simple percussion tube or primer, D, with a metallic cartridge by passing it through an aperture in the side thereof, as seen in Figs. 1, 2, and 3 of the drawings. This combination is not in itself novel, but my invention consists in a novel mode of confining and supporting the inner end of the priming-tube D, by causing it to pass through an annular wad, B, of papier-maché, or other suitable material, fixed in the bottom of the cartridge.

This wad B is made in the shape of a ring, so that the knotted end of the retracting-cord used in connection with all cartridges to be fired from my improved breech-loading gun, as patented October 30, 1860, may be inclosed in the center thereof. To receive this cord, an aperture is pierced through the wad from its inner to its outer periphery, and the outer end of the cord extending from the wad is then passed through an aperture in the side of the cartridge and depends therefrom, as seen in Figs. 1 and 2. The wad, when slipped into its place in the interior of the cartridge, thus incloses the knotted end of the cord, which is thereby prevented from slipping out. When the wad is thus fixed in the bottom of the cartridge and held by the retracting-cord, it forms a solid support for the inner end of the long, slim percussion-tube or hollow wire D, Fig. 2, which contains the fulminate. This priming-tube is driven into an aperture pierced in the side of the cartridge-case, and through the wad to its inner central cavity, until its inner end reaches this cavity. Its outer end will then project slightly beyond the exterior surface of the cartridge, as seen in the drawings, to receive the blow of the gun-hammer or hammer-rod. By dipping the tube D in a varnish or cement before driving it into place, the joint between the tube and cartridge may be made water-proof.

I contemplate the use of a thick annular layer of cement in place of the wad B, the cement to be so molded as to have substantially the form of said wad B, as described. In this case the percussion-tube D is to be placed in the cartridge and pressed through the cement before the latter has become fully hardened.

Having thus fully described my improvement in primed metallic cartridges, what I claim therein as new, and desire to secure by Letters Patent, is—

The combination of a wad, B, of any suitable material, with the interior of a cartridge, A, so as to form a firm support for the inner end of a priming-tube, D, passing out through the side of said cartridge, substantially in the manner herein set forth.

EDWARD MAYNARD.

Signed in presence of—
GEO. W. MAYNARD,
J. H. H. DOTY.

E. Maynard.

Cartridge.

No 45,120. Patented. Dec. 13. 1864.

Fig: 1.

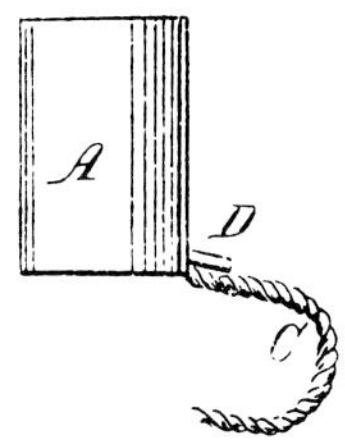

Fig: 2.

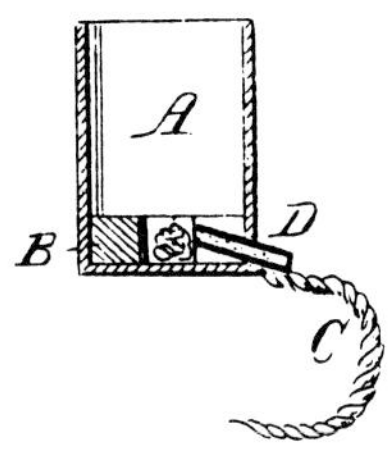

Fig: 3.

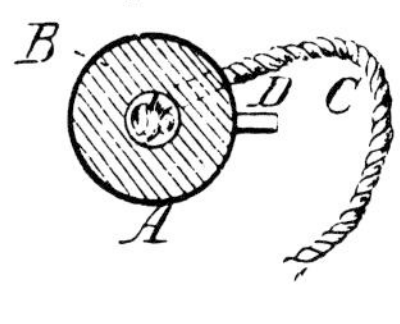

Witnesses { Chas H Anderson / Randolph Coyle Jr }

Edward Maynard
By Robbins & Burr
Attorneys

This is a transcription of the handwritten specification from Dr. Edward Maynard's own copy of his Letters Patent for "Improvement in Breech Loading Fire Arms" number 48,423 granted June 27, 1865, "and of the INDEPENDENCE of the United States of America the eighty ninth."

To All whom it may concern.

Be it known that I, Edward Maynard, of the City of Washington, the District of Columbia, have invented a new and useful improvement in the solid hinged breech-block of breech loading muskets, and do hereby declare that the following is a full, clear and exact description of the reference being had to the accompanying drawings, forming a part of this specification, in which

Fig 1, is a side elevation of a portion of a breech loading musket , furnished with my improved breech block.

Fig. 2 is a top, or plain view of the same.

Fig. 3, a longitudinal section thereof on the line x, x, of figure 2, showing the relative arrangement of the hammer of the gun and the plunger in the breech block.

Fig. 4, a transverse section in the line y, y, of figure, 3, showing the form given to the hinged breech block, and also the diagonal hole in the same, in which the retaining spring of the plunger operates.

Fig. 5, a transverse section in the line z, z, of figure 3, showing the form of the base of a primed metallic cartridge of peculiar construction, which may be advantageously employed in connection with my present improvement in the breech block, and-

Fig's 6, 7 & 8, respectively an elevation, central vertical section and a transverse section of this cartridge.

Fig. 9, a longitudinal section of the breech block in the line of w, w, of fig. 2 &

Fig. 10 & 11, respectively a central vertical section and a bottom view of another form of cartridge.

Like letters denote like parts in the several figures.

The musket now in use in the United States serfvice, recognized and accepted as the regulation musket, is well known to be satisfactory to the Government in the matter of size, shape and weight. Considering the fact that its harmonious proportions, and the result of many experiments, and that they cannot be advantageously varied, I have endeavored to give to this finished arm all the advantages which are formed in the modern breech loading weapons, without imparing its established proportions, by the addition of an improved solid hinged cone seat to such muskets.

This improvement is described in the letters, patent of the United States granted to me on the 30th day of October A.D., 1860. Since that date the advantages attending the use of primed cartridges, in breech loading firearms, have become so decidedly apparent as to lead me to seek, so to improve the regulation musket, altered or manufactured in accordance with my patent of October 30th, 1860 as to adapt it to the use of such cartridges.

I have therefore substituted, for the cone, in the hinged breech block A, of such a musket, a plunger B, playing freely in an aperture, pierced obliquely through the breech block A, on a distinct line from a point in the seat of the usual cone socket, to a point a little above the notch cut in the side of the barrel of the gun, to receive the retracting thong or arm of the cartridge inserted in its chamber; this point when the breech block is closed, being also in opposition to a primed flange or offset formed upon the said cartridge, all as clearly shown in Fig. 3.

The outer end of this rod, or plunger, projects from the sear end of the breech block, in the place of the usual cone, so that it may be struck by the hammer of the gun, as ordinarily arranged, without any change in its position. It is of such a length as that, when its outer end projects, as described, its inner or lower end will be flush with the front face of the breech block and will rest loosely against the primed flange or offset of the cartridge, when the breechblock is closed. The plunger B is retained in this position secured from accidental violent contact with the cartridge, and is prevented from turning by means of a spring s, projecting up through a hole, extending from the underside of the breech block, to the aperture in which the plunger works..Fig's, 3 & 4. The lower end of the spring is secured in a recess formed upon the under side of the breech block, as seen in Fig. 9, and its upper or vibrating end is received into a hole formed in the underside of the plunger B. A simple set screw inserted through the side of the breech block and received into a notch or recess in the plunger, or any other equivalent device may be substituted for he spring s., but I deem he employment of the spring, as described peculiarly desirable, for the hole in which it operates may be advantageously employed as a channel through which to convey oil to the plunger, for the purpose of facilitating its operation.

The end of the plunger is bevelled off, or slightly rounded, as shown in Fig. 3, of the drawings, so that if, from the possible defective action of the spring s, or from any other cause it should accidentally project beyond the face of the breech block, it will be forced back to its proper position by contact with the barrel of the gun.

The nose of the ordinary hammer used upon the regulation musket of most nations, needs but to be cut off, as shown in the drawings, and the hammer perhaps slightly bent, in order to adapt it to use in connection with the improved plunger, and the lower portion of that end of the recess, or loading aperture in the gun, which is next to the chamber, needs to be slightly enlarged, or cut down to admit of the introduction of a flanched (sic) cartridge. These are the only alterations required in the musket proper, in addition to those set forth in my patent of October 30th, 1860, to attain the desired improvement.

Any cartridge having an encircling priming flange at its base or which is primed at a single point (so arranged and devised as to fall in line with the end of the plunger, when the cartridge is inserted in the gun and the breech block is in its closed position) may be used in the musket improved as described, but the cartridge clearly represented in Fig's 6, 7 & 8 and 10 & 11, of the drawings obviate the necessity of an enlargement of the loading aperture, and as they otherwise present peculiar advantages for use in connection with my present improvement, I have made them the subject of a separate application for letters patent.

By cutting down the nose of the hammer, as represented in the drawings, it is adapted to strike a direct blow upon the plunger. I hereby, not only avoid the expense of a new hammer in altering the regulation musket of most nations, to the use of primed cartridges, but also perfect the action of the old hammer; for as it is made lighter there is less weight and consequent inertia to be overcome by the spring when it is released by the trigger, so that the movement of the hammer is quicker whilst the blow is lighter, and therefore better suited to the primed cartridge.

The operation of the gun, improved as hereinbefore described, is simple. When loaded in the usual manner and ready for firing, the plunger of the gun, under the control of the spring s, and its inner end touching the flange or point of the cartridge, in which the fulminate is placed. When the hammer is released by the trigger it falls directly upon the end of the plunger, and the blow, overcoming the influence of the spring s, acts immediately upon the priming of the cartridge, causing the discharge of the load from the gun.

Having fully described my improvement in breech loading muskets, what I claim therein as new and desire to secure by Letters Patent, is

The continuation of the retaining spring s, with the pluinger B in the hinged breech block of the breech loading musket, or other fire arm, when the said spring is contained in a hole extending from the underside of the breech block to the aperture in which the plunger works, substantially in the manner and for the purpose herein set forth.

The foregoing specification of my improvements in breech loading fire arm signed this______day of September 1864.

Edward Maynard

Witnesses.

George W. Maynard,
J. H. H. Doty.

(Note: On the back of the last sheet of the patent are handwritten notes, apparently those of Dr. Maynard, as follows: In the upper right-hand corner--June 27-65- (date this patent was awarded).

Centered, at the top of the sheet:

J. Davis July 7, 1863, Nr. 39198
B.F. Joslyn Aug 4 1863 - " 39407
Jas. Nemer February 23, 1864 " 41732
J. Goulding May 3 1864 " 42573
G. Millers June 8 1864 "20503
Smith & Wesson July 14 1854 "10535

These are related United States patents which the inventor apparently consulted).

№ 48.423.

Edward Maynard's Imp^ts in Breech-loading Muskets.

Fig. 1.

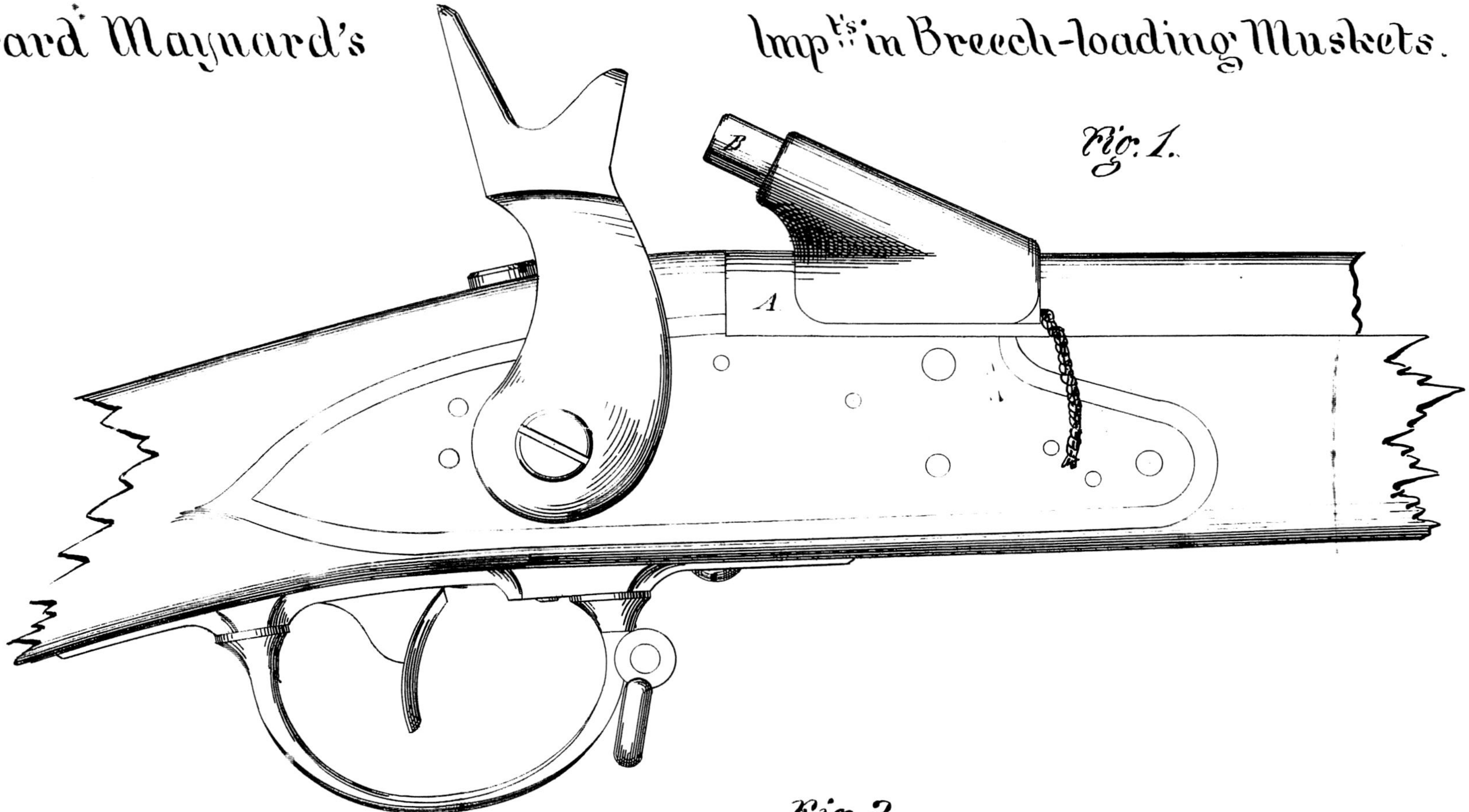

Fig. 2.

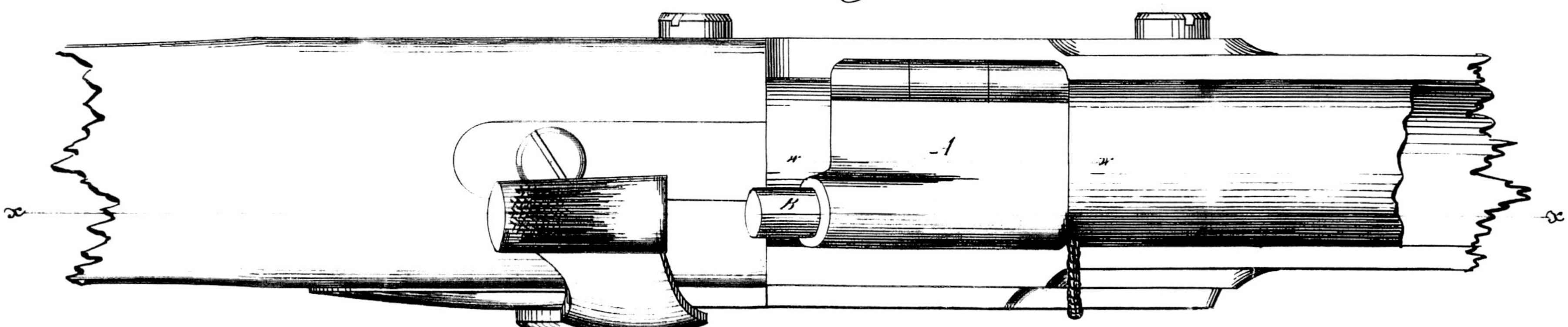

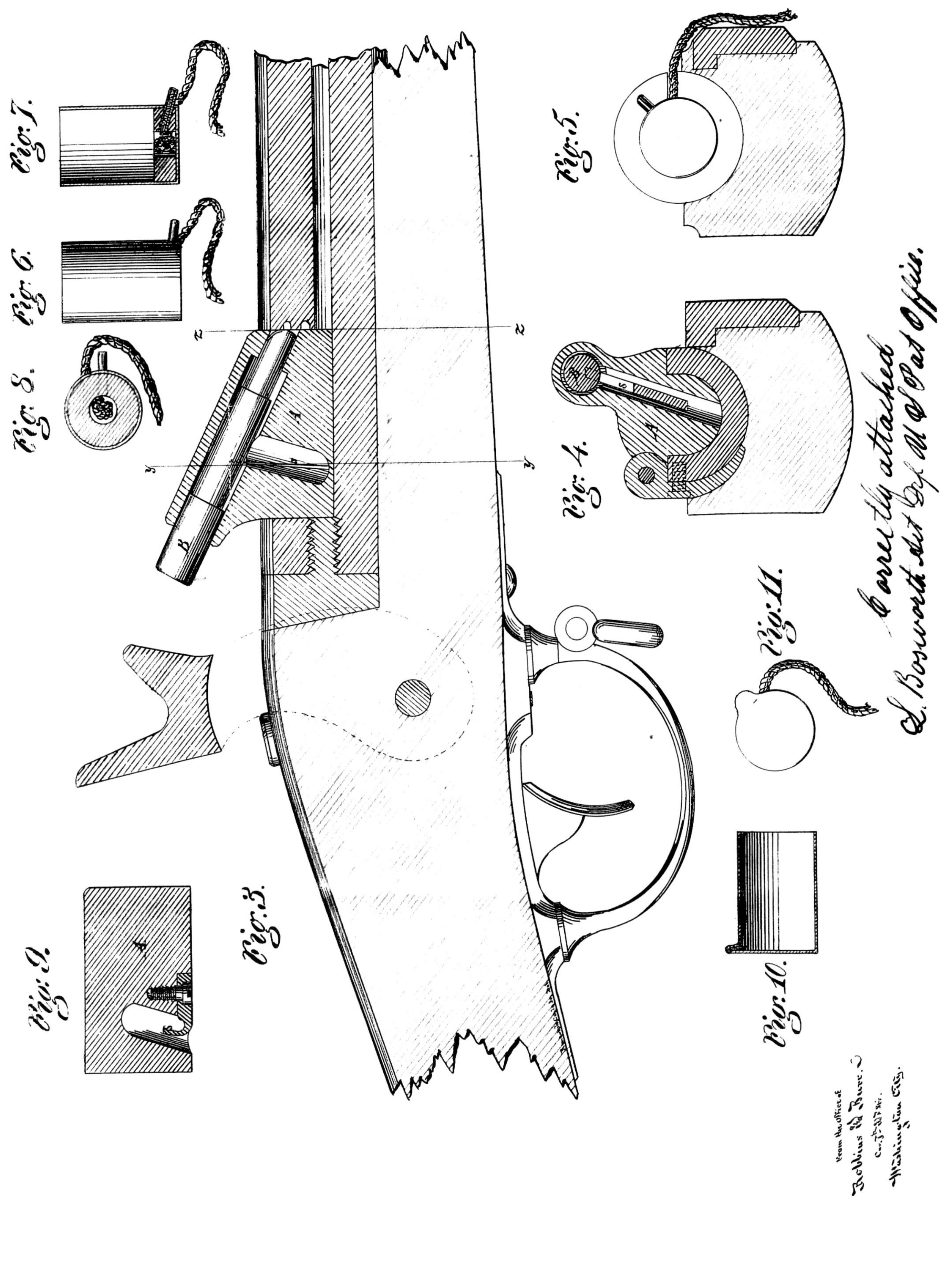
Fig: 7.
Fig. 6.
Fig. 8.
Fig: 5.
Fig. 4.
Fig: 3.
Fig: 9.
Fig: 10.
Fig: 11.
Correctly attached

UNITED STATES PATENT OFFICE.

EDWARD MAYNARD, OF TARRYTOWN, NEW YORK.

IMPROVEMENT IN PRIMING METALLIC CARTRIDGES.

Specification forming part of Letters Patent No. 59,044, *dated October* 23, 1866; *anted ated October* 3, 1866.

To all whom it may concern:

Be it known that I, EDWARD MAYNARD, of Tarrytown, in the county of Westchester and State of New York, have invented a new and useful Improvement in Primed Cartridges; and I do hereby declare the following to be a full, clear, and exact description thereof, reference being had to the accompanying drawings, forming a part of this specification, in which—

Figure 1 is a central longitudinal section of the simple form of my improved primed cartridges; Figs. 2 and 3, similar sections, respectively, of modifications thereof; Fig. 4, a transverse section through the base of the cartridge illustrated by Fig. 3 in the line *x x* thereof; Fig. 5, a bottom view of a second form of retracting-arm cartridge as improved by my invention, and Fig. 6 a transverse section through the bottom thereof, showing the position of the priming-cavity therein with reference to the arm and supporting-lug.

Similar letters indicate like parts in all of the figures.

The nature of my invention consists in the combination of an outer cap or cover of thin metal with a priming-recess formed exteriorly in the base of a cartridge, when said base is provided with a projecting rim or flange, or with pins or arms, so as to be firmly supported when placed in the gun, and is made of such strength as to afford proper resistance to the blow of the hammer in firing the charge.

Fig. 1 illustrates my invention as applied to the metallic cartridges for which I obtained Letters Patent on the 28th day of May, 1861.

These cartridges are constructed of a thin metallic cup, *a*, secured, by solder or otherwise, upon a solid disk, *b*, of a diameter so far greater than that of the cup as to project and form a supporting-flange around the bottom thereof, as seen in the drawings.

In the center of the thick double bottom of this cartridge I form a small priming-cavity, *c*, of any suitable form, having a vent communicating therefrom with the interior of the cartridge. When this cavity has been filled with fulminate, I protect it by a cap, *e*, of thin metal made to fit closely over the entire disk, forming the base of the cartridge. This cap *e* shields completely the fulminate from exposure, and, as it extends to the outer circumference of the rim of the cartridge, will, when it is discharged, protect the firing-pin and the lock from the gases leaking back from the charge by conveying them outwardly.

This metallic cap may be combined with the base of any cartridge where said base is firmly supported in the gun, either by means of a projecting flange, as in Figs. 1, 2, and 3, or by supporting lugs or pins, as seen in Figs. 5 and 6, and to any cartridge having such a solid supported base, whatever the composition of the cup or casing may be, whether of papier-maché, metal, or otherwise.

In Fig. 2 a modification of the base is illustrated, in which, although a sufficiently solid supporting-flange is retained, the central part only of the disk is covered by the shielding-cap *c*.

In Fig. 3 its application to such forms of cartridges as are provided with a solid retracting-arm, as set forth in my Letters Patent of September 8, 1863, is shown. In applying my improvement to these cartridges, the solid base must be either slightly enlarged (see *m* in Fig. 3) to provide a supporting-flange thereto, or a projecting lug, *g*, must be formed, (as seen in Figs. 5 and 6,) in addition to the arm, for the same purpose.

When a lug is formed upon the base of the cartridge in relation to the retracting-arm, it becomes necessary in such cartridges as are to be used in my improved fire-arms having a hinged breech-block, or others of a similar pattern, to place such lug *g* at the end of the chord of an arc less than a semicircle, and in order to obtain proper resistance to the blow of the hammer upon the base when thus supported, I place the priming-recess in such cartridges in a line between the arm *f* and lug *g*, as illustrated in Figs. 5 and 6, rather than in the center thereof.

In fitting the shielding-cap *e* over and upon the base *b* of such cartridges, I cause its rim to embrace the retracting-arm and the lug by cutting away said rim at these points.

Having thus fully described my improvement in primed cartridges, what I claim therein as new, and desire to secure by Letters Patent, is—

A closely-fitting metallic shield or cover, in combination with the base of a cartridge, when said base is so constructed as to receive solid support from the gun and is primed exteriorly, substantially in the manner and for the purpose herein set forth.

EDWARD MAYNARD.

Witnesses:
H. H. YOUNG,
DAVID A. BURR.

E. Maynard,

Cartridge.

No. 59044. Patented Oct. 23 1866.

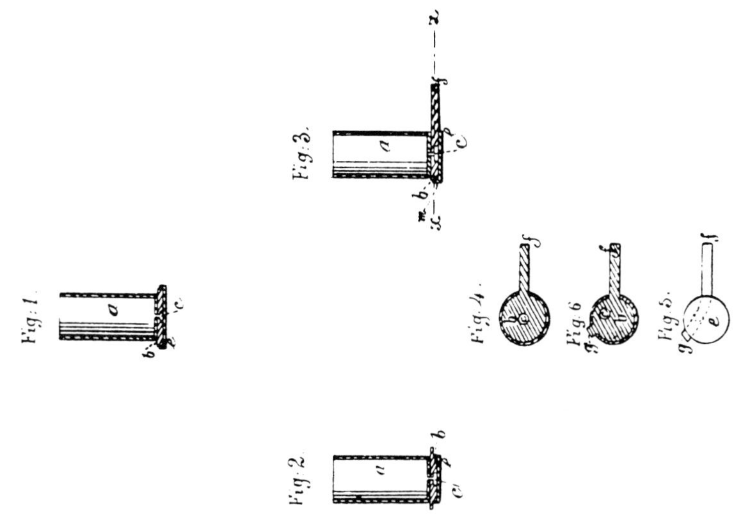

Edward Maynard

Inventor

Witnesses

W. Burris

M. Elliott Burns

David A. Burr

Atty.

UNITED STATES PATENT OFFICE.

HIRAM BERDAN, OF NEW YORK, N. Y.

IMPROVEMENT IN METALLIC CARTRIDGES.

Specification forming part of Letters Patent No. 52,818, dated February 27, 1866.

To all whom it may concern:

Be it known that I, HIRAM BERDAN, of New York city, in the county of New York and State of New York, have invented a new and useful Improvement in Center-Fire Metallic Cartridges for Breech-Loading Arms; and I do hereby declare that the following is a full, clear, and exact description of the same, reference being had to the accompanying drawings, forming part of this specification, in which—

Figure 1 is a longitudinal section of one of my improved center-fire metallic cartridges complete. Fig. 2 is a view of the rear end of the same. Fig. 3 is a view of the front end of the cartridge. Fig. 4 is a section view of a partly-made metallic cartridge. Fig. 5 is a view of the rear end of the center-pin and ball.

Similar letters of reference in the several figures indicate corresponding parts.

The object of my invention is to protect metallic cartridges against accidental explosion, and also to combine in one cartridge the advantages of a "center-fire" and the taper form in such manner as to render their manufacture less expensive and dangerous.

In the construction of metallic cartridges with a center-fire it has never been deemed practical, on account of the danger of premature explosion, to make the cases of a taper form by the drawing or swaging process after the fulminate, powder, and ball are inserted; that is, at one blow, by means of dies. The premature discharge would occur in the swaging process with other cartridges than mine, on account of the fulminate being placed in the hollow rim of the metallic case or between disks at the center of the head of such a case, and, thus located, is subjected to the force of the blow upon the bullet and powder necessary to give the required taper form to the case and press the bullet to its proper position.

The nature of my invention consists, first, in a cavity or fulminate-chamber in the end of the anvil or centering-rod, in combination with the head of a cartridge-case; second, in a tapering cartridge-case drawn or swaged from sheet metal, with the smallest end forward and with a center fire, the same constituting an improved new article of manufacture.

To enable others skilled in the art to make and use my invention, I will proceed to describe the same.

To make my improved metallic cartridge, I take sheet metal and first bring it into a cylindric form, the cylinder being open at one end. The head of the case is formed in the usual manner, but without a fulminate-chamber; the corrugation *a* in the head is formed by the same die which forms the head. A conic ball, with grooves in its circumference and a cavity in the base of the ball for the anvil or centering-stem *b*, is now brought into use. The stem of this ball has in its rear end a small cavity or chamber, *c*, in which the fulminate is placed so as to be free from the inside of the head of the cartridge-case when the ball and case are put together, as illustrated in Fig. 1 of the drawings. The case, charge of powder, and the ball and stem having all been brought together in the usual mode, the swaging or drawing die is brought down over the ball and case with sufficient force to change the case from the form shown in Fig. 4 into the form shown in Fig. 1. The ball is held true or in center during the swaging or tapering process by means of the stem and the die and the milled ring-shoulder of the cartridge-case head. The same operation that produces the taper form of the case also fastens the ball, stem, and case firmly together, and insures uniformity in the external surface.

It will be observed that the fulminate of the cartridge is entirely isolated from the head of the cartridge, and is so located that it cannot possibly be ignited by a square contact of the head with any object, or vice versa; nor by a contact of the flange with any object, or vice versa. In a word, it is necessary, in order to ignite the fulminate, to indent the head of the cartridge-case directly at the center, and to effect this a very small implement must come in contact with the head at that point.

It is a very important matter to have a strong flange to the cartridge-case, and this I secure by making the head solid and of thicker metal than can be employed when the fulminate-recess is formed in the head.

It is also important to have the cartridge-case of taper form, in order that it may be readily withdrawn from its seat after it has been exploded in a gun. The taper form also gives a close fit between the cartridge-case and the breech of the barrel, as it enters on the wedge principle.

It is also important to ignite the charge at the center of the cartridge, for reasons well understood.

It is also important to make the taper cartridge-case of sheet metal, as well as unite the pin, ball, and case together at one operation by the swaging or drawing process, as by this means a much firmer anvil to detonate against is formed, and also much time and expense are saved, and, withal, a more uniform external surface is secured.

It is further important to have the fulminate within the compass of the circumference of the pin in order that the swaging or drawing die may be forced down to the flange upon the ball, powder, and case with the necessary power and rapidity without liability of igniting or detonating the fulminate.

And, finally, by combining a taper-form case and a center fire the force of the blow which is necessary to indent the head of the cartridge-case or produce the detonation of the fulminate or agent employed to fire the charge is to a great degree received by the inclined shoulder presented by the taper cartridge to the counter-bore of the gun, and thus any tendency to strip or force back the flange and press the case too far into the gun-barrel or over the ball is overcome.

I believe I am the first one who has ever combined any description of center fire and a swaged or drawn tapering sheet-metal cartridge-case, whether the fulminate is contained in a cavity of the stem or between disks, or in a cap placed on a cone which is in communication with the charge of powder in the cartridge, or with any description of center fire.

What I claim as my invention, and desire to secure by Letters Patent, is—

As a new article of manufacture, a metallic cartridge, having a stem projecting from the base of the bullet with a cavity at the rear end for the fulminate, the case being made with a flange at the rear end and tapering toward the forward end, and drawn down upon the bullet by means of a swage or die, whereby the external surface of the cartridge can be made more uniform, less liable to explode in the act of construction or transportation, and the shell more readily extracted after being fired.

HIRAM BERDAN.

Witnesses:

R. T. CAMPBELL,

EDW. SCHAFER.

Fig. 2

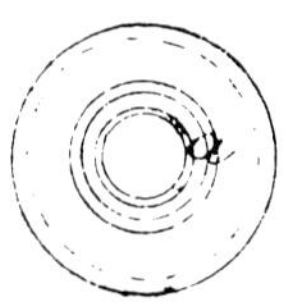

Fig. 1.

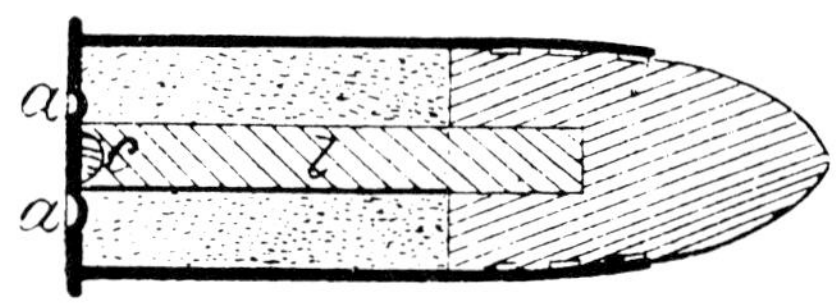

Fig. 3

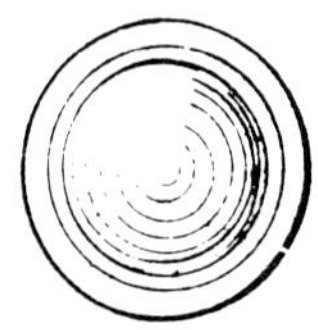

Fig 4

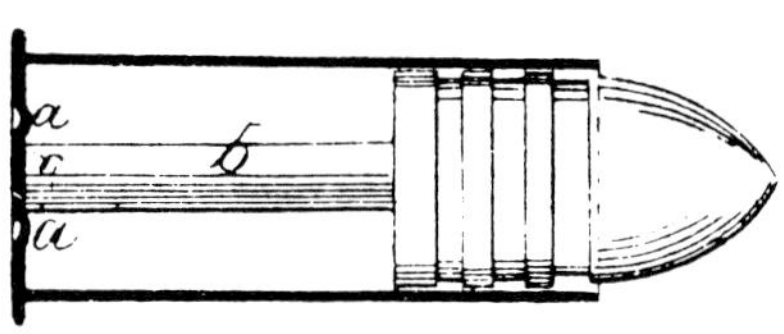

Fig. 5

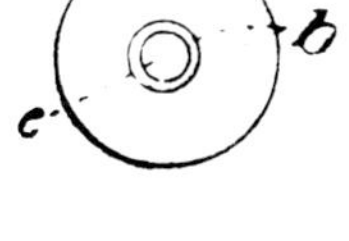

UNITED STATES PATENT OFFICE.

HIRAM BERDAN, OF NEW YORK, N. Y., ASSIGNOR TO THE BERDAN FIRE-ARMS MANUFACTURING COMPANY, OF SAME PLACE.

IMPROVEMENT IN PRIMING METALLIC CARTRIDGES.

Specification forming part of Letters Patent No. 53,388, dated March 20, 1866.

To all whom it may concern:

Be it known that I, HIRAM BERDAN, of the city, county, and State of New York, have invented a new and useful Improvement in Cartridges for Breech-Loading Fire-Arms; and I do hereby declare that the following is a full, clear, and exact description of the same, reference being had to the accompanying drawings.

This invention relates to an improvement in that class of cartridges known as center-fire cartridges, whether such cartridges have their entire shells or their heads or bottoms only made of metal. The object of this improvement is partly to insure the prevention of the accidental ignition of the fulminate-priming to which other center-fire cartridges are liable in the manufacture, or when packed and transported, a number together; partly to better insure the ignition of said priming by the blow of the hammer; and partly to enable the shell to be reloaded, reprimed, and used over again, if desired, for target practice, or in any case in which it might be convenient to save the shell, without renewal or replacement of any of its parts but the cap, pellet, or other primer. The head or bottom of the cartridge is constructed or fitted with a cup, into which the fulminate-priming, in the form of a cap, pellet, or otherwise, is inserted in such manner as to be entirely contained some distance within the outer face of the head or bottom of the shell; and in the bottom of the said cup there is a permanently-fixed projection in a rearward direction, against which the fulminate-priming is driven by the hammer, and which presents a point or so small a surface that the entire force of the blow of the hammer is so concentrated on a small portion of the priming as to insure its ignition.

Figure 1 in the accompanying drawings is a central longitudinal section of a ball-cartridge constructed according to this invention, on a scale larger than the real size. Fig. 2 is a rear end view or the same.

Similar letters of reference indicate corresponding parts in both figures.

The shell of the cartridge is represented in the drawing as made entirely of copper; but it may be made of paper, with a copper or other metal head. The cup N, which receives the priming, is represented as made in a separate piece, inserted tightly into a hole in the center of the head, and with a flange, *n*, countersunk into the head; but it might be practicable to make it of the same piece with the head. In the bottom of this cup, which projects into the cartridge, there is a vent-hole, *r*, for the passage of the fire from the priming to the charge, and there is also an outward teat-like projection, *s*, or projecting point, which is represented as formed by indenting the bottom of the cup on the inner side, but which may be formed by depressing the metal around it, or by making a hole and inserting and securing a small pin or teat of solid metal; but in any case it must constitute a permanent portion of the cup which will not be detached during the use of the cartridge. The priming is represented as being in the form of a shallow pellet-like cap, *t*, containing the fulminate *u*; but it may be in any other form that will permit its insertion loosely into the cup N, in which it is secured by means of water-proof varnish or other adhesive material, at such distance from the outside of the head that it may be protected from the points of the conical bullets or any other bodies, the contact or concussion of which might be liable to produce the ignition of the fulminate. The fulminate is toward the teat or point *s*.

The fulminate is to be fired by means of a firing-pin struck by the hammer or by the point of a hammer, which will enter the cup N, and drive the priming forward against the teat or point *s*, which presents so small a surface that the force of the blow is concentrated on an exceedingly small surface of the priming and will insure its explosion. The rim of the cap *t* expands within the cup N in firing, in such manner as to form a gas-check.

In the operation of this cartridge no portion of the metal is displaced, the pellet, cap, or priming alone being driven forward against the fixed point *s*; and when the discharged cartridge-shell or its metal head has been withdrawn from the fire-arm, the cap *t* or residual portion of the priming will drop out from the cup, in which a new priming can be inserted if desired to reload the cartridge.

One advantage of this cartridge is that it may be loaded before the application of the priming, thus not only rendering its manufacture less dangerous than that of other primed cartridges, but permitting the cartridges to be stored in an unprimed condition, and the priming to be applied just before they are required to be used, thereby not only rendering the storage safer, but preventing the deterioration to which the fulminate-priming is subject in cartridges when kept for a long time.

What I claim as my invention, and desire to secure by Letters Patent, is—

The fixed permanent teat-like projection *s* at the bottom of the cup N in the head of a cartridge, in combination with the pellet or other priming inserted into and protected within the said cup, substantially as herein set forth.

H. BERDAN.

Witnesses:
HENRY T. BROWN,
J. W. COOMBS.

H. Berdan.

Cartridge.

No. 53388. Patented Mar. 20. 1866.

Fig:1

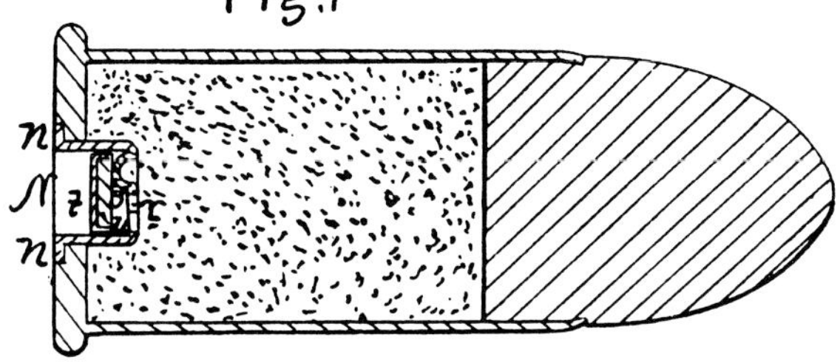

Fig:2

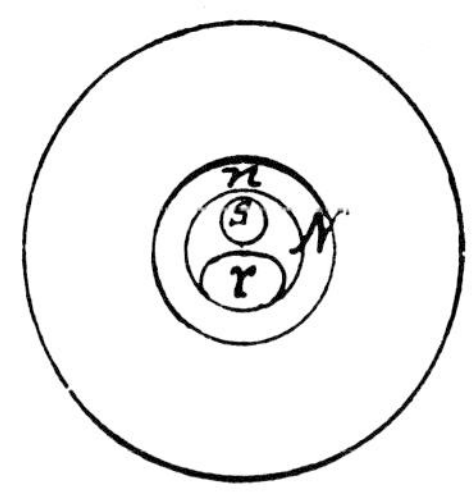

Witnesses:
J. W. Coombs
Henry T. Brown

Inventor:
H Berdan

UNITED STATES PATENT OFFICE.

O. F. WINCHESTER, OF NEW HAVEN, CONNECTICUT.

IMPROVEMENT IN METALLIC CARTRIDGES.

Specification forming part of Letters Patent No. 60,814, *dated January* 1, 1867.

To all whom it may concern :

Be it known that I, O. F. WINCHESTER, of New Haven, in the county of New Haven and State of Connecticut, have invented certain new and useful Improvements in Metallic Cartridge-Cases; and I do hereby declare that the following is a clear, full, and exact description thereof, reference being had to the accompanying drawings, making part of this specification, and to the letters of reference marked thereon, like letters indicating like parts wherever they occur.

Figure 1 is a longitudinal sectional view of my improved case complete. Figs. 2, 3, and 4 are views of the separate parts, shown mostly in section.

The object of my invention is threefold: first, to produce a cartridge that shall be less liable to accidental explosion, both in handling and also in the magazine of the gun; second, to so strengthen the base or head as to prevent the breaking or pulling off of the flange or head; and third, to produce a central-fire cartridge so constructed as to act as a cushion, for the purpose of relieving the breech-pin more or less from the sudden shock or strain caused by the explosion of the charge.

To accomplish these objects I first construct the copper case or tube A in the usual manner prior to having the flange formed thereon, and then form a small recess or depression externally in the center of the base, as shown in Fig. 2, *n* representing the recess.

I then form a shorter tube or cap, B, having a corresponding recess, *a*, in the center of its head or base, as shown in Fig. 3, this cap B being of sufficient diameter to fit over the rear end of the case A.

I next make a circular disk, of rubber, felt, or other soft and elastic material, of a diameter corresponding with case A, as represented in Fig. 4. This disk has a small hole at its center, and this hole should be slightly less in size than the cup or recess in the base of the cases A and B.

I then place the fulminate in the recess on the outside of the base of case A, drop the disk or cushion *b* into the cap B, and slip the latter over the base of case A, and then subject the cases thus united to a compression that shall reduce them in diameter at the rear, until the exterior of cap B shall be of a size uniform with case A, as shown in Fig. 1.

It will be seen that with a cartridge-case thus constructed there is no possibility of exploding it except by driving in the base of the cup or recess in the head of cap B until it impinges against the fulminate *c*, and thus ignites it by compression between the two cases A and B, the inner one serving as an anvil to resist the blow of the hammer or firing-bolt. It will also be apparent that if, by any means, particles of the fulminate should get located at any other point than the recess on the base of the case, it will not be accidentally ignited, as the soft disk *b*, intervening between the two metallic surfaces, will prevent the latter from coming in contact. This feature not only renders the cartridge more safe to handle and transport, but especially fits it for use in a magazine-arm in which the cartridges are placed lengthwise, and where the point of the ball in one necessarily presses against the base of another. By making the point of the ball flat, and of a diameter slightly larger than the recess *a* in the cap B, it will be seen that any concussion produced by a sudden jar or shock of the gun and its magazine will expend its force on the rubber disk or cushion *b*, and hence will not affect the fulminate.

It will be further apparent that, with the addition of the cap or re-enforce B, the head of the cartridge will be so strengthened as to obviate entirely all danger of rupture and consequent escape of the gas at that point—a difficulty of frequent occurrence with the ordinary single case, in which the case is greatly weakened, and frequently cracked, by the crushing down of the head to form the flange. By this method of construction the inner case A is not thus weakened, as no flange is formed thereon, and the metal is consequently left of its full thickness and strength at that point. By inserting the cushion or disk *b* between the cases, as described, it will receive the force of the explosion of the charge, and by means of its elasticity will thus relieve the breech-pin somewhat from the sudden shock and strain that would otherwise be exerted upon it, and hence would be less liable to injure the arm.

Having thus described my invention, what I claim is—

A cartridge-case consisting of the case A and cap B, with the intervening disk or cushion *b*, arranged to operate substantially as and for the purpose set forth.

O. F. WINCHESTER.

Witnesses:
W. W. WINCHESTER,
GEO. SPENCER VIBBERT.

O. F. Winchester,

Cartridge.

No 60814. Patented. Jan 1. 1867.

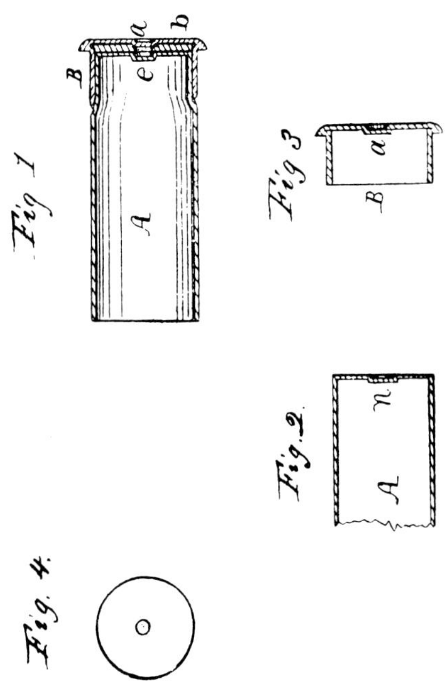

Witnesses.
P. T. Dodge
J. Case

Inventor,
O. F. Winchester
By W. C. Dodge
Atty

UNITED STATES PATENT OFFICE.

EDWARD MAYNARD, OF TARRYTOWN, NEW YORK.

IMPROVEMENT IN PRIMING METALLIC CARTRIDGES.

Specification forming part of Letters Patent No. 61,225, *dated January* 15, 1867; *antedated December* 5, 1866.

To all whom it may concern:

Be it known that I, EDWARD MAYNARD, of Tarrytown, in the county of Westchester and State of New York, have invented a new and useful Improvement in Metallic Cartridge-Caps; and I do hereby declare the following to be a full, clear, and exact description thereof, reference being had to the accompanying drawings, forming a part of this specification, in which—

Figure 1 is an elevation of a metallic cartridge primed in accordance with my invention by the combination therewith of a primed detachable base-cap. Fig. 2 is a central vertical section in the line *x x* of Fig. 1. Fig. 3 is a similar section on an enlarged scale of the detachable primed base-cap; Fig. 4, an enlarged top or inner view of the same; and Figs. 5 and 6, similar enlarged sectional views of the detachable base-cap, illustrating modes of combining additional priming-receptacles therewith by riveting and soldering.

Similar letters indicate like parts in all of the figures.

The nature of my invention consists in priming a metallic cartridge, or a cartridge having a metallic bottom, with fulminate placed upon the inner side of a detachable cap, made to cover the base of the cartridge or a portion thereof.

The detachable cap may be carried apart from the cartridge already primed, or the fulminate may itself be carried separately from either cartridge or cap, and be placed in the cap as occasion may require, a receptacle being formed upon the inner side of the cap to receive and secure the fulminate.

When a cap so formed is placed upon the cartridge the fulminate will be thereby confined against a suitable vent in the bottom of the cartridge, and also perfectly protected from moisture or dampness.

The object of my invention is to provide such ready means of priming loaded metallic cartridges as that they may be transported for security without such priming, free from the liability of accidental explosion, and yet be with perfect ease and dispatch at any moment made ready for firing.

The thin metallic cup or cap *a* which I use is made to fit closely over and upon the base of a cartridge having a metallic bottom, as seen in the drawings, or over a projection therefrom, as illustrated in the specification of improvements in metallic cartridges which I filed in the Patent Office on the 30th day of March, 1866, preparatory to obtaining Letters Patent thereon. With this cap I combine a charge of fulminate by placing it at such a point as shall correspond and coincide with a vent in the base of the cartridge, but which I prefer to make central, as illustrated in the drawings.

To secure, confine, and protect the fulminate when placed upon the inner side of the cap or base-cup *a*, I form a small receptacle, *c*, therefor. This receptacle may be formed in many ways. I contemplate all of them, provided that the receptacle shall not project externally beyond the base of the cap, but prefer to make it either by striking or stamping up an annular crease or encircling projection from the outer side to project from the inner side and inclose a small space for the fulminate, as illustrated in Figs. 3 and 4 of the drawings, or by fastening upon the inner side of said base-cup *a* a small, distinct priming-cup, riveted as illustrated in Fig. 5, or soldered in place, as seen in Fig. 6.

Instead of forming the receptacle in the manner indicated, the fulminate may be attached at a single point on the interior of the metallic base-cap by means of adhesive cement, glue, or varnish; or a film of fine gauze (either of metal wire or of linen or other fiber) or of any thin fabric or material which will allow the spark from the fulminate to pass freely through it may be employed for the purpose of retaining the fulminate in place, either of these methods being regarded as equivalents for the receptacle above described.

The cap or base-cups *a*, furnished with a priming-receptacle upon their inner side, may be provided and sold as a distinct article of manufacture apart from the cartridges with which they are to be used.

Having thus fully described my invention, I claim as a new article of manufacture, and desire to secure by Letters Patent—

A primed metallic cap for the base of a cartridge, when the fulminate is secured at a single point on the inner side of said cap, and the priming point or receptacle does not project externally therefrom beyond its base, all substantially in the manner and for the purpose herein set forth.

The foregoing specification of my improvement in metallic cartridge-caps signed by me this 2d day of May, A. D. 1866.

EDWARD MAYNARD.

In the presence of—
DAVID A. BURR,
H. H. YOUNG.

E. Maynard.
Cartridge.

No 61225. Patented Jan 15 1867

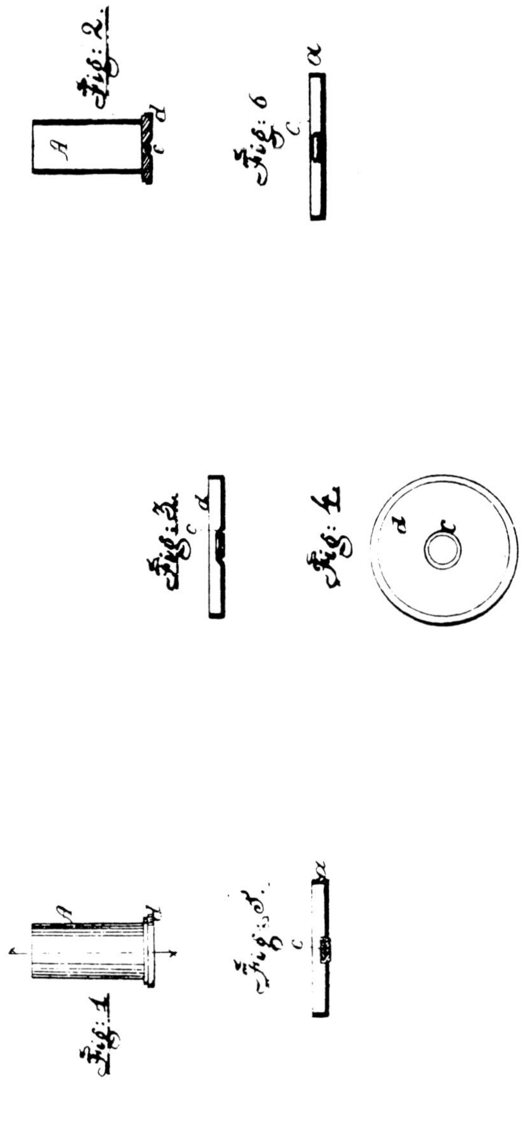

UNITED STATES PATENT OFFICE.

ISAAC M. MILBANK, OF GREENFIELD HILL, CONNECTICUT.

IMPROVEMENT IN PRIMING METALLIC CARTRIDGES.

Specification forming part of Letters Patent No. 62,283, *dated February* 19, 1867.

To all whom it may concern:

Be it known that I, ISAAC M. MILBANK, of Greenfield Hill, in the county of Fairfield and State of Connecticut, have invented and made a certain new and useful Improvement in Cartridges; and I do hereby declare the following to be a full, clear, and exact description of the said invention, reference being had to the annexed drawing, making part of this specification, wherein—

Figure 1 is a longitudinal section of a cartridge fitted with my improvement. Fig. 2 is a section of the case and its metallic base; and Fig. 3 is a section of my fulminate-tube.

Similar marks of reference denote the same parts.

The object of my invention is to provide a metallic cartridge-case that can be loaded and used as fixed ammunition, or may be loaded several times with powder and ball, so as to avoid carrying so much weight of cartridge-cases, and save their cost.

The nature of my said invention consists in a metallic base introduced within the sheet-metal case and soldered permanently in place, so as to afford the necessary strength to resist the blow of the hammer upon the fulminating material, which fulminate I introduce in a hollow cap having a tube projecting into the powder of the cartridge. These tubes and caps are changeable, so that when one has been fired it can be withdrawn from the case and another introduced and the case reloaded. By using these tubular caps the fire from the fulminate communicates with the gunpowder near the center of the charge, rendering the same much more effective than when the fire is only at the rear end.

In the drawing, *a* is the sheet-metal cartridge-case or shell, of any usual size or shape; *b*, the ball. *c* is a metallic base sufficiently strong to resist the force of the hammer striking against the rear of the cartridge. This base *c* is permanently attached within the case *a* by soldering the same into its place, and the base *e* of the case *a* covers up the base *c*, so that said base *c* may be made of iron or other cheap metal.

Through the center of the base *c* is an opening, with a countersink at the rear portion, and into this I introduce the tubular fulminate-cap, formed with a hollow head, *o*, and a tube, *i*, extending any required distance into the powder-space of the cartridge.

The hammer of the fire-arm is to strike the cap *o* in the rear, to explode the fulminate contained therein.

What I claim, and desire to secure by Letters Patent, is—

The base *c*, soldered or brazed into the cartridge-case *a*, and covered by the base *e*, in combination with the fulminate-tube *i* and head *o*, setting within a countersink in the base *c*, as and for the purposes set forth.

In witness whereof I have hereunto set my signature this 16th day of October, A. D. 1866.

I. M. MILBANK.

Witnesses:

M. A. P. WARNER,

GEO. W. WARNER.

I. M. Millbank,
Cartridge.

No. 62283. Patented Feb. 19. 1867.

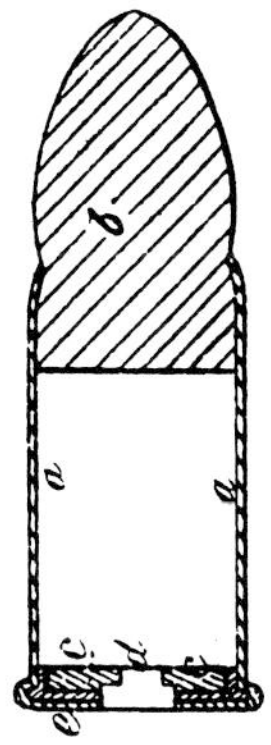

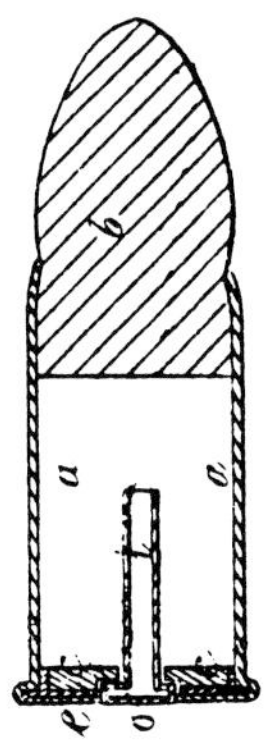

Witnesses:

Geo. D. Walker

Chas. H. Smith

I. M. Millbank

by L. W. Sewell

Atty.

UNITED STATES PATENT OFFICE.

BETHEL BURTON, OF BROOKLYN, NEW YORK.

IMPROVEMENT IN PRIMING METALLIC CARTRIDGES.

Specification forming part of Letters Patent No. 81,058, *dated August* 11, 1868.

To all whom it may concern:

Be it known that I, BETHEL BURTON, of the city of Brooklyn, in the county of Kings and State of New York, have invented a new and useful Improvement in Metallic Cartridges; and I do hereby declare that the following is a full, clear, and exact description of the construction and operation of the same, reference being had to the annexed drawings, making part of this specification, in which—

Figures 1 and 2 are sectional views of the cartridge and bullet with my improvement attached. Fig. 3ª is a perspective view of the cup through which the firing-pin passes, as prepared to enter the cartridge. Fig. 3ᵇ is a top view of the same. Figs. 4, 5, and 6 are paper, cloth, or other suitable materials. Fig. 4 represents the paper disk, coated on one side with amorphous phosphorus; Fig. 5, a similar disk, coated with chloride of potash. Fig. 6 represents an annular washer or disk, having a circular hole in the center.

In making cartridges on my plan, I take paper, cloth, or other suitable materials, and coat them with gum or glue. While wet, I sprinkle one with amorphous phosphorus, and the other with chloride of potash, which may be mixed or not with other materials, if desired. When dry they are cut to the required size. The cartridge-shell is first made in the usual way. A hole is made in the head, corresponding in size to the cup, Fig. 3, the base of which is made in the form of a disk. It is dropped down into the cartridge. When pressed down it flattens out the disk, which enters the rim of the cartridge. The firing-pin is then introduced through the base of the cartridge, passing through the hole in the cup, Fig. 3, at *c*. The cap *f*, made of brass or other metal, is then placed in the cup, which prevents any gas escaping that may pass through the hole *c*, in which the firing-pin *h* enters. The head of the firing-pin *h* is made to fit closely in the cap *f*, so that any gas escaping serves to drive back the pin, leaving the cartridge ready for reloading.

When the pin and cap are in place the cartridge is ready to receive the powder, which is pressed tightly in the cartridge. The disks or wafers, Figs. 4 and 5, with the prepared surfaces facing each other, separated by the annular washer, Fig. 6, are then placed on the powder. A greased wad may then be placed thereon for lubrication, and the bullet inserted, when the cartridge is ready for firing.

Another mode of priming may be used, as shown at *k*, Fig. 1—a cap primed with the before-mentioned materials, Figs. 4 and 5, or other fulminate, the cap being longer than common, for the purpose of conveying the fire through the lubricating material to the powder when ignited by the firing-pin. The lubricant may be grease, or, if a wad, as before, a hole must be made to allow the cap to pass through.

The use of heavy charges of powder increases the length of the cartridge, which makes it impossible for some guns to eject the shell from the chamber of the barrel, and to overcome this difficulty the cartridge is shortened and enlarged, as shown at *m*, Figs. 1 and 2.

The shoulder thus formed creates resistance to the free passage of the unburned powder passing out of the cartridge, and hence increased recoil.

Several ineffectual attempts have been made to fire the powder in front instead of at the base, but failed in consequence of the liability to premature discharge by the cartridge being struck a blow or falling on the ground. To overcome this difficulty the cup, Fig. 3, is elongated; the point of the firing-pin is kept at a safe distance from the explosive matter. The length of the cup gives space for the pin being driven forward into the fulminate. The amorphous phosphorus and chloride of potash, being separated, cannot ignite by the bullet being pressed against them. The desired security is here given in making and using cartridges on my plan, as well as the diminished recoil, by firing the powder in front.

The cup *a*, Fig. 3, and cartridge may be one piece, if preferred, instead of as described.

When the charge in the cartridge is exploded the pressure of the gas within the cap *f* forcibly presses it outward against the internal surface of the cup *a*, forming a gastight joint and preventing its escape.

What I claim as my invention, and desire to secure by Letters Patent, is—

1. The firing-pin *h*, when made with a piston on its rear end working in the cylindrical cap *f*, which cap also works in the cylindrical cup *a*, all as shown and described, and for the purpose specified.

2. The application and arrangement of the prepared fulminate on paper, cloth, or other suitable materials, which may be separately cut with safety and combined for use in cartridges, substantially as set forth.

3. The mode of separating the two prepared disks or wafers by means of an annular washer, substantially as set forth.

In presence of— BETHEL BURTON.

A. H. SIDELL,

DAN. VAN VOORHIS.

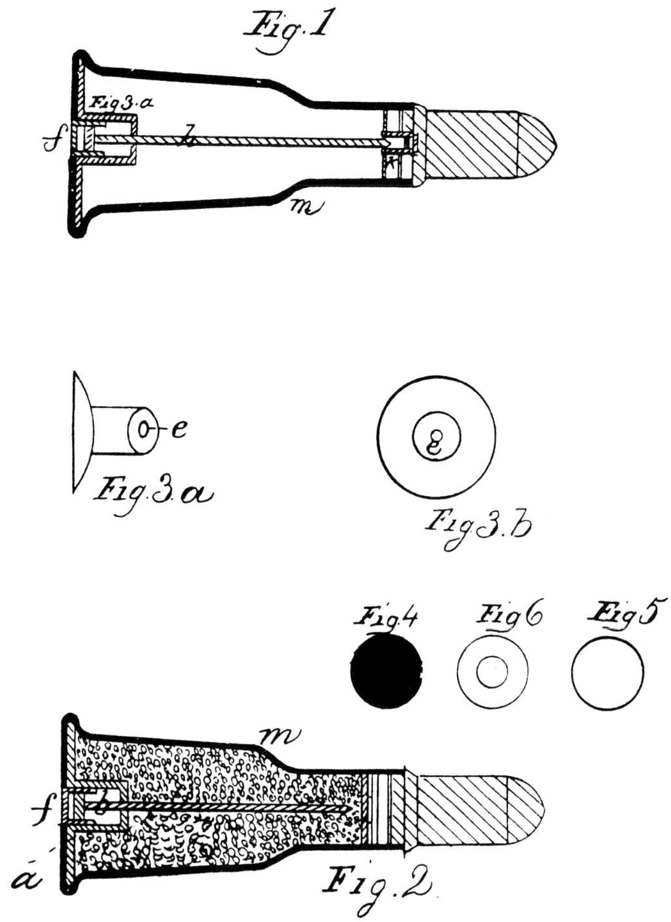

Witnesses

A. H. Sidell

Dan. Van Voorhis

Inventor.

Bethel Burton

UNITED STATES PATENT OFFICE.

HIRAM BERDAN, OF NEW YORK, N. Y., ASSIGNOR TO THE BERDAN FIRE-ARMS MANUFACTURING COMPANY, OF SAME PLACE.

IMPROVEMENT IN METALLIC CARTRIDGES.

Specification forming part of Letters Patent No. 82,587, dated September 29, 1868.

To all whom it may concern:

Be it known that I, HIRAM BERDAN, of the city, county, and State of New York, have invented certain new and useful Improvements in Cartridges for Breech Loading Fire-Arms; and I do hereby declare that the following is a full, clear, and exact description of the same, reference being had to the accompanying drawings, forming part of this specification, in which—

Figure 1 is a longitudinal outside view, on an enlarged scale, of a cartridge with my improvements. Fig. 2 is a central longitudinal section of the same, the charge of powder being omitted to enable the re-enforcement of the head of the shell to be more plainly seen.

Similar letters of reference indicate corresponding parts in both figures.

The first part of this invention relates to the re-enforcement of the head of the metal shell of a cartridge by means of an internal lining-cup of metal, and its object is to prevent the swelling of the rear portion of the shell by the insertion of such cup, which is liable to occur when the said cup is secured in place by being driven tightly into the shell, as heretofore has been commonly done; and to this end it consists in securing the said cup in place by providing in the bottom of the said cup an opening or cavity, and providing the head of the shell with an internal projection which fits tightly into said opening or cavity, and upon which the said cup is driven tightly, the exterior of the said cup being made to fit easily into the interior of the shell.

The second part of the invention relates to the use of patched bullets in cartridges with metal shells.

The advantage of patching bullets, when accurate shooting is desired, is too well known to require explanation here. Hitherto, however, it has not been considered practicable to use patched bullets in metal-cased cartridges for breech-loaders, and with the copper shells heretofore commonly used it has been totally impracticable; for, owing to the inelasticity of copper, when the shell has been made of that metal of ordinary thickness, the tightness with which it has been necessary to drive in the bullet to insure its retention therein has caused the part of the shell which receives the bullet to be bulged or swelled out, and the bullet or patch has had to be driven in so far that there would not be left protruding from the shell a sufficient length of patch to enter as far as desirable into the bore of the barrel beyond the shell in loading, and if the metal has been made thicker, the tightness above mentioned has caused the patch, when made of paper, which is the only material practicable for patching the bullets of such cartridges, to be torn from the bullet in inserting the latter into the shell.

After many experiments I have found that by using a patch of thin but very tough paper, and a cartridge-shell drawn from rolled sheet-brass, which is very elastic, the patched bullet may be inserted into the shell so easily, and yet be held by the elasticity of the shell with sufficient security, without either bulging the shell perceptibly or tearing the patch; and this part of my invention consists in the combination of a patched bullet with a brass cartridge-shell.

To enable others to understand fully the construction and operation of my invention, I will proceed to describe it with reference to the drawings.

A is the metal shell of the cartridge, made substantially of the usual form, except that it is contracted from *e* forward, and that the center of the head is depressed inward from the exterior in such a manner as to form a cylindrical or very slightly conical projection, *c*, in the interior of the shell. The contraction above mentioned is not, however, essential to the carrying out of the present invention, and the depression in the head is not of itself a part of the said invention, as it serves the purpose for which such a depression has been previously used—namely, as a receptacle for the percussion primer *b*. B is a re-enforce, made of brass or other metal or material, and fitted snugly, but quite easily, into the part of the shell in front of the head, and having a central opening or cavity, *e*, in its bottom, fitting tightly onto the projection *c* in the interior of the shell. This cup is inserted into the shell before the forward part is contracted, and is driven down tightly onto the projection *c* with a punch of suitable character, and thereby secured in the shell without any reliance upon the fit of its exterior to the interior of the shell, thereby preventing any bulging of the shell in front of the head, which might interfere with loading, but yet fitting so snugly to the sides of the shell that when the cartridge is fired the sides of the cup will at once expand against the shell, and so insure perfect re-enforcement and protection. This method of securing the re enforce cup is applicable to shells of brass, copper, or any other metal. C is the bullet. D is the patch, made of thin linen paper, such as is used for bank-notes, rolled tightly round the cylindrical portion, to the extent of about two thicknesses, and then drawn over its base and tied or otherwise secured.

To prepare the brass shell for the reception of the bullet thus patched, the mouth of the shell has its edge slightly beveled inward, as shown at *i i*, by means of a punch, which enters the shell while the latter is held in a die. The shell having been thus prepared, the patched bullet is inserted by any suitable means whereby its axis is kept in line with the axis of the bullet. The elastic character of the brass of which the shell is composed causes the bullet to be held with sufficient tightness to prevent its accidental displacement, without inserting the patched portion into the shell so far as not to leave enough of it protruding beyond the mouth of the shell to insure the proper entrance of the patch and bullet into the barrel of fire-arm.

What I claim as my invention, and desire to secure by Letters Patent, is—

1. Securing the re-enforce cup in the cartridge-shell by means of a projection on the interior of the head of the shell, and an opening or cavity in the bottom of the cup, fitting tightly on the said projection, substantially as and for the purpose herein described.

2. The combination of the patched bullet and the brass cartridge-shell drawn from sheet metal, substantially as and for the purpose herein set forth.

H. BERDAN.

Witnesses:
A. LE CLERC,
J. W. COOMBS.

Fig. 1.

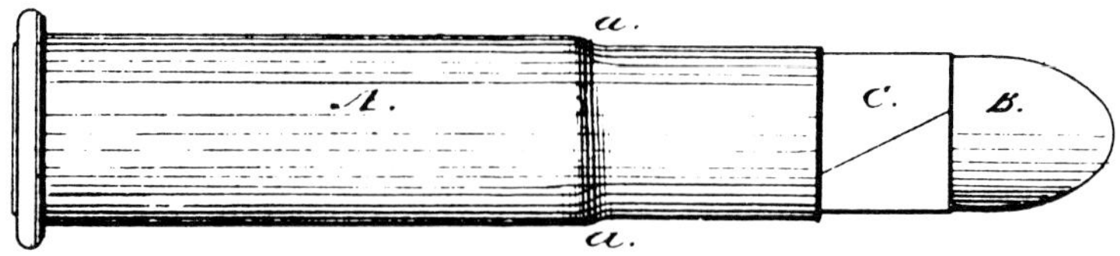

Fig. 2.

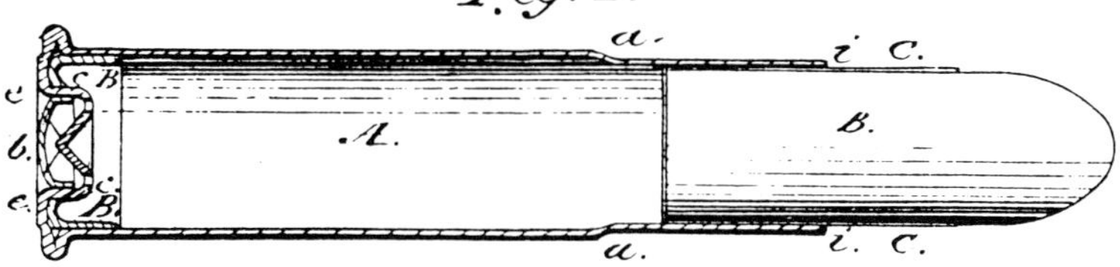

Inventor;
H Berdan

Witnesses;
A LeClerc

UNITED STATES PATENT OFFICE.

W. F. PARKER, OF MERIDEN, CONNECTICUT.

IMPROVEMENT IN CARTRIDGES.

Specification forming part of Letters Patent No. 88,202, *dated March* 23, 1869.

To all whom it may concern:

Be it known that I, W. F. PARKER, of Meriden, in the county of New Haven and State of Connecticut, have invented a new Improvement in Cartridges; and I do hereby declare the following, when taken in connection with the accompanying drawings and the letters of reference marked thereon, to be a full, clear, and exact description of the same and which said drawings constitute part of this specification, and represent, in—

Figure 1, a perspective view of the cartridge, looking toward the rear end; Fig. 2, the percussion-cap; Fig. 3, a longitudinal section of the cartridge-shell; Fig. 4, the same with the cap inserted; Fig. 5, a rear-end view of the same; and in Figs. 6 and 7, a rear and sectional view of the head with the cap inserted, as before my improvement.

This invention relates to an improvement in cartridge-shells in breech-loading fire-arms, and especially to that class of shells which are used for shot for sporting purposes, the shell being removed from the barrel after discharge and reloaded, so that the same shell answers for many discharges, the object being to adapt the shell to the use of a percussion-cap formed with a ring around its base or head, against which it bears when placed in the head. A perspective view of the cap is shown in Fig. 4, which is in form a miniature copper shell.

Heretofore the head has been formed, as seen in Figs. 6 and 7, with an opening through the head of a diameter to receive the body of the shell, and the recess, so that the flange will set in flush and fill the recess. This cap, after discharge, is removed with difficulty, it being necessary to punch it out from the opposite side, and the opening through the shell, being necessarily large, compels the necessity of placing the cap into the head before loading, to remain there until the shell is discharged, and therefore be liable to discharge by accident. In my invention these difficulties are entirely removed; and it consists in forming a recess in the rear end of the cartridge-head, so as to expose the flange to the action of any instrument which may be applied thereto upon the outside, the said recess being of sufficient depth to carry the cap into the shell flush with or lower than the surface of the head.

In order to the clear understanding of my invention, I will fully describe the same as illustrated in the following drawings:

A is the shell; B, the head, joined to the shell in the usual manner. Centrally in the head I bore out a chamber, *a*, of proper diameter and depth to receive the percussion-cap, and through the remainder of the head in the recess I make a small perforation, *d*, of sufficient size only to make a communication from the percussion-cap to the powder for the purpose of ignition, so that the powder in the shell will not escape therethrough. The rear end of the shell I form concave, or make a recess of much larger diameter than the head of the cap, as seen in Figs. 1 and 3, so that when the cap is set into the recess *a*, as seen in Fig. 4, it will be flush with or below the surface or edge of the head, and yet so that the rim will be entirely exposed above the recessed or concave surface. Thus constructed, the cap is easily removed by the thumb and finger nail, or by any sharp instrument, as the blade of a knife, and is not of necessity placed into the head until the shell is to be placed in the arm. Thus I facilitate the removal of a percussion-cap, avoid the risk of accident by the necessary presence of the cap in the loaded cartridge, and prevent the escape of powder through the opening in the head.

I find it advantageous to construct the cartridge so that the cap may be removed by an instrument having a small point, which may be made for the purpose. To this end I form in the concave surface of the head, down into the recess formed by the cap, one or more channels, *f*, (see Fig. 5,) of sufficient size to permit the insertion of a small point beneath the head or rim of the cap, which, used as a lever, easily throws the cap out of the shell.

Having fully described my invention, what I claim as new and useful, and desire to secure by Letters Patent, is—

1. A cartridge-shell the head of which is constructed with a recess, *a*, for the percussion-cap, its head recessed so as to expose the rim of the cap, substantially in the manner and for the purpose set forth.

2. The channels *f* in the head of a cartridge, with or without the recess *a*, substantially in the manner and for the purpose set forth.

W. F. PARKER.

Witnesses:

A. J. TIBBITS,

J. H. SHUMWAY.

W. F. Parker.

Cartridges.

No 88,202. Patented. Mar. 23. 1869.

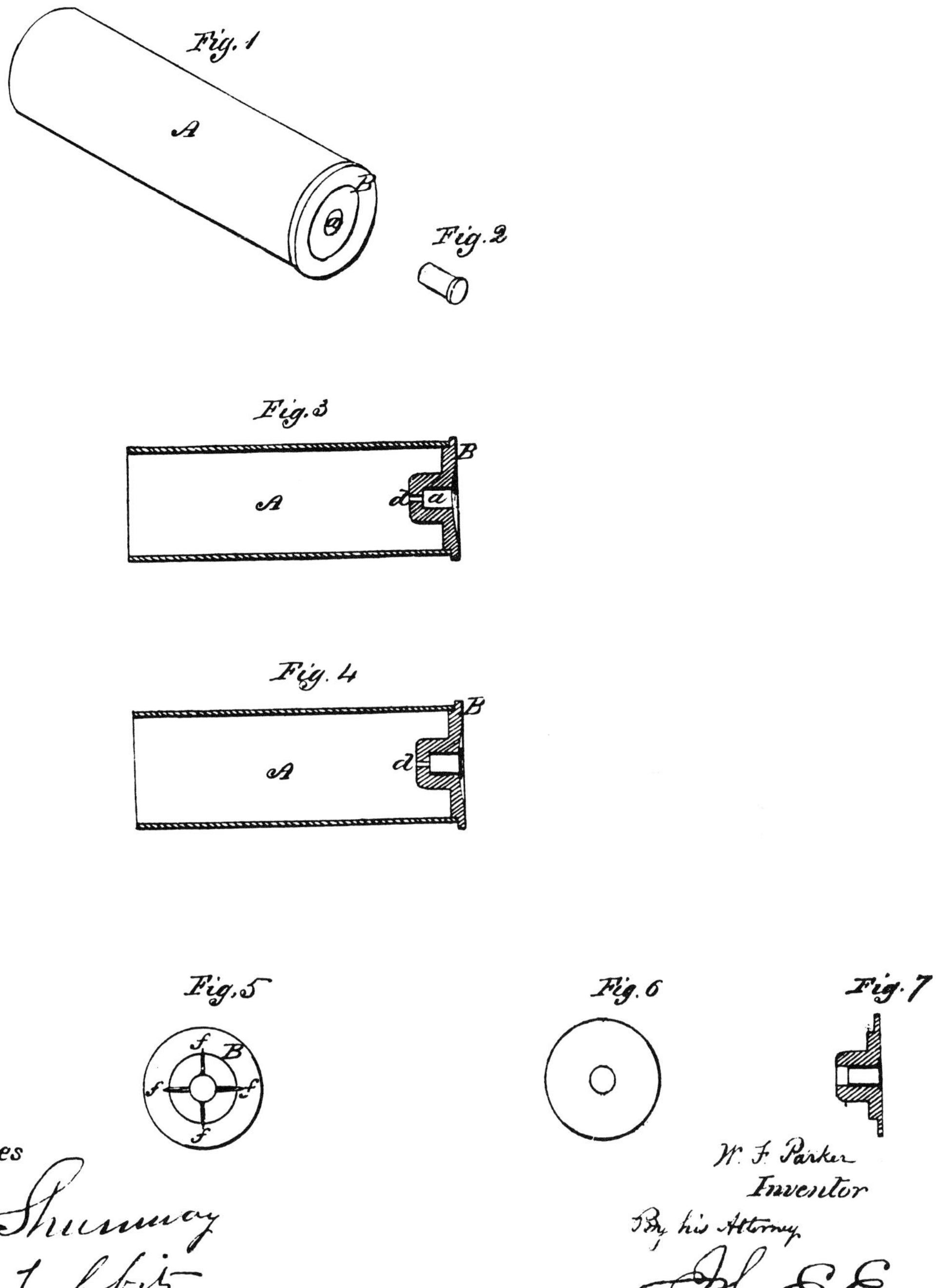

Witnesses
J. H. Shumway
A. J. Tibbits

W. F. Parker
Inventor
By his Attorney
John E. Earle

UNITED STATES PATENT OFFICE.

E. M. BOXER, OF THE ROYAL ARSENAL, WOOLWICH, ENGLAND.

IMPROVEMENT IN METALLIC CARTRIDGES.

Specification forming part of Letters Patent No. 91,818, dated June 29, 1869; patented in England October 13, 1866.

To all whom it may concern:

Be it known that I, EDWARD MOUNIER BOXER, of the Royal Arsenal, Woolwich, in the county of Kent, England, Colonel Royal Artillery, have invented an Improved Cartridge for Breech-Loading Fire-Arms; and I do hereby declare that the following is a full and exact description of the said invention, reference being had to the accompanying drawing, and to the figures and letters marked thereon; that is to say—

My invention has reference, first, to an improved form and arrangement of the anvil to central-fire breech-loading cartridges generally, whereby the liability of the cartridge to explode when accidentally dropped, or subjected to any other blow than by the striker of the fire-arm, is obviated. In the arrangements heretofore employed, the anvil has been made of such a form that if the cartridge was subjected to a sufficiently sharp blow it could pass more or less entirely into the percussion-cap, so as to strike against the fulminating compound and thus explode the cartridge.

Now, my improvement consists in forming shoulders or projections of such a form on the sides of the anvil that when the anvil is inserted into the cap they rest against the edge of the latter, and thus prevent the anvil from passing further into the cap under any circumstances, the point of the anvil being then in such a position that it is impossible for it to cause the ignition of the fulminating compound, except when the back of the cap is indented by the striker on firing.

My invention consists, secondly, in forming the base of the cartridge of a solid metal disk, and in securing the body of the case to this disk in the following manner: The flanged cap chamber, that is to say, the flanged tube which contains the cap and anvil, or other means of ignition, has one or more cannelures or flanges formed upon its exterior surface. The rear end of the body of the cartridge, whether of paper alone or of metal alone, or of paper, calico, linen, or other similar material combined with metal, is turned over upon the wad which forms the bottom of the body. This wad, when condensed, enters into the cannelures of the cap-chamber, and at the same time jams the turned-over portion of the body between its bottom surface and the surface of the metal disk, thus securely fastening the metal disk, the cap-chamber, and the body of the cartridge together. The wad I sometimes make of an alloy of lead and antimony, or lead and tin, or an alloy of a similar character, and of a cup-shape, so that the explosion of the charge readily expands the edges of the cup, and thus prevents the gas from passing down the sides of the case.

The above-described arrangement is applicable to blank ammunition and to cartridges which have not to sustain any severe strain. In ball-ammunition, I prefer to strengthen the body of the cartridge by means of a thin metal cup, which fits over the end of the body. In cartridges which have to withstand a severe strain, I further stren then the bottom of the body by means of one or more strips of thin sheet metal coiled on the exterior of the body, but inside the cup, which strip extends beyond the cup, and in some cases I employ a second cup in place of or in addition to the strips of metal. The strip or strips of coiled metal or cup prevent the edge of the outer cup from cutting through the body of the cartridge when it is expanded on firing.

This form of charge is applicable to every description of fire-arm, and when properly arranged it increases the force of the charge from ten to fifteen per cent.

Having thus set forth the nature of my invention, I will now proceed more fully to describe the manner of performing the same.

Figure 1, on the accompanying drawing, shows a sectional elevation to an enlarged scale of the before-described shoulder-anvil A, inserted in the percussion-cap B, against the edge of which the shoulders *a a* of the anvil abut, thus preventing it from passing far enough into the cap to cause the ignition of the fulminate, which can therefore only be effected by indenting the cap by the action of the striker of the gun.

Figs. 2 to 9 show one arrangement of my improved mode of forming the base of the cartridge. This consists of the following parts, namely: First, a metal washer, C, with a cen-

91,818

tral hole, *c*, shown detached in section and plan at Fig. 3. Second, a thin metal cup, D, also having a central hole, *d*, shown detached in section and plan at Fig. 4. Third, a cap-chamber, E, shown detachedin section at Fig. 5.

These parts are fitted together as shown in elevation at Fig. 6, the cap-chamber E being passed through the holes *c* and *d* in the parts C and D. The body of the cartridge-case F, formed either of paper alone, as at Fig. 7, or of tubular metal, or of paper, calico, linen, or other similar suitable material, combined with metal, as at Fig. 8, is then inserted into the base, the rear end of the case having first been turned over a wad, G, shown detached in elevation and plan at Fig. 9, as indicated at Figs. 7 and 8. The case is then secured to the base, as shown in the section at Fig. 2, by forming a cannelure, *e*, on the cap-chamber E by upsetting the end thereof, into which cannelure the wad G fits, the latter being made to press firmly against the cap-chamber, and against the case F by being condensed by pressure. The upsetting of the end of the cap-chamber may be effected either before or after the wad is condensed.

Figure 2A shows a section of a cartridge consisting of the same parts described with reference to Fig. 5, with the addition of a coil of thin metal, H, shown detached at Fig. 10, turned over the rear end, which coil is inserted between the case F and the cup D, and is made to extend beyond the edge of the latter, as shown in the detached elevation of these parts at Fig. 11.

The base of the cartridge, shown in section at Fig. 2B, is formed of the same parts as the cartridge, Fig. 2A, with the further addition of another coil of thin metal, I, extending beyond the first coil H, as shown in the detached elevation at Fig. 12. In this arrangement I sometimes substitute a very thin cup for the first coil, as shown at H′, Fig. 13, in which case the base-cup D is made thinner than when a coil is employed. The object of these base-coils in the cartridges, Figs. 2A and 2B, being principally to prevent the cutting through of the body of the case at the time of firing, (which sometimes occurs from the body expanding more readily than the base-cup,) the relative thicknesses and resistance to expansion of the cups and coils must be regulated accordingly.

When the case of the cartridge consists of a solid metal tube, and where the strain is not great, as in pistol-charges, the base cups may be dispensed with, as before stated, and the case be secured by simply being turned over at the lower end and jammed between the disk C and the wad G, as shown at Figs. 14 and 15, of which Fig. 14 shows a section of a pistol-cartridge and Fig. 15 shows a section of the case for the same.

Figs. 16 and 17 show my improved form of compressed and recessed powder-charge for cartridges. Fig. 16 shows a section of a rifle-cartridge of 0.577 bore, with the charge applied, and Fig. 17 shows a section of a cartridge of 0.5 bore with the same. The construction of the cases of the cartridges is the same as before described with reference to Fig. 2A. K is the powder-charge, compressed, as before described, and having a deep recess, *k*, formed at its rear end, and a shallow one, *k*′, at its front end.

On firing the cartridge the charge is ignited on the entire surface of the recess *k*.

Having now described the nature of my invention, and in what manner the same is to be performed, what I claim is—

1. Constructing the anvil to cartridges for breech-loading fire-arms with shoulders or projections so formed as to abut against the edge of the percussion-cap, and thus to prevent its entering the cap to such an extent as to cause the ignition of the fulminate, except when the cap is acted upon by the striker of the gun, substantially as hereinbefore described with reference to Fig. 1 on the accompanying drawing.

2. The separate-flanged cap-chamber E, so made as to inclose the cap and anvil, or other means of ignition, and at the same time by compression of wad G around the same to secure the body of the cartridge-case, to a solid metallic disk, C, with or without the intervention of a base-cup, D, substantially as hereinbefore set forth.

3. Interposing one or more pieces of coiled metal or cups between the base-cup and the cartridge-case, which coiled metal or cups extend beyond the edge of the base-cup, substantially as and for the purposes hereinbefore described with reference to Figs. 10 to 13 on the accompanying drawing.

In testimony whereof I have signed my name to this specification in the presence of two subscribing witnesses this 17th day of December, 1867.

E. M. BOXER.

Witnesses:
C. D. ABEL,
THOS. TAYLOR.

E. M. Boxer.

No. 91818

PATENTED
JUN 29 1869

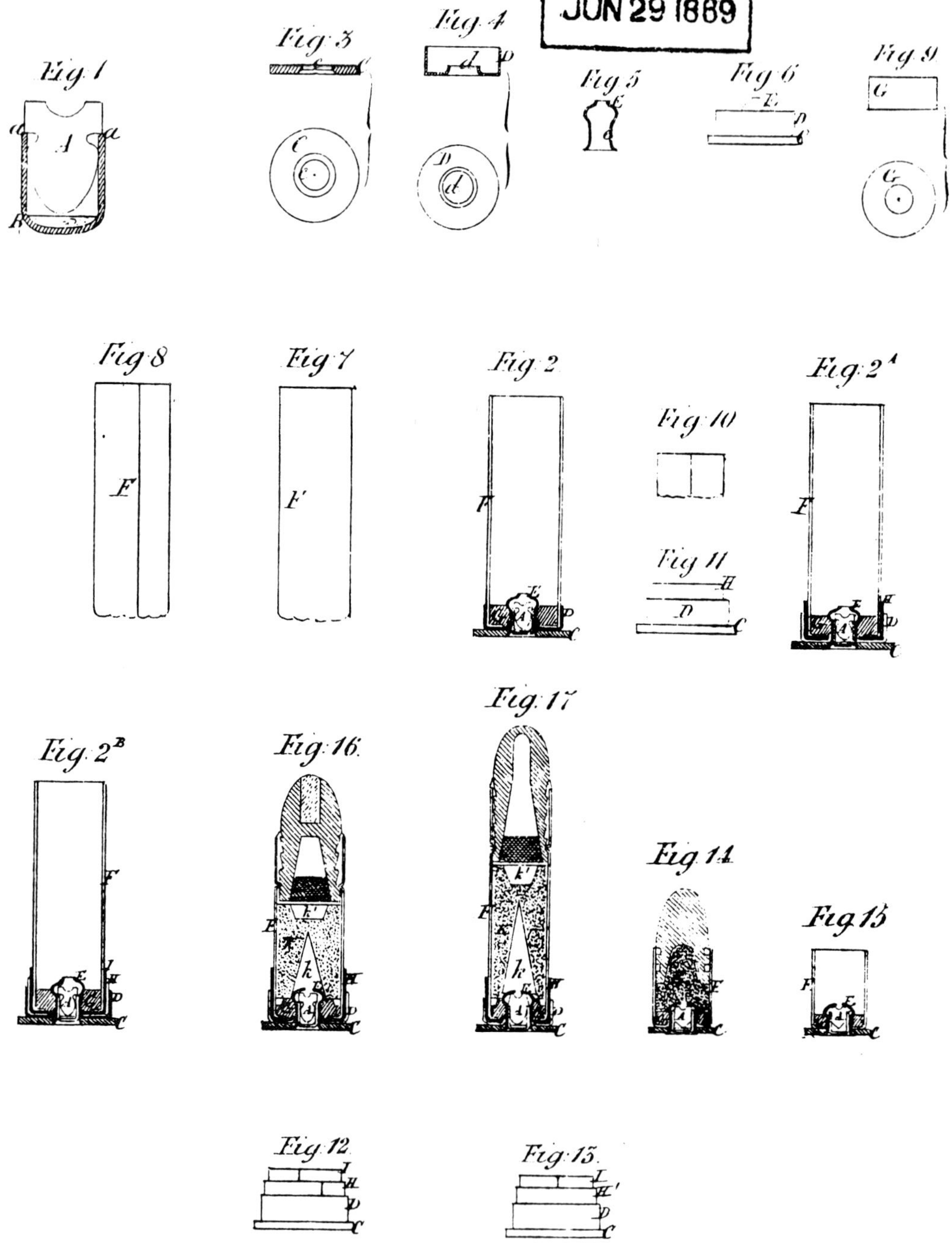

Witnesses

E M Boxer

UNITED STATES PATENT OFFICE.

ISAAC M. MILBANK, OF GREENFIELD HILL, CONNECTICUT.

IMPROVEMENT IN METALLIC CARTRIDGES.

Specification forming part of Letters Patent No. 93,546, *dated August* 10, 1869.

To all whom it may concern:

Be it known that I, ISAAC M. MILBANK, of Greenfield Hill, in the county of Fairfield and State of Connecticut, have invented and made a new and useful Improvement in Cartridges for Breech-Loading Fire-Arms; and I do hereby declare the following to be a full, clear, and exact description of the said invention, reference being had to the annexed drawing, making part of this specification, wherein—

Figure 1 is a vertical section of the cartridge-case complete. Fig. 2 is the ring of solder or alloy separately, and Fig. 3 is a section and end view of the primer-socket.

Similar letters apply to the same parts.

In Letters Patent granted to me August 6, 1867, a cartridge-case is set forth with an interior metallic base soldered into the sheet-metal case. In my present invention the base of the sheet-metal cartridge-case is strengthened by solder, or similar alloy or metal, melted into the said case, so as to flow into intimate contact with the same and form an interior support; and I also make use of a socket for receiving the fulminate or cap, said socket being a separate piece secured into the case by the metal melted around it.

In the drawing, *a* represents the cartridge-case of sheet metal, 2 being the cylindrical portion, 3 the base, and 4 the flange.

The flange 4 is filled with an alloy or metal, and the case strengthened in the flat base and around the rim by such metal or alloy being melted within said case in the presence of muriate of zinc or any other flux that will cause the proper union of the metals, and I find it most convenient to cut out disks or rings *b* of solder from sheet metal, and drop them into the case in the presence of a flux, and then set the shells upon a heated plate until the metal melts and unites with the interior of the shell *a*, strengthening the same at the base, so that the base will not be liable to split by the force of the explosion, and the base will be supported against the blow of the hammer on the fulminate.

The socket *c* is made of a short piece of tubing, with an inward flange or contraction at 5, against which the open end of the primer or cap rests, so that said primer will not be driven into the cartridge by the blow of the hammer. This socket *c* is introduced through an opening made in the base of the cartridge-case *a*, the metal being pressed inward so as to form a short cylinder for receiving the said socket *c*, and the solder or alloy melted into the base also unites with the exterior of the socket *c*, holding the same firmly to place.

This manner of strengthening the base by the metal melted within the case may be employed in cartridges where the cavity or space for receiving the fulminate is made in some other way than by the socket *c*, and said socket *c* might be employed with a metal base introduced within the case, as in my aforesaid patent.

It will be evident that the sheet metal forming the base of the cartridge-case might be pressed inward to form a teat or cylinder for receiving the primer, and the base strengthened by a ring of solder or other alloy melted into the space between said teat and the sides of the shell.

I do not herein claim soldering in a separate disk of metal within the cartridge-case, the same being the subject set forth in my patent before named; neither do I claim a metal base connected to a sheet-metal cylinder, and having a cavity for a primer.

What I claim, and desire to secure by Letters Patent, is—

1. A sheet-metal cartridge-case in which the base is strengthened and stiffened by the metal itself—that is, melted to cause it to flow into intimate contact with the sheet metal of said case, as set forth.

2. A tubular socket for the fulminate or cap, extending inward from the center of the sheet-metal base of the case, when combined with the metal melted around such socket, and within the base of the metallic case, for the purposes set forth.

In witness whereof I have hereunto set my signature this 16th day of April, 1869.

I. M. MILBANK.

Witnesses:

CHAS. H. SMITH,
GEO. T. PINCKNEY.

I. M. Millbank.

Cartridge.

No. 93.546. Patented. Aug 10.1869.

Fig. 1.

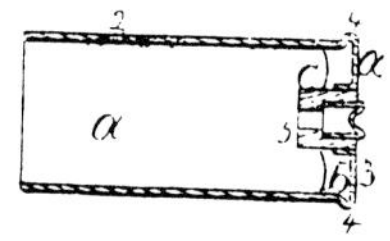

Fig. 2.

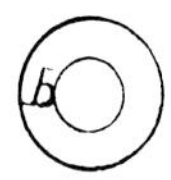

Fig. 3.

Isaac M. Millbank

per L. W. Serrell Atty

Chas H Smith

Geo. S Walker

UNITED STATES PATENT OFFICE.

J. V. MEIGS, OF WASHINGTON, DISTRICT OF COLUMBIA.

IMPROVEMENT IN METALLIC CARTRIDGES.

Specification forming part of Letters Patent No. 87,352, dated March 2, 1869.

To all whom it may concern:

Be it known that I, JOE V. MEIGS, of Washington city, in the District of Columbia, have invented a new and useful Method of Constructing Metallic Cartridges, of which the following is a full, clear, and exact description:

The objects of my invention are to secure in a metallic cartridge certainty in firing, simplicity and cheapness in construction, and immunity from accidental explosions; and to these ends the improvements herein claimed consist in a novel mode of constructing cartridge-shells with a central chamber to receive the fulminate formed in the base of the cartridge, this chamber being so formed as not to extend to the edge of the base of the shell, by which means the cartridge can only be exploded by a sharp punching stroke in the line of the axis, as otherwise the metal would yield and widen the space in which the fulminate is contained instead of compressing it, which compression is necessary to produce an explosion.

To carry out these objects I stamp out or spin up the case A in any well-known way, into the shape shown in Fig. 1, and then compress the central cavity B into the form shown in Fig. 2. The fulminate C is then placed around the neck or flange, as shown in Fig. 3. In order to insure the instantaneous explosion of the charge, I propose to place around the fulminate and within the shell a fuse, D, which will burn even more rapidly than powder, and thus accelerate the explosion of the entire charge; or the fuse may be so constructed as to ignite the charge at its forward end. The bending or folding of the metal renders the case rigid enough to resist the blow of the hammer without an anvil, and it will be seen that the bottom of the case is concave; which form prevents the fulminate from accidental explosion by dropping on a flat surface.

The operation of my improved cartridge is illustrated by Sheet 2 of the accompanying drawings, in which—

Figure $B^{\times}$ represents the effect of an axial crushing force exerted by a blunt-pointed instrument larger than the base of the shell. Fig. $C^{\times}$ represents the effect of a diagonal inward blow from a similar instrument. Fig. $D^{\times}$ represents the effect of an axial punching stroke from a firing-pin, such as I propose to use. Fig. $E^{\times}$ represents the effect of a diagonal outward blow from a bullet. Fig. $H^{\times}$ represents the effect of a direct blow from the point of a bullet. Fig. $A^{\times}$ represents a circular receptacle inclosing the fulminate. Fig. $d^{\times\prime}$ represents a chamber riveted in the shell. Fig. $d^{\times\prime\prime}$ represents a similar one soldered to the base of the shell. Fig. $b^{\times\prime}$ represents a riveted button-shaped fulminate-chamber, and Fig $b^{\times\prime\prime}$ a similar soldered one. Fig. $F^{\times}$ represents a single indented overlapping straight flange; Fig. $G^{\times}$, a similar double overlapping flange, and Fig. $I^{\times}$, still another form of the same.

From these diagrams it will be seen that any blow (except that of a sharp punching instrument as in Fig. $D^{\times}$) on the shell will tend to separate the walls of the fulminate-chamber instead of compressing them, and thus prevent the ignition of the fulminate, and consequently greatly diminish the liability to accidents, which is so great among cartridges as at present constructed.

In Fig. 4 I have shown the fulminate as inclosed in the overlapping flanges of a straight central groove, by which the same result is attained in a different manner from that hereinbefore described.

Figs. 5 and 6 show another mode of accomplishing the same result by a cross-shaped groove.

By the modes of construction above shown the walls of the fulminate-chamber are rendered sufficiently rigid to explode the fulminate by the blow of the hammer upon the firing-pin without using a separate anvil, and the fulmitate chamber itself forms the anvil on which to explode the fulminate by compressing it in its groove, but when the cartridge is struck in any save the proper direction and by the proper point, the groove in which the fulminate is held will open instead of being compressed, and thus prevent the explosion.

What I claim as my invention, and desire to secure by Letters Patent, is—

The method, herein described, of preventing accidental explosions in metallic cartridges by placing the fulminate in such relation to the base and to the anvil formed by the walls of the fulminate-chamber that it can only be exploded by a blow in a given direction, substantially as set forth.

In testimony whereof I have hereunto subscribed my name.

JOE V. MEIGS.

Witnesses:
WM. D. BALDWIN,
BALTIS DELONG.

J. V. Meigs,

Cartridge.

2 Sheets, Sheet 1.

No 87352.

Patented Mar 2 1869

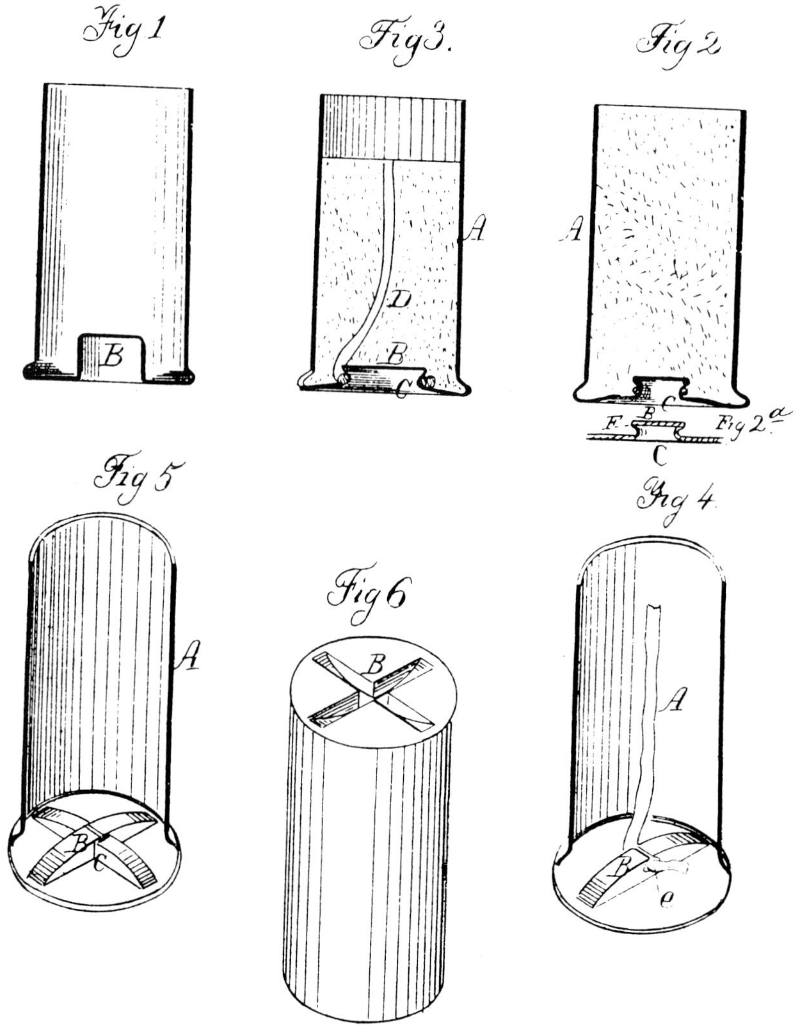

Witnesses
Baltis De Long

Inventor
J. V. Meigs
by his Atty's

2. Sheets Sheet 2.

J. V. Meigs,
Cartridge.

No. 87.352. Patented Mar. 2.1869.

Witnesses
M. M. Betts
Ballis DeLong

Inventor
J. V. Meigs
by his Attys
Baldwin & ...

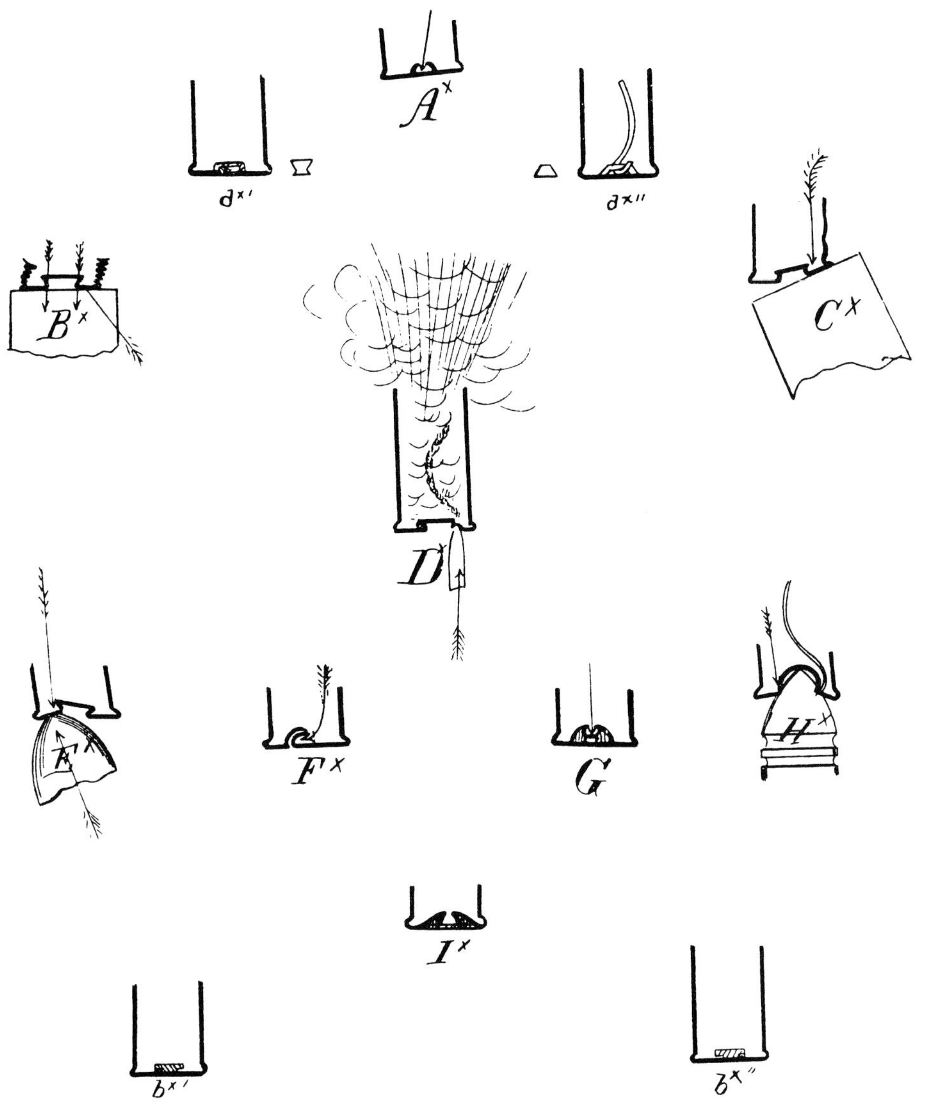

UNITED STATES PATENT OFFICE.

J. V. MEIGS, OF WASHINGTON, DISTRICT OF COLUMBIA.

IMPROVEMENT IN METALLIC CARTRIDGES.

Specification forming part of Letters Patent No. 90,951, dated June 8, 1869.

To all whom it may concern:

Be it known that I, J. V. MEIGS, of the city and county of Washington, in the District of Columbia, have invented a new and useful Improvement in Metallic Cartridges, of which the following is a full, clear, and exact description, reference being had to the annexed drawings, which make part of this specification, in which—

Figure 1 represents a view of the interior of the cartridge-case, a part being broken away; Fig. 2, a vertical central section through a filled cartridge without a flange; Fig. 3, a view of the outside of several cases, showing the depressions in the base; Fig. 4, a vertical central section through a flanged filled cartridge; and Fig. 5, a broken cartridge-case, seen from the outside, showing the anvil in part.

It is the object of my invention to render metal cartridges secure against explosion in transportation, or accidental discharge in magazine-guns; and to this end my invention consists in a novel method, hereinafter described, of constructing a metallic cartridge by placing the fulminate centrally in a bar-anvil, securing the anvil in the base of the cartridge-case, and grooving the base from edge to edge, or nearly so.

To carry out the object of my invention, I form the cup A of soft metal, with a flange at the base, as shown in Fig. 4, or without a flange, but with an annular groove, as in Figs. 2 and 3, in any of the most approved modes of spinning or striking up from a disk. The bar-anvil K, of hard metal, is curved so that its ends, which are thinner than it is in the middle, will rest in the flange, as in Fig. 4, or beneath the annular groove, as in Fig. 3, and be held securely. The anvil is punched on its under side to form a cup, *r*, in which the fulminate *s* is deposited. The bottom of the cartridge-case A is a plane surface, and on the outside of this bottom I sink a depression, *q*, or two depressions, *q* and *t*, the one being at a right angle to the other, as shown in Fig. 2 at C′ and A′, and reversed or seen from the inside, as at *q* and *t* in Fig. 1. These depressions are made so as to be left square at their edges, and may extend from one side of the cup to the other, or they may be made so as not to extend entirely across the bottom. In either case, however, they are to be made deeper at the center than at the extremities of the depressions.

When the cartridge is to be fired, a rod to be struck by the hammer enters the point at which the depressions cross, or the center of the single depression, and, by its percussion on the fulminate in the center of the anvil, produces the desired explosion of the cartridge.

Now, when the anvil containing its fulminate is placed within the flange, it is manifest that the case or cup may be charged with ammunition, and that a considerable lateral pressure may be applied to close the upper edge of the cup around the ball without any danger of exploding the fulminate; and it is further obvious that, as the grooves in the bottom of the cup will present a rigid resistance to any body the cartridge may encounter, without striking the fulminate, the danger of explosion in the cartridge from transportation in boxes or otherwise will be almost entirely removed. It is further obvious that when the cartridge so constructed shall be used in magazine-guns, even such as carry one cartridge above another in a tube, the rigid corners of the depressions will prevent the contact of the point of a ball from pressing against the fulminate of the cartridge next it, and thus render this cartridge peculiarly safe for such guns.

I do not, under this application, claim the bar-anvil *per se*.

What I claim as my invention, and desire to secure by Letters Patent, is—

The combination, as set forth, of the bar-anvil with the square-edged depression in the bottom of the cartridge case, constructed, arranged, and operating as and for the purpose set forth.

In testimony whereof I have hereunto subscribed my name.

JOSIAH V. MEIGS.

Witnesses:

WM. D. BALDWIN,

JOS. I. PEYTON.

J. V. Meigs,
Cartridge.

No. 90.951. Patented. June. 8. 1869.

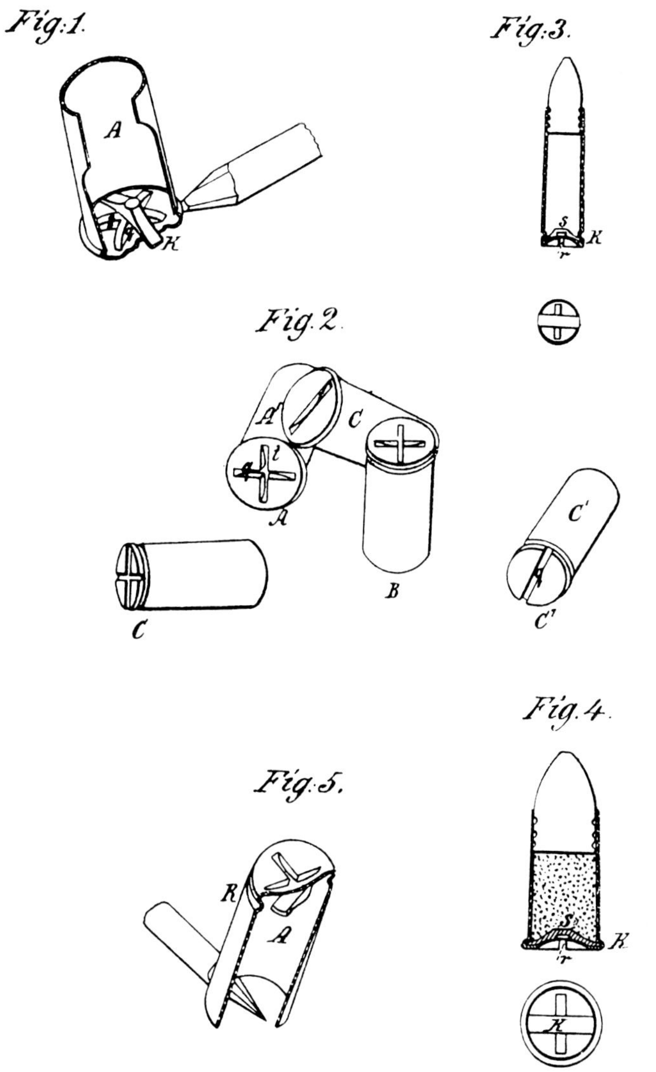

Witnesses
J. J. Peyton
Theodore Lang

J. V. Meigs.
by his Attys
Baldwin & Lan

UNITED STATES PATENT OFFICE.

EDWIN MARTIN, OF SPRINGFIELD, MASSACHUSETTS.

IMPROVEMENT IN METALLIC CARTRIDGES.

Specification forming part of Letters Patent No. 105,348, dated July 12, 1870.

To all whom it may concern:

Be it known that I, EDWIN MARTIN, of Springfield, in the county of Hampden and State of Massachusetts, have invented a new and useful Improvement in Metallic Cartridges; and I do hereby declare the following, when taken in connection with the accompanying drawings and the letters of reference marked thereon, to be a full, clear, and exact description of the said invention, and which said drawings constitute part of this specification, and represent, in—

Figure 1, a plan view of the inside of a cartridge shell with my invention applied. Fig. 2 is a vertical longitudinal section of the same through line A B of Fig. 1. Fig. 3 is a plan view of the base or cone. Fig. 4 is a side elevation of the same. Fig. 5 is a plan view of another modification of my invention. Fig. 6 is a vertical longitudinal section through line C D of Fig. 5.

My invention relates to that class of metallic cartridges known as center-fire cartridges; and it consists in forming a metallic base of sufficiently hard metal, somewhat conical in form, and having any desirable number of grooves around its periphery set into a pocket or recess in the outside of the head of the shell, and securing it therein by forcing down the said pocket upon and around the said conical base, by which process the metal is forced into the annular grooves, thus securing the base within the recess. The base may be made of steel or iron, and upon the outer or large end is a recess, at the center of which recess is a nipple upon which to place a percussion-cap; and a perforation is made through the base at the center of the nipple, and into the interior of the shell through the bottom of the pocket; and my invention also consists in re-enforcing the interior of the head of the shell with graphite or plumbago, pressed in firmly around the pocket which holds the conical base at the head of the cartridge, for the purpose of strengthening it, and preventing the gas from blowing through the head of the cartridge when it is exploded.

That others skilled in the art may be able to make and use my invention, I will proceed to describe its construction and the mode of its operation.

In the drawings, *a* represents the shell, in the head of which a recess or pocket is formed by forcing in the metal from the outside by means of a proper punch or tool, *r* being the extreme inner base of said recess, and *f* the ordinary flange at the head.

Fig. 3 represents a plan view of the outer end of the conical base *m*, in which *o* is a center projection or nipple, a perforation, *n*, at the center of said nipple, communicating with the interior of the shell, as shown in section in Fig. 2. The outer projection *m′* projects beyond the plane of the center projection or nipple *o*, so that the outer part *m′* serves as a guard to prevent the cap, when placed upon the nipple, from being exploded by any accidental blow.

The metallic base *m* has one or more annular grooves or channels *i* thereon, and is of a proper form to fit, in a measure, the recess in the head of the shell. The base *m* is placed within the pocket or recess, and a tool, having a cavity in the end, of a corresponding form to the exterior of the pocket or recess *r*, but a trifle smaller, is forced down upon the inside of the shell, so that the sides of the cavity in the tool are pressed hard against the exterior of the recess *r*, so that the metal at the side of the recess all around is forced hard against the metallic base *m*. This operation forces the metal at the sides of the recess, or a portion of it, into the annular grooves *i* firmly, so that the metallic base *m* cannot, by any ordinary process, be withdrawn from the recess. This forms a shell which may be loaded and fired as often as desirable.

I strengthen this shell, as well as any other having a projection, or a projecting pocket upon the interior of the head of the shell, in the following manner: I prepare graphite, by reducing it to powder, or grinding it, and place a suitable quantity within the shell, and then press it in firmly by means of a suitable tool, having an annular projection upon the end of sufficient face to cover the graphite at *c*, and force it in firmly.

Fig. 1 shows a shell having a pocket formed as described in Letters Patent granted to me, and dated March 23, A. D. 1869, numbered 88,191, but re-enforced with graphite, as herein described. This method of re-enforcing a shell gives great strength to the head of the shell, and forms a most perfect gas-check, as the graphite, after being pressed in, is perfectly hard, and, as heat has no effect upon it whatever, it may be fired any number of times without deteriorating in the least.

It is evident that a shell having a metallic base, *m*, secured therein, as herein described, is operative as a shell, without the re-enforcement of graphite, which is represented in black at *c*, but this said re-enforcement is a valuable addition, as it renders any shell much more perfect and durable.

Whenever it is desirable to remelt the copper of which shells are made, having a graphite packing, it may be done, and, as heat will have no effect upon the graphite, it will of itself separate from the molten metal and remain on top, and may be easily removed. This is a great advantage over the use of different metals in the manufacture of shells, as the remelting of the same would form an alloy which injures its ductility in its manufacture into shells again.

I am aware that shells have been heretofore made so that they could be fired and reloaded at pleasure; but never to my knowledge has the device herein shown and described been before known or used; and I am aware, also, that shells have heretofore been re-enforced or strengthened by various metallic devices; and I disclaim any and every device in which graphite is not used as a strengthening material; but—

Having described my invention, what I claim as new, and desire to secure by Letters Patent, is—

1. The metallic perforated base *m*, having the nipple *o* thereon, and one or more annular grooves, *i*, when secured within the recess *r* at the head of the cartridge-shell, as described.

2. The re-enforcement of graphite, pressed firmly in at the interior of the head of the cartridge-shell, for the purpose and in the manner substantially as described.

EDWIN MARTIN.

Witnesses:

T. A. CURTIS,

OGDEN GRISWOLD.

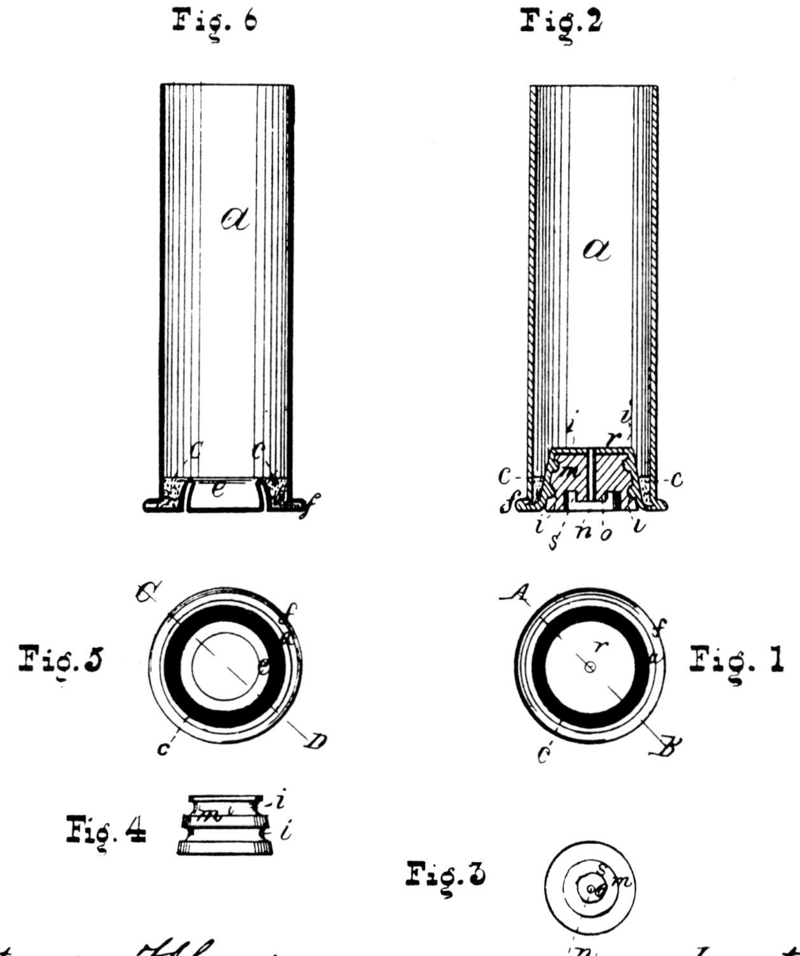

Witnesses. T. A. Curtis
Ogden Griswold

Inventor:
Edwin Martin

FIG. 3.

FIG. 5.

FIG. 6.

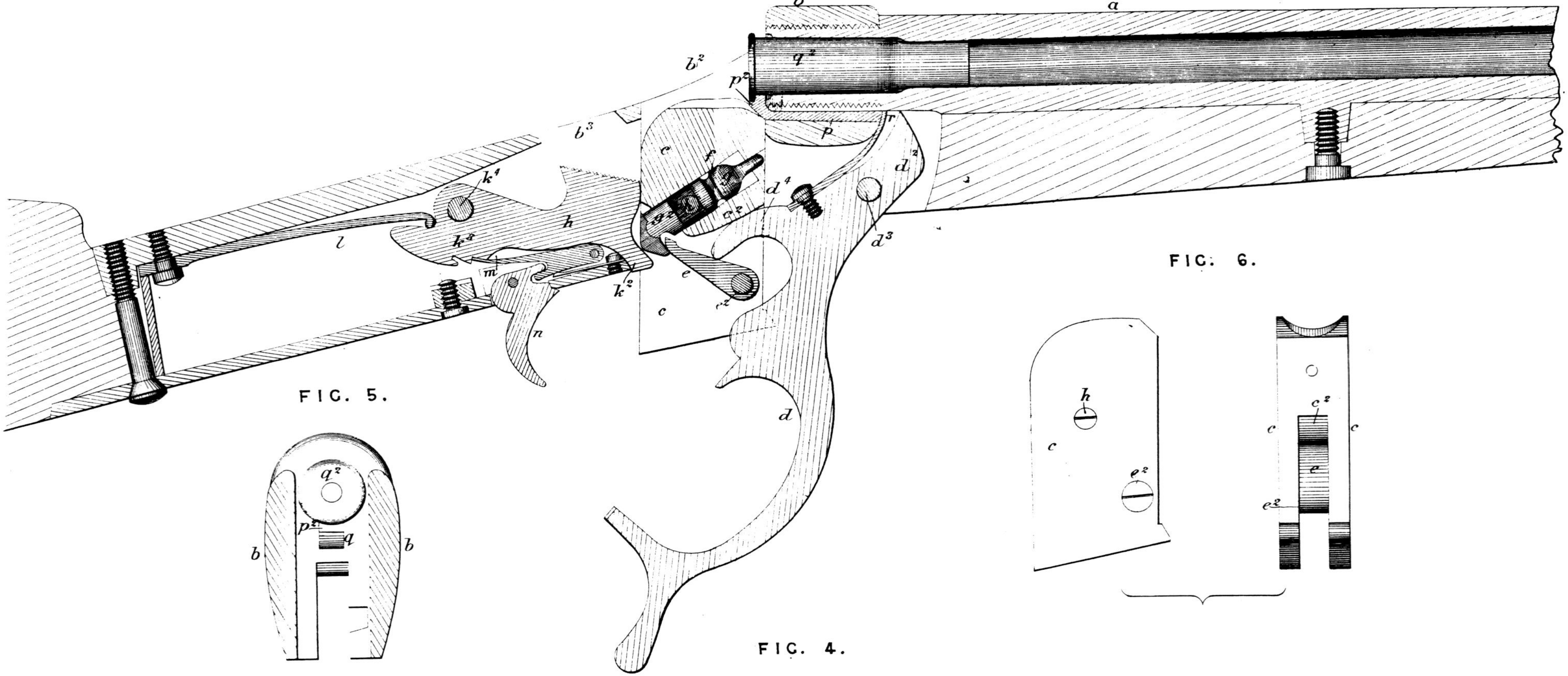

FIG. 4.

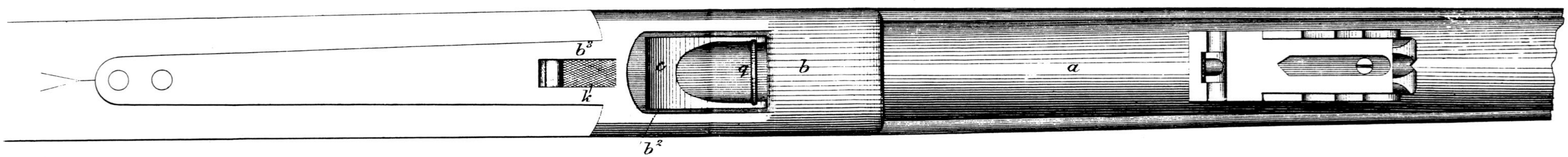

Among papers in the Sharps Rifle Company's archives were a number of patents for single shot, falling block breech-loading rifles. This is one of the drawings from British patent No 2574 granted September 28, 1870 for a hammerless rifle designed by Olof Nilson of Redwing, Minnesota.

Mr. Nilson's rifle must have been a contender in the trials that resulted in British adoption of the Martini-Henry. Sharps Rifle Company ordered copies of his and the other patents to determine the possibility of building a hammerless rifle. Courtesy of John Hintlian.

UNITED STATES PATENT OFFICE.

EDWIN MARTIN, OF SPRINGFIELD, MASS., ASSIGNOR TO HIMSELF, SAMUEL W. PORTER, AND JAMES F. CRANSTON, OF SAME PLACE.

IMPROVEMENT IN METALLIC CARTRIDGES.

Specification forming part of Letters Patent No. 111,856, *dated February* 14, 1871.

To all whom it may concern:

Be it known that I, EDWIN MARTIN, of Springfield, in the county of Hampden and State of Massachusetts, have invented a new and useful Improvement in Metallic Cartridges; and I do declare that the following is a full, clear, and exact description thereof, reference being had to the accompanying drawing, making a part of this specification, and to the letters of reference marked thereon, in which—

Figure 1 is a longitudinal section through the center of the shell, die, and the inside and outside heading-tool, showing the process of forming the head and the fold around the head. Fig. 2 is a longitudinal section of the shell after the head is formed. Fig. 3 is a similar section, showing the shell after the pocket for the fulminate has been formed in the head. Fig. 4 is a longitudinal section, showing an anvil secured within the shell by means of the fold at the head. Fig. 5 represents a shell with a re enforcing cup secured therein, showing the use of the fold in filling up the cavity between the re enforcing cup around the bottom and the head of the shell. Fig. 6 is a longitudinal section of a shell, with the priming and reducing tool and the crimping-tool, showing the shell as it is primed and crimped. Fig. 7 is a similar section of a shell with the priming and reducing tool, showing the operation of introducing and securing the conical anvil within the pocket in the head of the shell. Fig. 8 is an end view of the priming and reducing tool. Fig. 9 is an end view of the crimping-tool, and Fig. 10 is a section and end view of the conical anvil.

My invention is an improvement upon the device for which Letters Patent were granted me dated March 23, 1869; and consists of a fold formed at the juncture of the head with the cylindrical part of the shell, said fold projecting into the interior of the shell and pressed firmly up against the head on the inside, thus making the head of the shell at this part of three distinct thicknesses of metal, giving it much additional strength, and also serving to prevent much more effectually any gas from being driven into the fold of the flange at the explosion.

That others skilled in the art may be able to make and use my invention, I will now proceed to describe its construction and mode of operation.

In the drawing, A, Fig. 1, represents the heading-die, and C, the inside heading-tool, having an annular shoulder at *s*, against which the end of the shell impinges when it is placed upon the tool C and driven against the tool B.

That part of the heading tool C below the shoulder *s* is made of a size to fit well the inside of the shell, which, being placed upon the tool C, fits well the interior of the die A.

The shell, after being drawn, is cut at the open end, so that the shell is somewhat longer than that part of the tool C below the shoulder *s*, and the tool C is tapered at the end and has a cavity, D, therein, as is seen in Fig. 1.

The outside heading-tool B has a projection, D′, upon the end, somewhat less in diameter than the cavity D, and, to form the head and fold, the drawn shell is placed upon the tool C, the open end of the shell abutting against the annular shoulder *s*, and the closed end of the shell projecting somewhat beyond the end of the tool C, which, with the shell thereon, is then inserted into the die A, so that the die covers the end at the shoulder *s*, and the end of the tool C lacking about the thickness of the shell, or a little less, of passing through the die.

The outside heading-tool B is then driven up smartly against the closed end of the shell, and the surplus metal is thus flattened upon the end and pressed out at the side, forming the flange *c′*. This part of the operation, however, I do not claim as new.

As the tool B, however, continues to press up the metal, the tool C, being tapered at the end, the surplus metal, which still remains to be disposed of, being prevented by the die A from being pressed outward, is forced inward, as shown at *c*, in Fig. 1, as there is nothing to prevent it, and this begins the formation of the fold *c*.

When the metal, in being pressed inward, reaches the tapered end of the tool C, it remains in that position, while that part of the fold *c* below the end of the tool C continues to be pressed in a little more, and is pressed firmly together, completing the fold, and, as the tool B reaches its nearest point to the die A, the outside fold c' is pressed firmly together, forming the flange, and, at the same time, the end of the tool C presses the inner fold *c* firmly and tightly against the interior of the head.

The projection of the fold *c* into the shell, whether more or less, may be regulated by the excess in the length of the drawn shell over that of the tool C below the shoulder *s*, whether more or less. This part of the process I claim as novel. At the same time this is done, the cup *a* in the head of the shell is formed by the projection D' forcing the metal into the cavity D in the tool C. This brings the shell into the condition shown in Fig. 2.

The pocket for the fulminate and anvil is then formed by driving back a portion of the metal at the base of the cup *a*, and reducing it, as explained in the before-mentioned Letters Patent granted to me, and as shown in Fig. 3. This inside fold *c* may also be used to hold in place a large anvil, as shown in Fig. 4, and is particularly useful in filling up the space between the corner of the re-enforcing-cup, when it is desirable to use one, and the flange c' of the head of the shell, although it is the object in using the fold *c* to obviate the necessity of using any re-enforcing cup in all ordinary gun-cartridges.

The anvils e', Figs. 6 and 7, may be punched from sheet metal of desirable thickness with a punch and die, and are inserted and secured in place by means of the tools shown in Figs. 6, 7, 8, and 9, and in the following manner: I' represents the inside priming and reducing tool, having an annular shoulder at *n* and a conical-shaped cavity, *m*, in the end.

The shell being left with the pocket *e* in a cylindrical form, as shown clearly in Fig. 7, the anvil e' being previously filled with fulminate, is placed, with its largest end upward, in the cavity *m* of the tool, and the tool is then forced into the shell, and, as it approaches the head of the shell, the anvil e' is first forced into the pocket *e*, and the tool I', being forced still farther in the conical form of the cavity *m* in said tool, gives a corresponding conical form to the walls of the pocket *e*, closing them firmly around and upon the conical anvil e', as shown more clearly in Fig. 6; and when the tool I' is in the position shown in Fig. 6, the tool I, having an annular fillet, *i*, upon its face, is driven up smartly against the outside of the head of the shell, and the fillet *i* forces in the metal, pressing it in firmly against the inner edge of the fold *c*.

The head of the shell is thus corrugated throughout most of its part, and very much greater strength or capacity of resisting the explosion is obtained, while, at the same time, the cavity in the flange or head is perfectly closed, and there is no possibility of the head of the shell being broken or burst off at the flange by the force of the gas at the explosion.

The greatest and most manifest advantage of this invention is that while these conical anvils may be made cheaper, taking much less stock in their manufacture, the whole process of forming the fold and introducing and securing the anvil within its pocket requires no extra operation over those gone through with in the manufacture of the shell or cartridge, as explained in the before-mentioned Letters Patent granted to me, dated March 23, 1866.

As the firing-pin of the arm is generally pointed or conical upon its end when a flat anvil is used, but a very small part of the fulminate is actually operated upon by the firing-pin in exploding it; but in this device, as the anvil is conical and the fulminate is compressed between the pin and the interior of the anvil upon all sides, or all around, the ignition of the fulminate is more full and thorough than when a common flat anvil is used.

Having thus explained my invention, what I claim as new, and desire to secure by Letters Patent, is—

1. A metallic cartridge-shell, having the fold *c* made therein, making the shell of three distinct thicknesses at the juncture of the head with the cylindrical part, all constructed substantially as and for the purposes herein described.

2. An annular fillet or corrugation upon the interior of the head of the shell, in combination with the fold *c*, all constructed substantially in the manner and for the purposes specified.

EDWIN MARTIN.

Witnesses:
T. A. CURTIS,
M. L. BOYNTON.

EDWIN MARTIN.

ASSIGNOR to self *Samuel W. Porter and James F. Cranston.*

IMPROVEMENT IN METALLIC CARTRIDGES.

111856 PATENTED FEB 14 1871

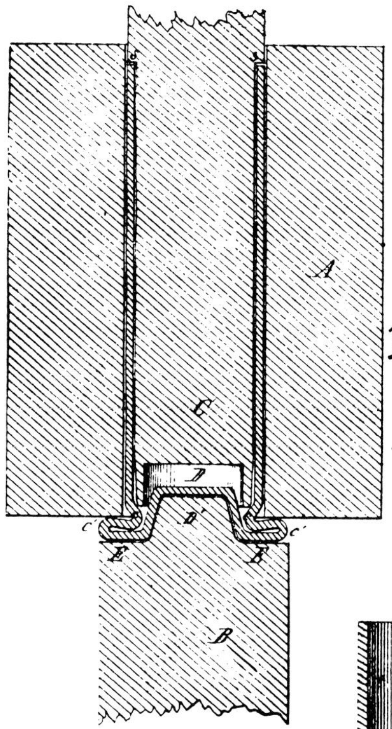

Fig. 1

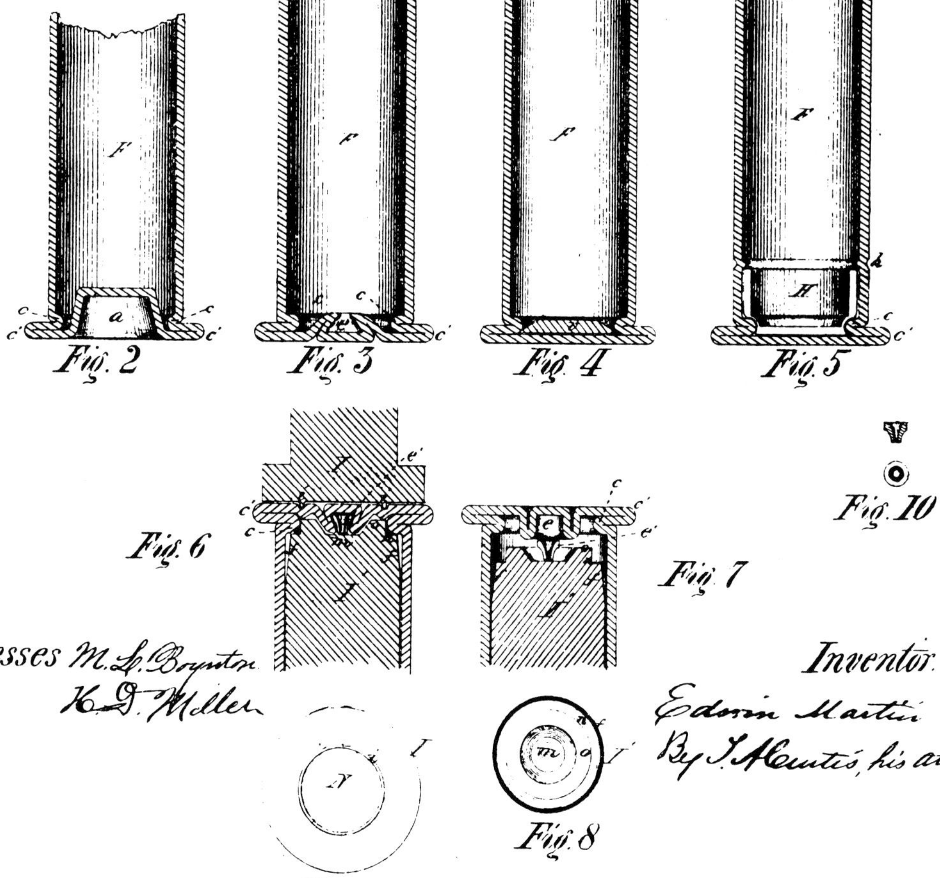

Fig. 2 Fig. 3 Fig. 4 Fig. 5

Fig. 6 Fig. 7 Fig. 8 Fig. 9 Fig. 10

Witnesses *M. L. Boynton*
K. D. Miller

Inventor.
Edwin Martin
By J. A. Curtis, his atty.

120,403

UNITED STATES PATENT OFFICE.

GEORGE R. STETSON, OF NEW HAVEN, CONNECTICUT, ASSIGNOR TO THE WINCHESTER REPEATING-ARMS COMPANY, OF SAME PLACE.

IMPROVEMENT IN METALLIC CARTRIDGES.

Specification forming part of Letters Patent No. 120,403, dated October 31, 1871.

To all whom it may concern:

Be it known that I, GEORGE R. STETSON, of New Haven, in the county of New Haven and State of Connecticut, have invented a new Improvement in Cartridges for Fire-Arms; and I do hereby declare the following, when taken in connection with the accompanying drawing and the letters of reference marked thereon, to be a full, clear, and exact description of the same, and which said drawing constitutes part of this specification, and represents in—

Figure 1 a side view; Fig. 2, a longitudinal central section; Fig. 3, the ball preparatory to the insertion of the cartridge; and Fig. 4, the shell inserted preparatory to forming the groove.

This invention relates to an improvement in that class of ball-cartridges in which the powder and ball are arranged in a metallic case, the object of the invention being to allow the use of swaged balls. Such balls, it is well known, are superior to molded balls, in consequence of greater perfection in shape and equal density of metal; but with such balls a patch has been required for lubricating purposes.

By my invention the use of the patch is avoided, and consists in forming the ball of the internal diameter of the shell—that is, the diameter of the bore of the arm in which they are to be used—then inserting the ball into the shell onto the wad or powder, and, after it is so inserted, by any suitable tool I form an annular groove in the ball near the end of the shell, pressing the lead to form this groove, which causes the lead to rise against the shell, the said groove being sufficient to retain the requisite quantity of lubrication; and the enlargement to so form the groove and shoulder is sufficient to take the grooves of the bore and give the requisite rotation to the ball.

A is the shell, of any known construction; B, the ball, (see Fig. 3,) is formed in a solid die into any desired shape, and the diameter of the ball should be the same as the bore of the barrel and the internal diameter of the shell. After the shell is charged this ball is inserted into the shell, as seen in Fig. 4, and then by any suitable device, by preference one which will cause the cartridge to revolve, I form an annular groove, *a*, in the ball forward of the end of the shell, the instrument used to form the groove being of such character as not to cut the metal, but to spin or throw it back against the end of the shell, as seen in Fig. 2. The groove is then filled with lubricating material in the usual manner.

In this construction it is not necessary to spin or turn the end of the shell down into the ball to retain it in position, as in the usual construction, the ball fitting sufficiently close in the shell to be retained; hence the ball will leave the shell more readily and in more perfect condition than in the common construction, and the labor of forming the groove in this manner is no more, if not less, than that required to turn the shell into the ball. The metal which is thrown up against the end of the shell may be to a greater or less extent; but as shown it is sufficient to take the grooves of the bore and give rotation to the ball, and possessing this advantage that, by the forcing of the lead against the end of the shell, it hermetically seals the shell against the admission of moisture.

I claim as my invention—

The combination, with a cartridge-shell, of a swaged projectile, in which an annular groove is formed and the metal displaced thereby is raised against the end of the shell after the insertion of said projectile, substantially as and for the purpose described.

GEORGE R. STETSON.

Witnesses:
A. J. TIBBITS,
J. H. SHUMWAY.

(32)

GEORGE R. STETSON.

Improvement in Metallic Cartridges.

No. 120,403. Patented Oct. 31, 1871.

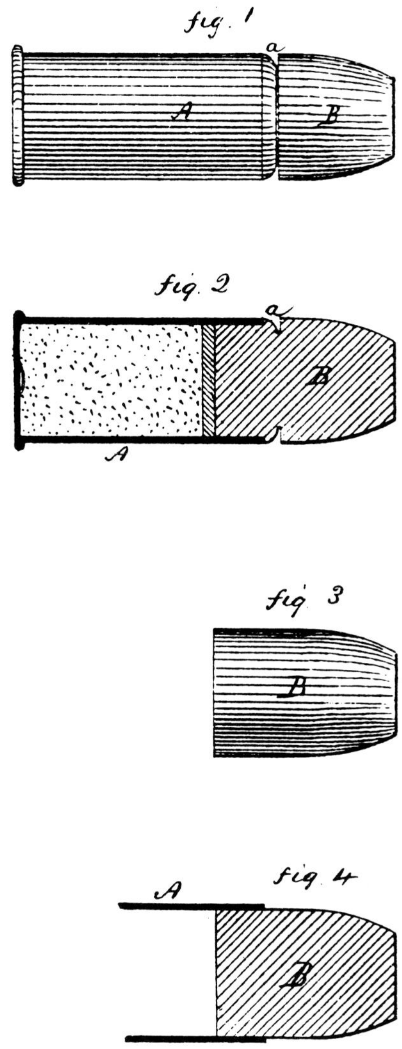

Witnesses:
J. H. Shumway
A. J. Tibbits

Geo. R. Stetson
Inventor
By his Atty
John E. Earle

123,622

UNITED STATES PATENT OFFICE.

GEORGE H. DUPEE, OF NEW HAVEN, CONNECTICUT, ASSIGNOR TO THE WINCHESTER REPEATING-ARMS COMPANY, OF SAME PLACE.

IMPROVEMENT IN CARTRIDGE-SHELLS FOR MAGAZINE FIRE-ARMS.

Specification forming part of Letters Patent No. 123,622, dated February 13, 1872.

To all whom it may concern:

Be it known that I, GEORGE H. DUPEE, of New Haven, in the county of New Haven and State of Connecticut, have invented a new Improvement in Cartridge-Shells; and I do hereby declare the following, when taken in connection with the accompanying drawing and the letters of reference marked thereon, to be a full, clear, and exact description of the same, and which said drawing constitutes part of this specification, and represents in—

Figures 1 and 2, side views of two classes of cartridges, in which the head or rim is of the same diameter as the shell; Fig. 3, the blank as struck, preparatory to forming the rim; Fig. 4, a longitudinal section of the same; Fig. 5, a longitudinal section completed.

This invention relates to an improvement in metallic shells for cartridges for fire-arms, the object of the invention being to strengthen the cartridge in the immediate vicinity of the rear or closed end, whereby I am enabled to produce shells having the head or rim in diameter equal to or less than the shell, the use of such shells being designed with special reference to that class of breech-loading fire-arms termed "repeating-arms," in which the cartridges are supplied to the barrel from a magazine or tube beneath the barrel, or in such position relative to the barrel that the cartridges may be readily transferred from the magazine to the barrel. In such arms it is necessary that the magazine be in diameter larger than the rim of the cartridge. Therefore, by constructing cartridges, as hereinafter described, with the head of no larger diameter than the shell, I am enabled to reduce the diameter of the magazine to that extent, and it consists in employing sheet metal thicker than that ordinarily used, as more fully hereinafter described, retaining nearly or the full thickness near the closed end, drawing it down to the usual thickness for the shell; the head formed by upsetting the metal instead of doubling the metal at the rim, as in the usual construction.

I will proceed to describe the manufacture of shells shown in Figs. 1, 2, 3, and 5, it being understood that this peculiar form of shell is not my invention.

I take, first, a blank from a metal which is in itself of sufficient thickness to resist the strain brought upon the rear end of the shell by the explosion—say about three times, little more or less, the thickness usually employed. I then, by suitable dies and punches, not altogether unlike the punches now in use, draw up the shell, retaining nearly or quite the full thickness at the rear end, as seen in Fig. 4, the shell contracted at the rear, the contracting of the shell commencing in the proper position to form the groove around the shell, the shell from this point forward being drawn down to about the usual thickness. This blank I then place in suitable dies, which will grasp around the shell at the contracted portion, and then a follower or other suitable instrument is forced down on the end, upsetting the metal to form the rim, as seen in Fig. 5, and leaving a groove around the shell forward of the rim, by means of which groove the retractor grasps the rim to withdraw the shell from the arm. Without the thick metal which is thus formed around the groove and neck of the cartridge the explosion would expand or tear the metal so much as to make the withdrawal uncertain or this class of cartridge ineffectual, and by this construction all the advantage of a reinforcement by introduction of other metals is avoided, and a shell produced at little if any more cost than ordinary shells. The fulminate or exploding device may be applied in any desirable manner. The rim is perfectly solid and therefore competent to sustain a much greater strain than the ordinary doubled rim.

I claim as my invention—

A cartridge-shell in which the rim is made by an annular groove formed in the shell, and the metal at the said groove and its vicinity in the process of manufacture left or made thicker for the purpose of strengthening the shell at that point, substantially as described.

G. H. DUPEE.

Witnesses:

O. F. WINCHESTER,

JOHN H. SHUMWAY.

GEORGE H. DUPEE.

Improvement in Cartridge Shell for Magazine Fire-arms.

No. 123,622. Patented Feb. 13, 1872.

fig 1

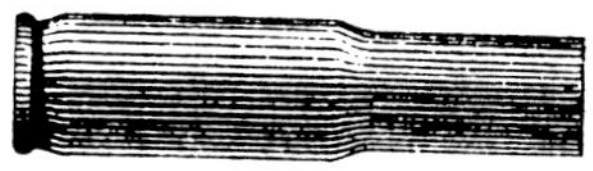

fig. 2

fig. 3

fig 4

fig 5.

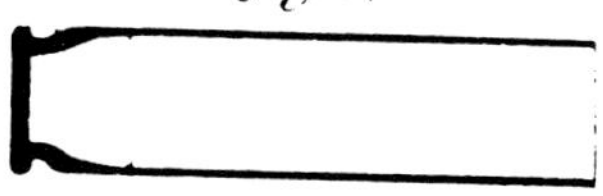

George H. Dupee
Inventor
By his Atty.
John E. Earle

Witnesses.
John H. Shumway
A. J. Tibbits

AM. PHOTO-LITHOGRAPHIC CO. N.Y. (OSBORNE'S PROCESS.)

125,640

UNITED STATES PATENT OFFICE.

FRANKLIN WESSON, OF WORCESTER, MASSACHUSETTS.

IMPROVEMENT IN GUN-LOCKS.

Specification forming part of Letters Patent No. 125,640, dated April 9, 1872.

Specification describing an Improved Hammer for Fire-Arms, invented by FRANKLIN WESSON, of Worcester, in the county of Worcester and State of Massachusetts.

The object of this invention is to enable both central and rim fire ammunition to be used in the same fire-arm and be exloded by the same hammer; to this end it consists in the novel construction hereinafter specified of a movable nose-piece and novel arrangement of the same within the head of the hammer, whereby it may be adjusted so that its upper portion will effect the discharge of rim-fire ammunition, or so that its lower portion will effect the discharge of center-fire ammunition.

In the accompanying drawing, Figure 1 is a side view of a portion of a breech-loading fire-arm fitted with my improved hammer. Fig. 2 is a partial section of the same cutting through a portion of the hammer and showing its nose-piece adjusted for rim-fire ammunition; and Fig. 3 is a similar view, showing the nose-piece adjusted for center-fire ammunition.

Similar letters of reference indicate corresponding parts in all the figures.

The fire-arm represented has a stationary breech, D, and movable barrel E, but the improvement may be applied in connection with other kinds of breech-loaders. Through the upper part of the breech there is an opening, *a*, for the nose of the hammer to strike directly upon the rim of a rim-fire cartridge, and opposite the center of the barrel there is inserted through the breech-piece a firing-pin, *b*, upon which the nose-piece of the hammer may strike to fire a center-fire cartridge. A is the hammer, which is pivoted and operated in the usual manner, but instead of having a solid head has the upper and front part of its head mortised, as shown at *e*, for the reception of the nose-piece B. This nose-piece consists of a flat piece of steel, having at the front of its upper portion a projecting nose, *c*, similar to that on the hammer of many fire-arms in which rim-fire cartridges are used, and just below this nose it is pivoted into a mortise, *e*, of the hammer, by a pin or screw, C, passing through it and the hammer. Below the pivot C there is in the front edge of this nose-piece a recess, *f*, so arranged as to come opposite the firing-pin *b*, and in its lower portion there are two holes, *i i*, one before the other, for the reception of a locking-screw, S, which passes through the hammer and one or the other of the said holes to secure the nose-piece in position.

To adjust the nose-piece B for rim-fire ammunition the forward of its two holes *i i* is brought opposite the corresponding hole in the hammer, as shown in Figs. 1 and 2, and the locking-screw or pin *s* inserted through the hammer and nose-piece, locking the latter with the point of its nose *c* so far forward and so depressed, and with the recessed lower portion *f* so far back, that in the action of the hammer the nose *c* will strike the rim of a rim-fire cartridge with which the fire-arm is loaded, but the lower portion of the nose-piece will not strike the firing-pin.

To adjust the nose-piece to operate the firing-pin the locking-screw is removed from the hammer and nose-piece and the nose-piece is shifted to bring the rear hole *i* opposite the corresponding hole in the hammer and the locking-screw is then reinserted and secures the nose-piece in such a position that its nose *c* is too high and far back to come in contact with the rim of the cartridge. The lower recessed portion *f* will come in contact with the firing-pin and operate it, thereby effecting the discharge of the cartridge.

In the front of the hammer-head there is a recess corresponding with the recess *f* in the nose-piece.

The projection *c* on the nose-piece passing through the breech D may be dispensed with and an additional pin, like *b*, may be arranged above the latter for firing rim-fire cartridges, such additional pin being operated upon by the upper part of the movable nose-pin B.

Claim.

The combination of the hammer A, the nose-piece B, the pivot C, the locking-pin S, and the firing-pin *b*, the whole arranged substantially as herein described.

FRANKLIN WESSON.

Witnesses:
EDWARD IRWIN,
WM. MCGRATH.

F. WESSON.

Improvement in Gun Locks.

No. 125,640. Patented April 9, 1872.

Fig. 1

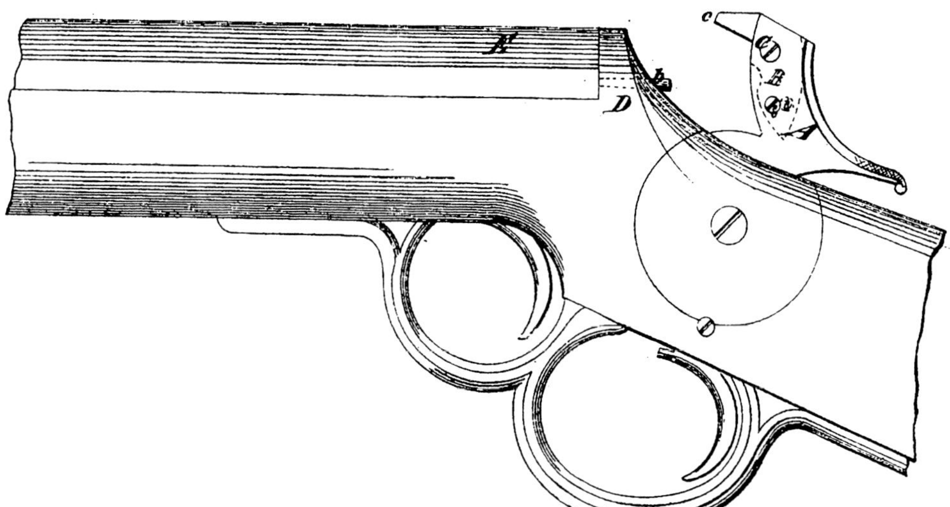

Fig. 2

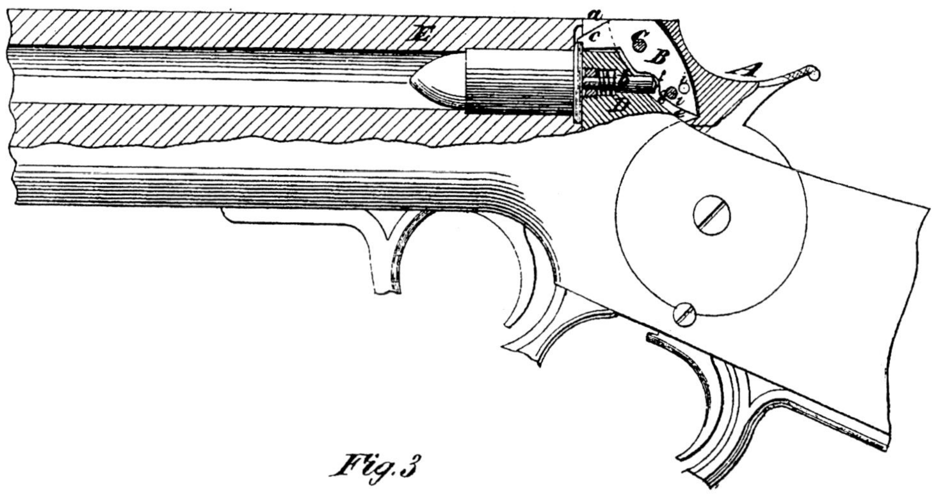

Fig. 3

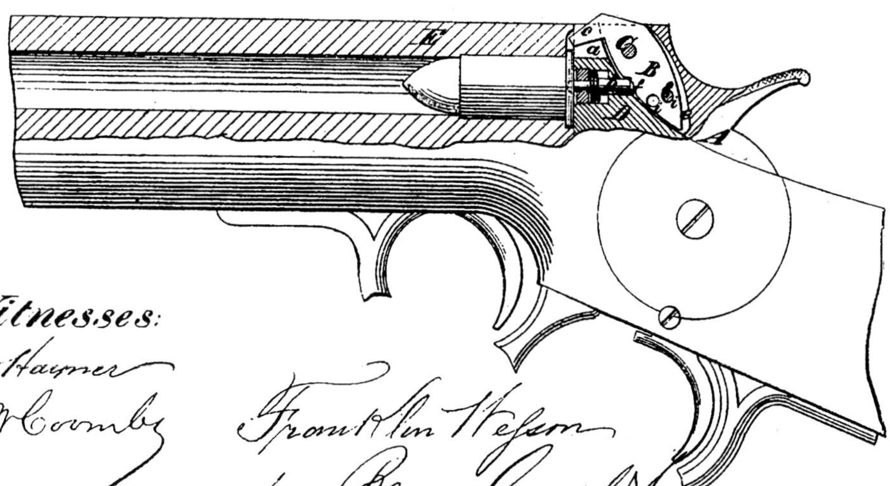

Witnesses:

Fred Haynes

J W Coombs

Franklin Wesson

per Brown Coombs & Co

Attorneys

UNITED STATES PATENT OFFICE.

BETHEL BURTON, OF BROOKLYN, NEW YORK, ASSIGNOR TO W. G. BURTON, OF SAME PLACE.

IMPROVEMENT IN METALLIC CARTRIDGES.

Specification forming part of Letters Patent No. **136,130**, dated February 25, 1873.

To all whom it may concern:

Be it known that I, BETHEL BURTON, of the city of Brooklyn, in the county of Kings and State of New York, have invented a new and useful Improvement in Cartridge-Shells; and I do hereby declare the following to be a full, clear, and exact description of the construction and operation of the same, reference being had to the annexed specification and to the figures and letters of reference marked thereon, and in which—

Figure I is a view of my cartridge prepared for use. Fig. II is a view of the shell as it appears, when drawn, before cut to length or cannelured. Fig. III is a section of the shell charged and primed with the bullet in its place.

As cartridges are now generally made from sheet metal, with a projecting flange and flat base, powerful machinery is required to upset and form the head, and, at the same time, the fiber of the metal, which runs in a general longitudinal direction of the case, is either obliterated or broken, and thereby weakened just at the fold of the flange, and where the greatest strength is required, as that is the only part of a cartridge which is wholly unsupported by the walls or breech of a gun when the same is in place. The projecting flange is also an objection when the cartridge is to be used in a magazine-gun, as it requires a tube of a diameter greater than that of the greatest diameter of the cartridge; and, beside that, the cartridges do not lie in a true line with each other, and have a tendency to rattle in the tube. The square end is objected to on the ground that if, by chance, there be a cartridge slightly shorter than the proper length it allows the next to project into the carrier-guide and over the carrier, and its square head prevents the carrier from moving up as required, when a round-headed cartridge would be pushed back into the magazine and the carrier operate equally as well.

The object of my invention is to obviate the above defects and difficulties: and to this end it consists in a cartridge drawn from a single piece of sheet metal having the fibers of the metal in a general longitudinal direction, gradually bent at and forming a rounded or spherical base, so that the fiber is neither broken or destroyed, and having a cannelure cut or made just in front of the rounded base for the extractor to catch behind to withdraw it from a breech-loading gun, and having a shoulder produced by the usual necking, which limits the extent to which it can be inserted into the gun.

Referring to the accompanying drawing, *a* is the shell; *b*, the bullet; *c*, the cannelure; *d*, the base or dome of the cartridge; and *e*, the primer.

The shell, as represented in Fig. II, is drawn from a blank in the usual way. It is then placed on a spindle revolving rapidly. On the same machine there are two cutters which move simultaneously, one of which cuts off the cartridge to length and the other cuts the cannelure the form required. When done the cutters withdraw and the shell is thrown off the spindle.

This operation is repeated as rapidly as one person can place the cartridges on the spindle; or they may be fed to the spindle from a hopper; whereas in making cartridges with a head or flange several machines are necessary to accomplish what I do with one.

The cartridge may be primed in the manner seen in Fig. III, or in any other desirable way.

By forming the base of the cartridge to that of a hemisphere I overcome the difficulty experienced with cartridges having a rim or flat base.

In the manufacture of these flat-based cartridges the greatest care is necessary to insure their being of equal length. Otherwise, when used in a magazine-gun, should one be too short, it allows the following cartridge to protrude from the magazine and prevent the carrier from raising by reason of the flat base. With the dome-shaped base of my cartridge, however, should it protrude by reason of a short cartridge, the carrier, in coming up, slides up along the arc, shoving back the cartridge into the magazine.

I do not confine myself to the particular shape of a hemisphere, or even to the frustum

B. BURTON.

Metallic Cartridges.

No. 136,130. Patented Feb. 25, 1873.

Fig I.

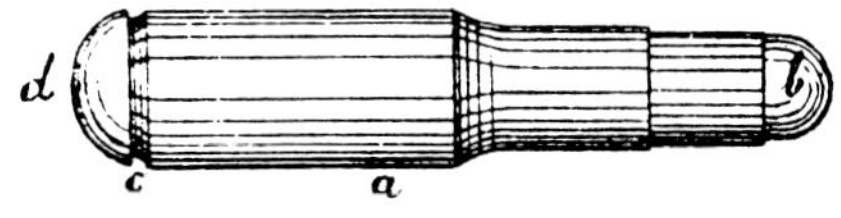

Fig II.

Fig III.

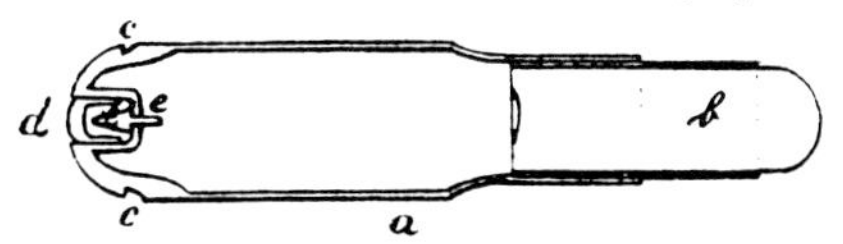

Witnesses.
Ell Clarkson
J. P. Murphy

Inventor.
Bethel Burton
per Brevoort & Bell
Attys

AM. PHOTO-LITHOGRAPHIC CO. N.Y. (OSBORNE'S PROCESS)

UNITED STATES PATENT OFFICE.

JOHN LOGAN, OF JERSEY CITY, AND GEORGE E. HART, OF NEWARK, N. J.

IMPROVEMENT IN CARTRIDGES.

Specification forming part of Letters Patent No. **142,924**, dated September 16, 1873; application filed January 21, 1873.

To all whom it may concern:

Be it known that we, JOHN LOGAN, of Jersey City, county of Hudson, in the State of New Jersey, and GEORGE E. HART, of the city of Newark, in the county of Essex and State of New Jersey, have invented new and useful Improvements in Metallic Cartridge-Shells; and we do hereby declare that the following specification, taken in connection with the drawings furnished, is a full, clear, and exact description thereof, which will enable others skilled in the art to make and use the same.

Our invention relates to improvements in cartridge-shells of the central-fire class, consisting chiefly in the assemblage and method of securing the parts together, as well as the formation of a furrow or recess across the center of the battery-cup, for the admission of an instrument for the purpose of removing the cap from the nipple when loaded or otherwise—a very desirable feature, as it enables the user to remove the cap, defective or otherwise, without removal of parts or unloading the shell.

Referring to the drawings, Figure 1 represents our improved shell with a part of its base cut away, exhibiting a sectional view of its parts. Fig. 2 represents a section of base, end view exhibiting battery-cup, groove, nipple, and slit or fire-passage.

To enable others skilled in the art to make and use our invention, we will describe in detail.

The cylinder part of the shell is made of suitable metal, generally brass, and in the usual manner. Referring to the drawings, it will be seen that the flange part of the shell, at its base, is depressed inwardly, thereby serving the double purpose of a seat for the base piece C, and to be engaged by the ring E within, when the parts are joined by riveting. To attain the greatest strength, the depression of the flange should be about an angle of forty-five degrees, as shown, although an angle varying from the above may be adopted. The base piece C, which is more clearly shown in Fig. 1, may be formed in dies, although it may be cast, and finished in the usual way. In diameter it should be made to fit exactly the depressed opening or seat at the base of the shell, extending within sufficiently to admit of being riveted, or of turning its edge over the ring E, which is placed in position within. Thus the flange of the shell proper is firmly embraced within the base piece at its center. The battery-cup is formed by drilling, or otherwise, sufficiently to admit of a nipple or anvil, with suitable surrounding space to admit of an ordinary percussion cap. This nipple or anvil may be fixed or movable. Holes or slits extend through the base piece at opposite sides of the nipple from its base, at H, for the passage of fire from the cap to the charge within the shell. A groove or furrow is formed across the center of the battery-cup, of a width and depth suitable for the reception of a suitable instrument for extracting the cap while the shell is loaded or otherwise. (See Figs. 1 and 2, at H.) The ring E may be formed in dies, or otherwise, of dimensions suitable to enter the space between the flange and cylinder within, as shown in Fig. 1, its edge admitting of being closed over by the part of the base piece provided for that purpose. (Shown at F.)

I would remark, when the slits are dispensed with at the base of the nipple a hole may be provided through the latter for the fire-passage.

Having thus set forth our invention, what we claim as new, and desire to secure Letters Patent of the United States for, is—

1. The combination of the cylinder A, breech piece C, ring E, and nipple D, formed and secured together substantially as herein set forth.

2. The breech piece C, provided with a channel through the battery-cup and a nipple, D, in combination with the cylinder A and fastening-ring E, for the purposes specified.

In testimony that we claim the foregoing we have hereunto signed our names and affixed our seals before two subscribing witnesses this the 4th day of January, 1873.

JOHN LOGAN. [L. S.]
GEO. E. HART. [L. S.]

Witnesses:
JOHN DANE, Jr.,
MANUEL W. COOKE.

J. LOGAN & G. E. HART.

Cartridges.

No. 142,924. Patented September 16, 1873.

Fig. 1.

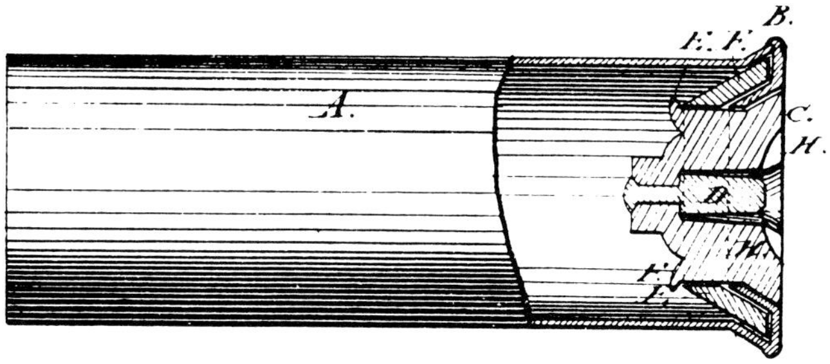

Fig. 2.

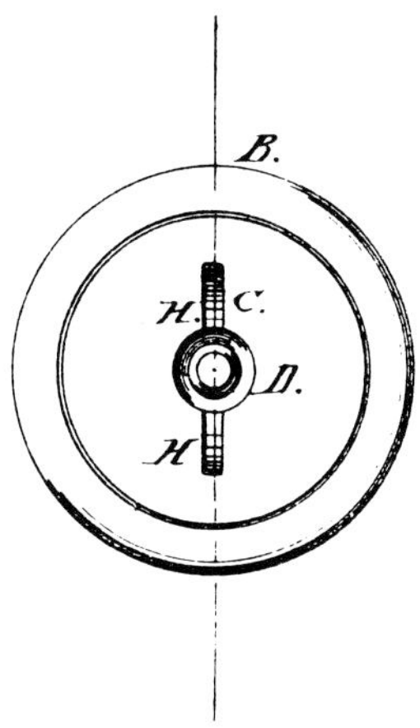

Attest:

Edw. W. Donn

M. Gardner

Inventor;

George E. Hart

John Logan

their atty J. E. Dame

UNITED STATES PATENT OFFICE.

FRANK W. FREUND, OF DENVER, COLORADO TERRITORY.

IMPROVEMENT IN METALLIC CARTRIDGES.

Specification forming part of Letters Patent No. **160,763**, dated March 16, 1875; application filed February 4, 1874.

To all whom it may concern:

Be it known that I, FRANK W. FREUND, of Denver, county of Arapahoe, Territory of Colorado, have invented a new and Improved Reloading Rifle-Cartridge; and I do hereby declare that the following is a full, clear, and exact description thereof, reference being had to the accompanying drawings, making part of this specification, in which—

Figure 1 is a longitudinal central section, and Fig. 2 a like section, showing the thinned portion of the cartridge screwed upon the thick rear portion, as at *y y*.

One general fault with cartridges of heavy charge for breech-loading guns has been that the bulk of powder has occupied the rear end of the cartridge, and consequently in the extreme rear end of the barrel, where the barrel is subjected to the heaviest strain of the explosion, so that when the cartridge is exploded such explosion simultaneously expands both the cartridge and the barrel, which, contracting unequally, makes it difficult to withdraw the cartridge, and still more difficult to reinsert the cartridge after it has been reloaded. This difficulty I obviate in my improved cartridge by making it considerably longer than the cartridge in common use, and by constructing it with a solid or thickened head of such length as will cause the bulk of the powder to be located (when the cartridge is inserted in the barrel) at a point considerably in advance of the extreme rear end of the barrel. By my invention I overcome or avoid the objectionable expansion both of the barrel and the cartridge, which is incident to the use of the cartridges as now commonly constructed.

In the drawings, D, Fig. 1, indicates a portion of my cartridge, which is made thinned from about the point *x x* to the mouth or muzzle, while in rear of the point *x x* the cartridge is made solid or thickened, as indicated at A B. From the point *x x* to the muzzle I harden the steel of which it is composed, making it of fine spring temper, and from said point *x x* to the rear termination of the shell I leave the steel soft or unhardened. I construct the part of my shell shown at A B, in rear of the point *x x*, in a solid or thick condition, for the reason, first, that I thus prevent an expansion of the shell against the barrel; and, second, I make it of metal which is soft or unhardened, because, from actual trial, I have discovered that practically such metal will not granulate and become weakened by the shock of explosion, and so split and break. The part D is made of spring temper, so that it will, if expanded by explosion, immediately return to its normal condition.

In the drawings, *a* represents the anvil for the percussion-cap, with a fine perforation passing through the solid part B, and making connection with a reduced powder-chamber, C, which in turn makes connection with the main powder-chamber D.

The object of constructing the cartridge with a reduced powder-chamber, C, in connection with a main powder-chamber, D, is, first, to so divide the charge of powder that the least shock of explosion shall fall upon the rear end of the cartridge as well as the extreme rear end of the gun, while at the same time the heaviest shock will fall upon the gun-barrel at a point considerably forward of its extreme rear end; second, by thus dividing the charge of powder the shock of explosion is divided and distributed in the cartridge, and the main chamber D relieved of a portion of the shock and power of the explosion.

The main powder-chamber terminates in an annular recess or gas-cup, *d d*, and the reduced powder-chamber C also terminates in a like recess or gas-cup, *c*, as shown in Figs. 1 and 2.

The recoil of an explosion is received in and checked by recesses *c d*, and thus the shock of discharge is lessened in its effect both upon the cartridge and the gun-barrel itself. Besides this the gas is more or less prevented from being forced through the small aperture in the anvil *a*, leading into the reduced chamber C. It is believed, also, that if the bulk of the powder lies close to the rear end of the barrel the danger from accident is greater than when the charge is located farther forward, as when at the rear the explosion is near the joint or fastening of the breech, and being near the fire-tube connecting with the powder the fire-tube becomes worn more easily than when the bulk of the powder is removed to a point forward in the barrel.

The advantages of my improved shell, it will be seen, are, first, that its rear portion is

F. W. FREUND.

Metallic-Cartridge.

No. 160,763 Patented March 16, 1875.

Fig. 1

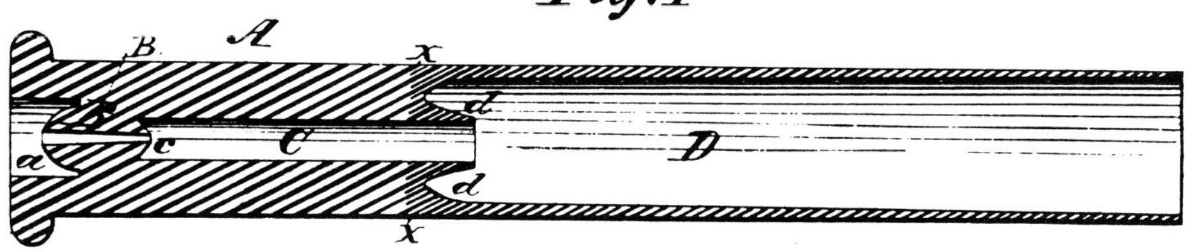

Fig. 2

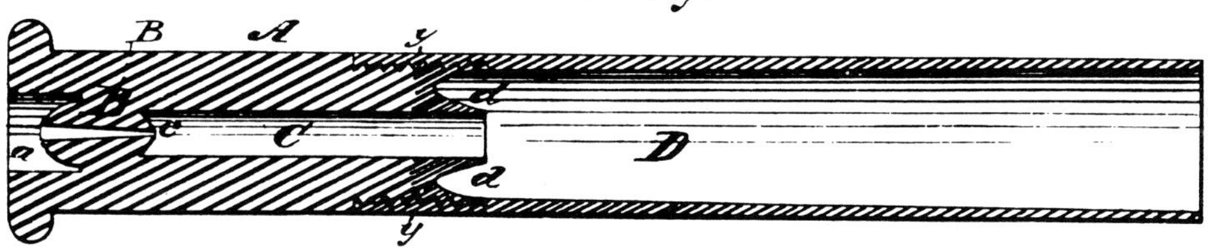

Witnesses.
James Martin Jr.
J. N. Campbell

Inventor:
Frank W. Freund
by his attorneys
Mason, Fenwick & Lawrence

THE GRAPHIC CO. PHOTO-LITH. 39 & 41 PARK PLACE, N.Y.

UNITED STATES PATENT OFFICE.

GEORGE E. HART AND JOHN LOGAN, OF NEWARK, NEW JERSEY; SAID LOGAN ASSIGNOR TO SAID HART.

IMPROVEMENT IN CARTRIDGES.

Specification forming part of Letters Patent No. **151,121**, dated May 19, 1874; application filed March 27, 1874.

To all whom it may concern:

Be it known that we, GEORGE E. HART and JOHN LOGAN, both residents of the city of Newark, in the county of Essex and State of New Jersey, have invented new and useful Improvements in Cartridge-Shells; and we do hereby declare that the following specification, taken in connection with the drawings furnished, is a full, clear, and exact description thereof, which will enable others skilled in the art to make and use the same.

The nature of our invention consists in the employment of a ring made to fit a recess in the rear part of the battery-cup, surrounding and covering a part of the top of the percussion-cap when the latter is placed in position in a cartridge-shell; the object of which is to provide a means for preventing the contact of the projectile of a following cartridge from engaging the cap of the preceding one when placed in numbers of two or more in a magazine-gun or other similar fire-arm, although equally serviceable when to be packed together in boxes or otherwise where their ends are likely to come in contact. This invention is particularly adapted for such shells as may be desirable to reload after firing, as the ring admits of being removed and replaced at will.

Referring to the drawings, Figure 1 represents sectional views of two cartridge-shells in close proximity to each other, with the end of the projectile resting against the ring, which forms the guard to protect the cap; Fig. 2, a shell with a part of the base cut away, exhibiting the ring in position, and a recess for the admission of a tool to remove the same, (it exhibits, also, a convenient tool for that purpose;) Fig. 3, an end view of shell, with ring in position, and the recess or slot for the admission of the instrument for removing said ring; Fig. 4, an enlarged view of the ring, (under side.)

The shell to which our invention applies may be of the usual kind, constructed with a centrally-arranged battery-cup, in which a percussion-cap is employed. In Fig. 1 it will be perceived that the seat for ring is made to conform somewhat to the shape of the latter, and is exhibited as being rounded, which may be accomplished by any of the known ways, and very conveniently done by being held in a chuck on a lathe and turned out by a suitable tool. The drawing represents suitable arrangements and proportions for its formation. The ring *c* is made with a recess therein, which partly closes over the top edges of the cap, as shown in Fig. 1, its outer diameter snugly filling the recess provided for it, which is slightly rounded; or it may be angular, with the mouth slightly smaller than the opening below it.

The ring is represented as open at one side, to admit of a slight compression to allow of being sprung into position, although it is not absolutely necessary, as a solid, or ends closely connecting, may be pushed into position readily, although the latter way has the effect to wear away the edges provided to retain it. Neither is it necessary to form the outer diameter of the ring by rounding, as any other formation that may be retained within the recess by a lip or similar means will serve the end desired.

The principal feature to be regarded is the partial covering of the top edges of the cap, so as to reduce the amount of surface which otherwise would be exposed; and, as the ring naturally projects above the cap, the end of the following projectile rests upon it instead of the cap.

We would remark that the slit in which the tool enters for removing the ring or cap is the same as covered by Letters Patent granted to us September 16, 1873.

Having thus set forth our invention, what we claim as new, and desire to secure Letters Patent of the United States for, is—

The combination of the ring *c* with the battery-cup of a center-fire cartridge, whereby the primer is protected and premature explosion obviated, substantially as set forth.

GEORGE E. HART.
JOHN LOGAN.

Witnesses:
ISAAC R. DENMAN,
JOHN DANE, Jr.

G. E. HART & J. LOGAN.

Cartridges.

No. 151,121. Patented May 19, 1874.

Fig. 1.

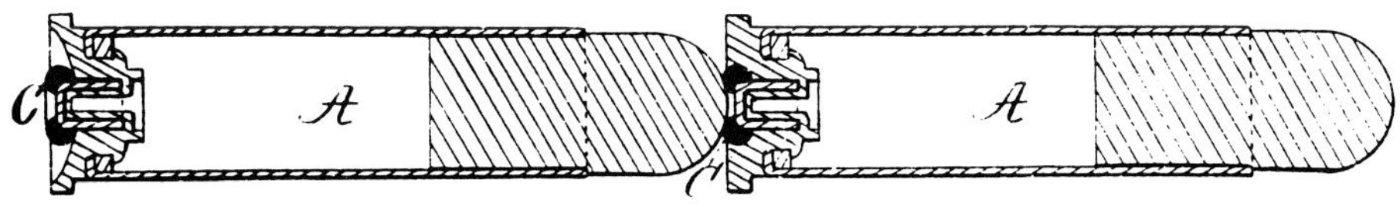

Fig. 2

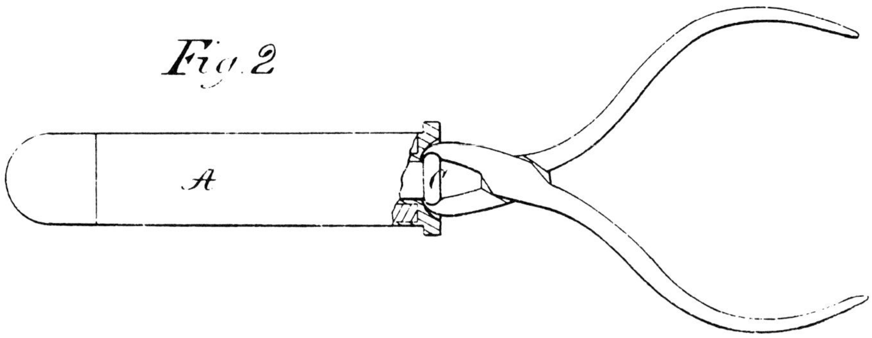

Fig. 3.

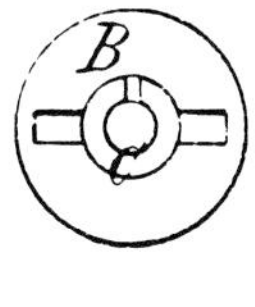

Fig. 4

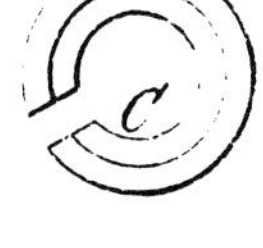

Witnesses.
Mary J. Uller
Edwin Harris

Inventors
George E. Hart
John Logan
by their atty John Dane

UNITED STATES PATENT OFFICE.

ROSWELL F. COOK, OF POTSDAM, NEW YORK.

IMPROVEMENT IN BALL-CARTRIDGES.

Specification forming part of Letters Patent No. **163.151**, dated May 11, 1875; application filed March 20, 1875.

To all whom it may concern:

Be it known that I, ROSWELL F. COOK, of Potsdam, in the county of St. Lawrence and State of New York, have invented a new and valuable Improvement in Cartridge-Shells; and I do hereby declare that the following is a full, clear, and exact description of the construction and operation of the same, reference being had to the annexed drawings making a part of this specification, and to the letters and figures of reference marked thereon.

Figures 1, 2, and 3 of the drawings are representations of longitudinal central sectional views of my shell; and Fig. 4 is an outside view of the same.

This invention has relation to improvements in metallic shells for cartridges.

The object of the invention is to devise a means whereby a patch may be used in breech-loading fire-arms as in muzzle-loaders, and the patch-covered ball deposited upon the grooves of the weapon.

To this end the nature of the invention consists in combining, with a flanged shell for receiving the powder, a second flanged shell for the ball, the latter being adapted to slide over and fit snugly on the latter, whereby a patch wrapped or otherwise secured upon a ball will be deposited upon the grooves of the weapon along with the ball, as will be hereinafter more fully explained.

In the annexed drawings, A designates a metallic cartridge-shell of the usual well-known form, and B is an open-ended tube or shell, having on one end a flange, *b*, in connection with which I propose to illustrate my invention. Shell A has in its closed end a percussion-cap, *c*, of the usual construction, and it is also provided with a flange, *a*, for the purpose of preventing it from penetrating too far into the barrel through tube B. I use these shells in the following manner, to wit: In accomplishing the result above mentioned, an oblong conoidal-ended ball, C, having been covered with a suitable patch, *d*, is inserted in the unflanged open end of tube B, which latter is then inserted in the open end of the breech of the barrel, and pushed in until its flange *b* rests against the end of the barrel. Shell A is then filled with powder, and is thrust into the rear open end of tube B until its flange *a* is brought in contact with the flange *b* of the latter, as shown in Fig. 3. Shell A, being preferably a little longer than tube B, will thrust the ball along with its patch out of the latter, depositing it upon the grooves of the barrel.

In breech-loading-arms heretofore constructed the ball is deposited in the barrel in rear of the grooves; consequently the butt or rear end of the ball will be upset upon the explosion of the powder before reaching the same, and when it does reach the grooves its rear end will be torn and abraded, thereby considerably lessening the range of the ball, and greatly impairing the accuracy of the firing.

By depositing the ball on the grooves, as above described, this defect is effectually done away with, and the necessity of wiping out the barrel after each discharge in test-firing obviated.

When the cartridge above described is put up as fixed ammunition, ready for use, it will present the appearance shown in Fig. 4, the ball resting upon the front edge of shell A. In this case the entire cartridge will be inserted into the breech of the weapon, with flange *b* resting against its rear. Shell A will then be thrust in, as above described, with a like useful result.

What I claim as new, and desire to secure by Letters Patent, is—

A cartridge for breech-loading fire-arms, consisting of the outer tube B, provided with flange *b*, powder-shell A, having flange *a*, and ball C, covered with a suitable patch, *d*, substantially as and for the purpose set forth.

In testimony that I claim the above I have hereunto subscribed my name in the presence of two witnesses.

ROSWELL F. COOK.

Witnesses:

A. P. REDWAY,

J. B. PELTON

R. F. COOK.

Ball-Cartridge.

No. 163,154. Patented May 11, 1875.

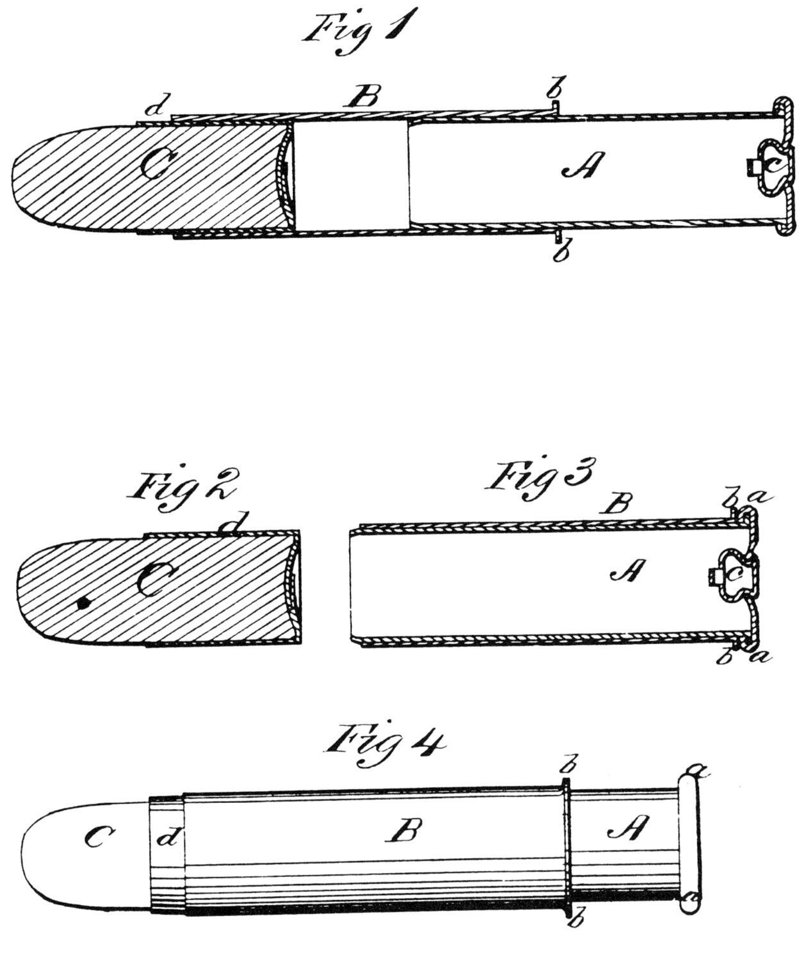

WITNESSES
Robert Everett
Francis J. Masi

INVENTOR
Roswell F. Cook
Chipman Hosmer & Co
ATTORNEYS

THE GRAPHIC CO. PHOTO-LITH. 39 & 41 PARK PLACE, N.Y.

UNITED STATES PATENT OFFICE.

GEORGE E. HART, OF NEWARK, NEW JERSEY.

IMPROVEMENT IN CARTRIDGES.

Specification forming part of Letters Patent No. **169,806**, dated November 9, 1875; application filed May 5, 1875.

CASE A.

To all whom it may concern:

Be it known that I, GEORGE E. HART, of the city of Newark, in the county of Essex and State of New Jersey, have invented new and useful Improvements in Metallic Cartridge-Shells; and I do hereby declare that the following specification, taken in connection with the drawings furnished, is a full, clear, and exact description thereof, sufficient to enable others skilled in the art to make and use the same.

My invention consists of certain improvements in the construction and mode of securing in position that part of a cartridge-shell which forms the base or breech piece in which the battery-cup is formed, having for its object simplicity, durability, and cheapness of construction, and such as to enable me to dispense with the usual number of parts as formerly employed.

Referring to the drawings, in which Figure 1 represents a section of a shell and its base-piece in position with shallow battery-cup. Fig. 2 represents same invention with the ordinary battery-cup, provided with an anvil or nipple. Figs. 3 and 4 represent the Berdan style of battery-cup, (sectional,) of the usual manufacture, with re-enforced back, exhibiting the weakened parts of the battery-cup, &c.

To enable others skilled in the art to make and practice my invention, I will describe it more fully in detail.

The cylindrical part of the shell is constructed in nearly the usual form, and of suitable metal, with an inverted flange at one end, extending inwardly about one-eighth of an inch toward its center of diameter, thereby forming a stiff rim suitable to be embraced by the edges of the base-piece, which, when riveted, or closed over the edge of said rim, as shown in Figs. 1 and 2, form lips of sufficient strength to hold the parts together without the aid of other pieces or means. This base-piece may be formed by any of the known methods by which such work is or may be accomplished, although I generally form those with shallow battery-cups in dies, drilling the hole for the fulminate or fire-passage, while those in which anvils are employed may be formed in the same manner, with the exceptions of the battery-cup, which may be more conveniently drilled in consequence of its depth and limited diameter.

As the drawings fully illustrate my improvement, further description would appear superfluous and unnecessary.

I am aware that metallic cartridge-shells with breech-blocks or base-pieces formed separately have been made heretofore in great varieties, yet for the most part complicated and difficult to properly construct. Even in their simplest form of construction the manner of uniting the parts together, particularly when solder is employed, proved defective and liable to soon get out of order. To obviate these objections and difficulties, and for the purpose of producing a cartridge-shell with a base-piece of a desired or proper thickness in the fewest number of parts and in the simplest form, is my purpose.

I am aware that rubber cartridge-cases have been provided with metallic vents to prevent the enlargement at this point that would otherwise occur in firing, &c. I am also aware that caps provided with an annular tongue or groove for insertion into the base of a shell have been in use prior to my said invention. Therefore, these I do not claim.

Having thus set forth my invention, what I claim as new, and desire to secure by Letters Patent of the United States of America, is—

A metallic cartridge-shell with a thickened breech-piece, constructed substantially as described, uniting with the inverted flange of the cylinder at its base, substantially in the manner and for the purposes set forth.

In testimony that I claim the foregoing I have hereunto subscribed my name before two subscribing witnesses on this 2d day of April, A. D. 1875.

GEORGE E. HART.

Witnesses:

JOHN DANE, Jr.,

S. E. HART.

G. E. HART.

CARTRIDGE.

No. 169,806. Patented Nov. 9, 1875.

Fig. 1.

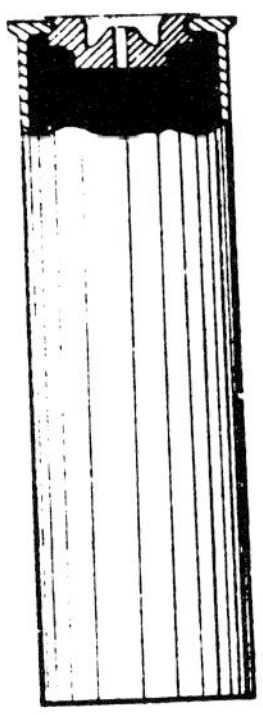

Fig. 2.

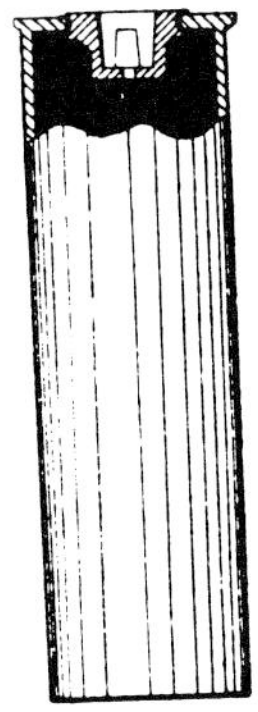

Fig. 3.

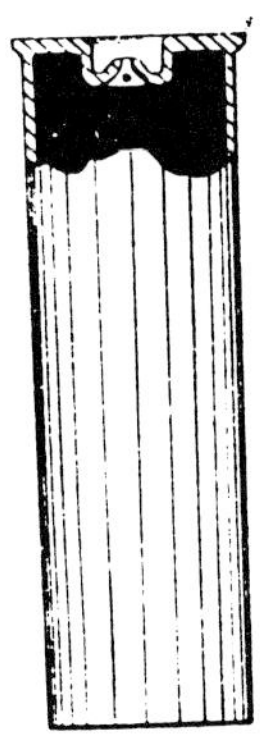

Fig. 4.

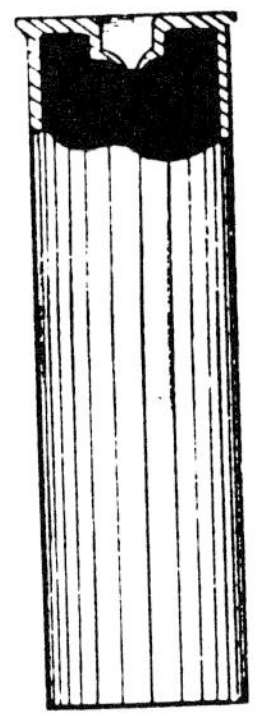

WITNESSES

Parker H. Sweet Jr.

C. S. Taylor

By

INVENTOR

George E. Hart

J. Wm. D. Dame Jr

Attorney

UNITED STATES PATENT OFFICE.

GEORGE E. HART, OF NEWARK, NEW JERSEY.

IMPROVEMENT IN CARTRIDGES.

Specification forming part of Letters Patent No. **169,807**, dated November 9, 1875; application filed May 5, 1875.

CASE B.

To all whom it may concern:

Be it known that I, GEORGE E. HART, of the city of Newark, in the county of Essex and State of New Jersey, have invented new and useful Improvements in Cartridge-Shells; and I do hereby declare that the following specification, taken in connection with the drawings furnished, is a full, clear, and exact description thereof, which will enable others skilled in the art to make and use the same.

The first part of my improvement consists in forming the base or breech piece of a cartridge-shell with a rounded or an ogee seat, into which the curved inverted flange (formed upon the cylindrical part) is held, the same being embraced between the latter and a ring, which forms a counterpart to said seat, the whole being formed into a unit by riveting, as shown in Fig. 1 of the drawings. The chief advantage in said formation and arrangement of parts is the great gain in strength at the points of union over the old or ordinary methods of construction.

The second part of my improvement consists in the employment of nickel or other similar non-corrosive metals for the nipple or anvil.

Referring to the drawings, in which Figure 1 represents a section of a shell exhibiting cylinder breech-piece, with ogee seat and ring in position; Fig. 2, the breech-piece (section;) Fig. 3, section of inner ring; Fig. 4, section of shell with curved flange; Fig. 5, section, showing ogee seat and parts joined together in a modified form.

To enable others skilled in the art to make my said improvement, I would first refer them more particularly to Figs. 2, 3, and 4 of the drawings for details; and for the arrangement of parts together as a unit to Figs. 1 and 5 of the drawings, which are such as to enable one skilled in the art to readily understand their construction.

It may be well to remark, however, that the cylindrical part is made by any of the known processes by which cartridge-shells are or may be made. The breech-piece may be formed by dies or by other suitable machinery. The important feature is the formation of the parts for the purpose of more perfectly securing the same together.

The anvil or nipple is constructed in the same manner as the ordinary nipple for such purposes.

I would remark that it is a well-known fact that steel and iron nipples for such purposes corrode and rust, which require much care and attention to prevent, particularly noticeable in damp weather, caused to a great extent by the action of the composition of which the fulminate is composed, and it is for the purpose of obviating these defects that I have adopted nickel and other non-corrosive metals.

Having thus set forth my invention, what I claim as new, and desire to secure by Letters Patent of the United States of America, is—

1. A cartridge-shell, provided with a curved flange, substantially as shown and described, and a thickened breech-piece, having a corresponding or an ogee seat, into which the said flange is compressed and held by an inner ring, substantially as and for the purposes described.

2. A cartridge shell, provided with a nickel or other non-corrosive metallic anvil, as and for the purpose set forth.

In testimony that I claim the foregoing I have hereunto signed my name before two subscribing witnesses.

GEO. E. HART.

Witnesses:
JOHN DANE, Jr.,
S. E. HART.

G. E. HART.

CARTRIDGE.

No. 169,807. Patented Nov. 9, 1875.

Fig. 1.

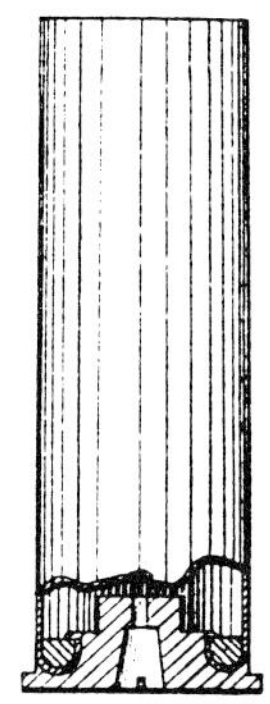

Fig. 2.

Fig. 3.

Fig. 4.

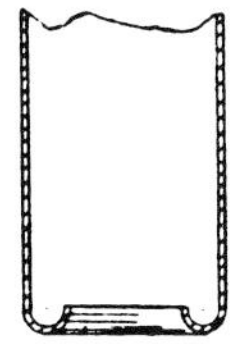

Fig. 5.

WITNESSES

Parker H. Sweet Jr.

James Boyd

By

INVENTOR

George E. Hart

Jno. D. Dane Jr.

Attorney

N. PETERS, PHOTO-LITHOGRAPHER, WASHINGTON, D. C.

UNITED STATES PATENT OFFICE.

LOUIS T. DE FROIDEVILLE, OF PARIS, FRANCE.

IMPROVEMENT IN CARTRIDGES.

Specification forming part of Letters Patent No. **172,714**, dated January 25, 1876; application filed August 4, 1875.

To all whom it may concern:

Be it known that I, LOUIS T. DE FROIDEVILLE, of Paris, France, have invented a new and Improved Cartridge; and I do hereby declare that the following is a full, clear, and exact description of the same, reference being had to the accompanying drawing, forming part of this specification, in which—

Figure 1 is a side elevation; Fig. 2, an end view; Fig. 3, a longitudinal section. Figs. 4, 5, 6, and 7 are different views of the cap; and Fig. 8 is the cylindrical cartridge-case, of tin-foil.

My invention relates to a novel construction of cartridge; and it consists in interposing between the powder and the bullet a layer of grease to keep out dampness, to operate as a gas-check, and lubricate the gun; and then, to prevent the grease from permeating the powder-grains, I place between the grease and the powder two wads, with a metallic plate placed between them to prevent the absorption and penetration of the grease through the wads, as hereinafter more fully described.

In the drawing, *a a* represent the cylindrical cartridge-case, which is made of pure unalloyed tin, in the shape of very thin leaves, known as tin-foil. The object of this form of case is to secure a metallic cartridge-case, which, while it effectually keeps out moisture, does not leave a shell in the gun, the tin-foil being volatilized and carried out with the discharged load.

In this case *a* I first insert the bullet *b*, and then place thereupon a layer of grease, *c*, to exclude dampness and lubricate the bullet in its passage out. Said layer of grease, being pushed out with the load, also operates as a gas-check to the burning powder in the rear, thereby increasing the trajectory impulse given to the bullet. I then place upon the grease the wad *d*, and, to prevent the grease absorbed by the wad from permeating the powder, I place upon the first wad, *d*, a metallic plate, *e*, and then cover the same with a second wad, *f*. I then pour in and ram the powder *g*, and afterward affix the cap *i* to the end piece *h*, and attach both to the cylindrical tin-foil cartridge-case, as shown in Fig. 3.

Having thus described my invention, what I claim as new is—

The metallic plate *e*, contained between wads *d f*, combined with the layer of grease *c*, and separating the same from the powder, substantially as and for the purpose described.

LOUIS THUILLARD DE FROIDEVILLE.

Witnesses:
ROBT. M. HUPER,
EMILE DUHAN.

L. T. De FROIDEVILLE.

CARTRIDGE.

No. 172,714. Patented Jan. 25, 1876.

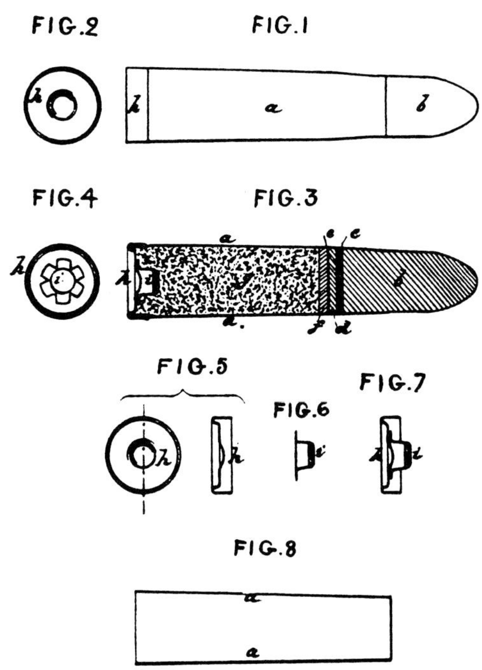

Witnesses:

Inventor:

N. PETERS, PHOTO-LITHOGRAPHER, WASHINGTON, D. C.

UNITED STATES PATENT OFFICE.

JAMES H. BULLARD, OF SPRINGFIELD, MASSACHUSETTS.

IMPROVEMENT IN METALLIC CARTRIDGES.

Specification forming part of Letters Patent No. **192,676,** dated July 3, 1877; application filed February 24, 1877.

To all whom it may concern:

Be it known that I, JAMES H. BULLARD, of Springfield, in the county of Hampden and State of Massachusetts, have invented a new and valuable Improvement in the Process of Manufacturing Cartridge-Shells; and I do hereby declare that the following is a full, clear, and exact description of the construction and operation of the same, reference being had to the annexed drawings, making a part of this specification, and to the letters and figures of reference marked thereon.

Figure 1 of the drawings is a representation of a plan view of the blank from which the shells are struck, and Fig. 2 is a side view of the same in the second stage. Fig. 3 is a sectional view of my cartridge-shell in the third state of construction; and Fig. 4 is a central sectional view in a further state of construction, showing the headed base. Fig. 5 is a front view of the headed base. Fig. 6 is a sectional view of a modification of my cartridge-shell; and Fig. 7 is a modification of Fig. 1, showing the different layers of metals.

This invention consists in the process of making cartridge-shells from two heterogeneous layers of metal, as will be hereinafter set forth.

My method or process of manufacturing is as follows: I take two loose thin sheets of metal and place one upon the other, without soldering, the lower being preferably brass, such as is ordinarily used for cartridge-shells, and the upper one being copper, or any metal which will not be injured by contact with gunpowder. I then place these sheets under a double-acting press, such as is in common use for making ordinary cartridge-shells, which cuts out the blanks and "cups" them at the same time, giving them the shape shown in Fig. 2. After this first operation the blank is drawn, in the usual manner, into the shape shown in Fig. 3. It is then "headed," so as to form the base, and the cartridge is complete, except such changes as may be necessary to adapt it to the chambers of certain fire-arms.

In the accompanying drawings, A designates the brass outer part of the shell, and B the protecting inner layer or lining, of material which is unaffected by gunpowder. C designates the base of the shell; and in heading or forming said base an annular rim, B′, of the copper or similar lining B is crimped within the annular expansion C′ of said headed base, which prevents it from any possibility of being detached or pulled out. This rim or bead B′, extending inward from the walls of the shell, acts as a re-enforce, which strengthens the base of the shell and prevents the same from being swollen or burst by the action of gases.

No permanent fastening is required to make heterogeneous layers A and B "draw" perfectly even. I prefer, however, to "tin," or coat with tin, either the upper side of the brass plate or sheet, or the lower part of the copper or similar plate or sheet, or the contiguous sides of such sheet of metal.

In the process of annealing the outer brass layer A and inner copper layer B will be soldered together by the action of the heat on said interposed layer of tin. This layer of tin (designated by D) is shown in Figs. 6 and 7.

The advantages of my method of construction above described are as follows: First, I am able to tell exactly how much of each metal is in each shell, and thereby to regulate the manufacture so as to obtain the greatest degree of elasticity, and also the requisite protection from the action of the powder. Secondly, my process does not require any special preparation of the metals, as in the case of soldering or other permanent fastening. Thirdly, I am enabled to separate the scrap without recourse to heat or any other agency. Fourthly, the only expense attending the manufacture, beyond that attending the manufacture of an ordinary brass cartridge-shell, is the cost of the copper or other material of which the lining is made, and that of the interposed tin when the latter is used. Finally, these cartridge-shells can be stored filled with powder for an indefinite length of time without injury.

What I claim as new, and desire to secure by Letters Patent, is—

1. An improvement in the art of manufacturing cartridge-shells, which consists in placing a sheet of copper upon a sheet of brass, cutting a blank from the double sheet, and drawing and heading the blank, substantially as described, and for the purpose set forth.

2. A cartridge-shell having a brass body, A, and headed base C, and a copper lining, B, provided with a re-enforcing rim or filling, B′, substantially as specified.

In testimony that I claim the above I have hereunto subscribed my name in the presence of three witnesses.

JAMES H. BULLARD.

Witnesses:

W. S. GREENE,
F. E. BECKWITH, Jr.,
M. G. BULLARD.

J. H. BULLARD.

METALLIC CARTRIDGE.

No. 192,676. Patented July 3, 1877.

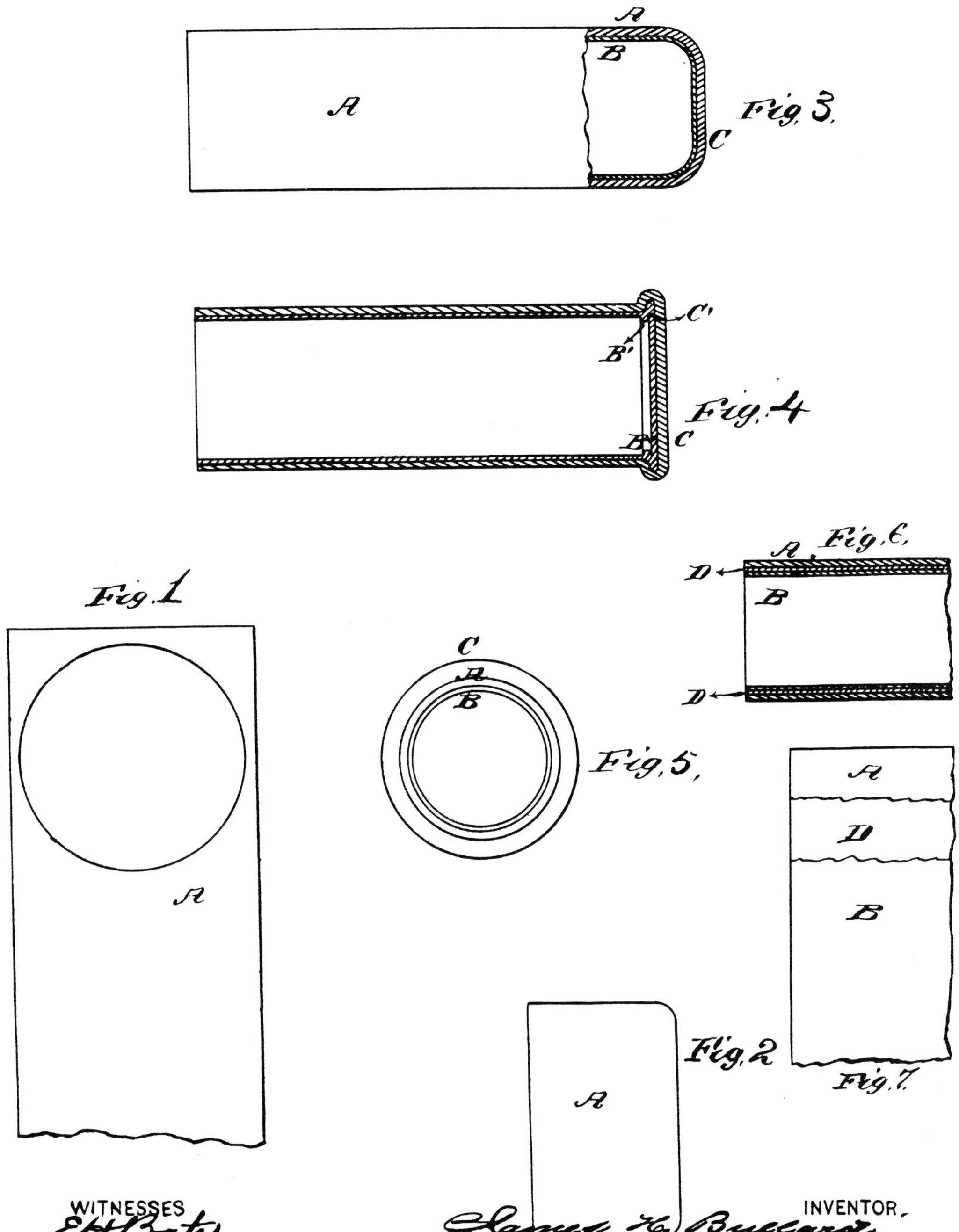

WITNESSES
E H Bates
George E. Upham

INVENTOR.
James H. Bullard.
Gilmore, Smith & Co.
ATTORNEYS

UNITED STATES PATENT OFFICE.

FRANKLIN WESSON AND CHARLES N. CUTTER, OF WORCESTER, MASS., ASSIGNORS TO SAID WESSON.

IMPROVEMENT IN BREECH-LOADING FIRE-ARMS.

Specification forming part of Letters Patent No. **193,060**, dated July 10, 1877; application filed May 7, 1877.

To all whom it may concern:

Be it known that we, FRANKLIN WESSON and CHARLES N. CUTTER, both of Worcester, in the county of Worcester and State of Massachusetts, have invented certain new and useful Improvements in Breech-Loading Fire-Arms, of which the following is a description, reference being had to the accompanying drawing, forming part of this specification.

The invention consists in a combination, with a side hammer, of a transverse lever in the breech-block, and a direct-acting center firing-pin, the whole being arranged so that the side hammer, in firing, will act through the interposition of the transverse lever on the center firing-pin, thereby doing away with the disadvantages which accrue to an oblique firing-pin, as heretofore used in fire-arms, having their hammers arranged on one side of the breech-piece; also, whereby the force of the blow of the hammer is increased, thus admitting of a less powerful mainspring being used. The invention likewise consists in a novel mode or means of fitting to its place the cartridge-ejector, whereby the latter may be readily entered and removed as required, also securely retained in position, and facility is afforded for attaching the barrel to the breech-receiver by screwing it therein.

Figure 1 represents a mainly sectional side view of a breech-loading fire-arm having our invention applied. Fig. 2 is a top view of the same. Fig. 3 is a transverse section on the line *x x*; and Fig. 4 a rear view of the breech-block, with its attached transverse lever for operating the firing-pin.

A is the hammer arranged on one side of the breech-receiver B. C is the breech-block, which is constructed to slide up and down within the breech-receiver to open and close the breech as required. D is the lever by which said breech-block is operated, said lever, which is pivoted at *b*, being connected with the breech-block by a link, *d*. E is the barrel of the arm, the rear end of which forms the cartridge-chamber. The cartridges *f* are inserted within, and when exploded ejected from said chamber in a straight direction through a groove in the breech-receiver when the breech-block is lowered for the purpose. Fig. 1 represents, by full lines, the breech-block as raised to close the breech, and, by dotted lines, as lowered to open the breech.

G is a transverse lever in the rear end of the breech-block, and arranged to work on a pivot, *s'*. Said lever occupies an oblique position across the back of said block, and crosses the axial line of the firing-pin *f** carried by said block, which pin is parallel with the barrel, and is a direct-acting center firing one. A spring, *g*, on the firing-pin serves to keep said lever pressed back or out, but not so as to interfere with the operation of the breech-block. When the side hammer A is operated to explode the cartridge, it acts, either directly or through the interposition of a pin, *h*, controlled by a spring, *i*, upon the transverse lever G, which in its turn, drives forward the center firing-pin *f** against the action of its returning-spring *g*. By this combination of the transverse firing-lever with the side hammer, and direct-acting center firing-pin, the force of the blow of the hammer is increased by said lever, so that a less powerful mainspring, H, suffices to control the hammer. Also, by using a direct-acting center firing-pin that objectionable opening in the face of the breech which is incidental to an oblique firing-pin is prevented, and the liability of the primer to catch in the breech-block is obviated, there being no sharp angle formed by the passage through said block for the primer to catch against, such as is unavoidable when an oblique firing-pin is used.

I is the cartridge-ejector, which is in the form of a slide or bar fitted to enter a longitudinal recess, *m*, in the under side of the barrel, and constructed with a rear turned-up lip, *n*, corresponding with the back end of the barrel, which receives the flange or head of the cartridge, said lip, which forms the acting portion of the ejector, fitting within a recess in the rear end of the barrel, when said ejector occupies its forward position. To provide for the entry or removal and retention in its working position of the ejector I, the breech-receiver B has a longitudinal groove, *r*, within it, back of the ejector, and within this groove a gib, J, is inserted. This gib, which, when in place, holds up the ejector to its working position, is secured or locked by a screw, *s*, entered from the under side of the breech-receiver. This forms a very simple and efficient mode of fitting the ejector, and admits of the barrel E being screwed at its rear end into the breech-receiver, as shown in Fig. 1.

The ejector I, which projects through the front of the breech-receiver, is operated to expel the exploded cartridge by means of a spur, *v*, on the breech-block operating lever D, said spur entering a notch or recess, *w*, in the ejector in front of the breech-piece or receiver B. A spring, M, fastened at one end to the ejector in front, and at its other end to the barrel on its under side, serves to return the ejector to its normal position. These constitute very simple and efficient means for operating the ejector.

I claim—

1. The vertically-sliding breech-block C, provided with the independent spring center firing-pin *f**, in combination with the lever G arranged in a diagonal recess in the back end of the breech-block, and pivoted at its lower end to the breech-block, substantially as and for the purpose described.

2. The combination, with the cartridge-ejector I, fitted within a longitudinal groove in the breech-receiver B, of the removable gib J, fitting in a groove, *r*, in the breech-receiver, to provide for holding the ejector in place, while at the same time both can be readily removed, substantially as and for the purpose described.

FRANKLIN WESSON.
CHARLES N. CUTTER.

Witnesses:
H. L. PARKER,
F. D. LEARY.

F. WESSON & C. N. CUTTER.
BREECH-LOADING FIRE-ARM.

No. 193,060. Patented July 10, 1877.

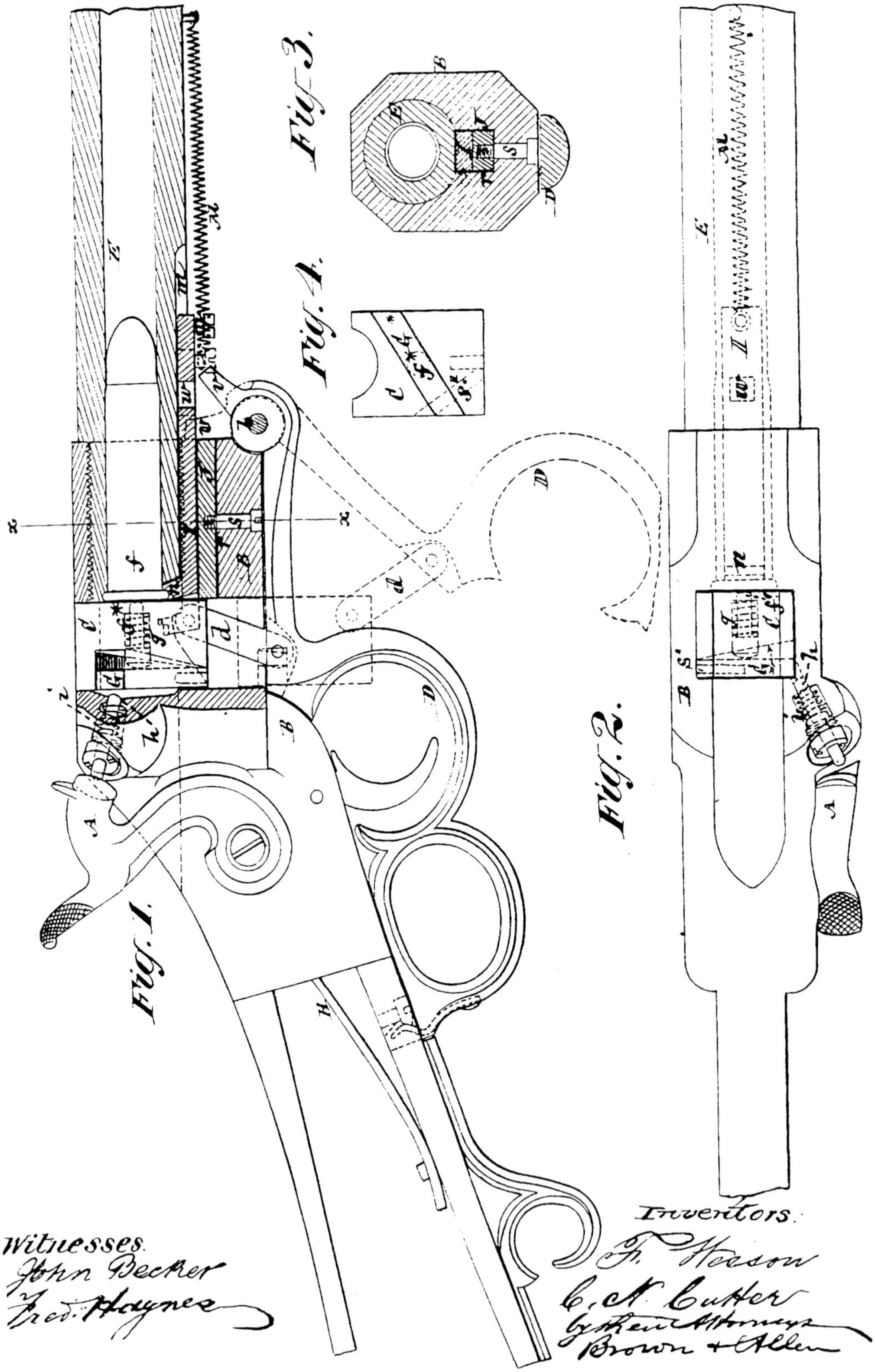

Witnesses.
John Becker
Fred. Haynes

Inventors.
F. Wesson
C. N. Cutter
by their Attorneys
Brown & Allen

N. PETERS, PHOTO-LITHOGRAPHER, WASHINGTON, D. C.

UNITED STATES PATENT OFFICE.

ISAAC DAVIS, OF FITZWILLIAM, NEW HAMPSHIRE.

IMPROVEMENT IN CARTRIDGES.

Specification forming part of Letters Patent No. **193,855**, dated August 7, 1877 ; application filed July 23, 1877.

To all whom it may concern:

Be it known that I, ISAAC DAVIS, of Fitzwilliam, county of Cheshire and State of New Hampshire, have invented a new and Improved Cartridge, of which the following is a specification:

Figure 1 is a longitudinal section of my improved cartridge. Fig. 2 is an end view of the sections.

Similar letters of reference indicate corresponding parts.

The object of my invention is to accelerate the motion of projectiles by dividing the charge of explosive, so that each successive portion of the charge increases the velocity of the projectile in arithmetical proportion.

In the drawing, *a b c* are sections of the cartridge-shell, which are connected by the plugs *d*, which are provided with nipples *e*, upon which percussion-caps *f* are placed. A ball, *g*, also having a nipple, *e*, and provided with a cap, *f*, is fitted to the last of the series of cartridge-shell sections.

The section *a* is provided with the ordinary percussion-head *h*, and when the cartridge is charged with powder and discharged in the usual way, the charge in the section *a* moves the plug between the sections *a b* forward, so that it strikes the percussion-cap *f* on the nipple of the second plug, which explodes the charge contained by the second section, *b*, and moves the second plug forward until it strikes the cap *f* on the nipple projecting from the ball *g*, and causes the third charge to explode.

The second plug moves with double the velocity of the first, and the ball moves with double the velocity of the second plug, or four times that of the first.

There are various ways of exploding the several charges, and the form of the cartridge may be varied from that shown and described; therefore I do not confine myself to the exact construction herein set forth.

Having thus fully described my invention, I claim as new and desire to secure by Letters Patent—

A cartridge having a shell made in two or more sections, connected by plugs, and each containing separate charges of explosive, the plugs and bullets being provided with nipples for receiving percussion-caps, substantially as shown and described.

ISAAC DAVIS.

Witnesses:

AMOS J. BLAKE,
A. R. GLEASON.

I. DAVIS.

CARTRIDGES.

No. 193,855. Patented Aug. 7, 1877.

Fig. 1.

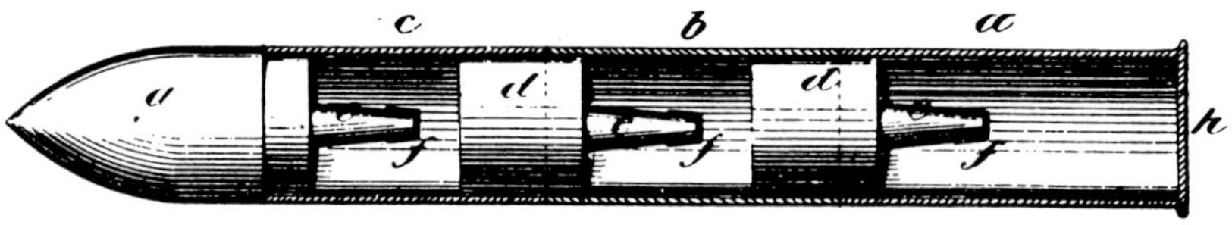

Fig. 2

WITNESSES:
Gustave Dieterich
J. H. Scarborough.

INVENTOR.
Isaac Davis.
BY Munn & Co
ATTORNEYS.

N. PETERS, PHOTO-LITHOGRAPHER, WASHINGTON, D. C.

BIBLIOGRAPHY

BARNES, FRANK C.

Cartridges of the World, Digest Book, Inc., Northfield, Illinois. Eight editions.

BUTTWEILER, ROBERT T.

Collector's Ammunition, RTB, Ltd., Houston, Texas. Volume 1 (1983) to Volume XII (1998). Three photo-illustrated ammunition auction catalogues per year.

DATIG, FRED A.

Cartridges For Collectors, Volumes One, Two, Three. Beverly Hills, California, 1956, 1958, 1962.

DE COUX, PETE.

Auction House For Ammunition & Related Items, Pete de Coux, Butler, Pennsylvania. Volume 1 (1997) to Volume 3 (November 1998). Photo-illustrated ammunition auction catalogues.

GRANT, JAMES J.

More Single Shot Rifles. Wm. Morrow & Co., NY 1959.

HOYEM, GEORGE A.

The History and Development of Small Arms Ammunition, Volumes One and Two. Armory Publications, Tacoma, WA 1981, 1982.

Cartridge Catalogues, Amory Publications, Coeur d'Alene, Idaho, 1997. Source of Winchester Repeating Arms Company, Union Metallic Cartridge Company and U. S. Cartridge Company catalogues.

HULL, EDWARD A.

Providence Tool Co. Military Arms, Edward A. Hull, 1978.

LOGAN, HERSCHEL C.

Cartridges. Huntington, West Virginia, 1948.

MARCOT, ROY M.

Remington. Primedia, Peoria, Illinois, 1998. Source for the Black Hills Rifle.

Spencer Repeating Firearms, Northwood Heritage Press, Irvine, California, 1993. Source of information on the Roper rifle and cartridge.

MARDEL, LUIZ.

Historia da Arma de Fogo Portatil, Imprensa Nacional, Lisbon, Portugal, 1887. Source for the needlefire rifle drawings.

OSBORNE, CHARLES & CO.

Patterns of Sporting, Muzzle, & Breech-loading Guns, Rifles &c. Birmingham, 1889. Source of the Union Metallic Cartridge Company ammunition list.

UNITED STATES PATENT OFFICE.

Cartridges For Small Arms, Patents, United States Patent Office, Washington, D. C. 1878. The first 286 patents for small arms ammunition granted in the United States from 1848 through 1877. Re-issued by Armory Publications, 1986.

CATALOGUES

WINCHESTER REPEATING ARMS COMPANY, New Haven, Connecticutt.

Winchester's Repeating Fire Arms, Single Shot Rifles, Rifled Muskets, Carbines, Hunting and Target Rifles, Etc., also Hotchkiss Magazine Fire Arms And Metallic Cartridges Of All Kinds. Various issues 1873 through 1911. Re-issued in 10 volumes 1993 by Armory Publications.

ALPHABETICAL INDEX

INDEX TO CARTRIDGES, AND DIMENSIONS

(Dimensions in inches. For metric equivalent, divide the inch factor by .03937)

MAYNARD

	Page	Rim	Head	Bullet	Case length
Experimental					
.35 Model 1867	9	.457	.358	.352	1.72
.40-40 Model 1867	9	.769	.452	.425	1.82
.46 rimless, 1846	4		.497	.460	1.12
.50 Cord extractor, 1863	5		.548	.525	1.215
.50 Wire Extractor, 1863	5		.548	.525	1.295
.50 capped, Model 1866	9	.64	.540	.520	1.30
.58 Mule Ear, 1866	5		.620	.580	1.40
.58 Lip Fire	9				
.64 Model 1867	9	.890	.680		2.52
.69 Thong Extractor	4		.725	.690	1.24
.69 Flop Ear	4		.730	.680	1.275
.69 Capsule	4		.680	1.25	1.25
.69 Ball, 1858	4		.728	.690	1.23
.69 Buck and Ball, 1858	4		.728	.680	2.08
.69 Tube Ignition, 1864	5				
1859/65					
.50-50	6	.766	.544	.520	1.21
.35-30	6	.760	.402	.372	1.522
.35-40	6	.762	.401	.375	2.01
.40-40	6	.767	.454	.420	2.310
.40-60	6	.770	.457	.429	2.03
.40-70	7	.768	.458	.428	2.36
.47	7	.750	.49		1.30
.50-100	7	.768	.550	.525	2.375
.55 Shot	7	.765	.585		2.27
.64 Shot	7	.885	.678		2.43
Stoppers	8				
1873					
.35-30	10	.765	.402	.375	2.13
.35-40	10	.760	.405	.375	2.58
.40-40 •	10	.750	.428	.403	2.074
.40-60	10	.759	.452	.425	2.27
.40-70	10	.757	.452	.425	2.45
.44-60	11	.762	.498	.458	1.750
.44-70	11	.754	.495	.457	2.190
.44-100	11	.755	.498	.455	2.88
.50-50	11	.765	.552	.525	1.56
.50-70	11	.757	.552	.527	1.84
.50-100	12	.757	.553	.525	2.39
.55-100	12	.768	.584	.553	2.14
.55 Shot	12	.768	.586		2.38
.64 Shot	12	.857	.685		2.50
1882					
.22-10	15	.307	.254	.225	1.17
.32-35	15	.510	.404	.315	1.86
.35-30	15	.483	.400	.370	1.56
.35-40	15	.505	.404	.360	2.058
.38-50	15	.506	.419	.378	2.85
.40-40	16	.530	.455	.416	1.77
.40-60	16	.525	.454	.429	2.20

	Page	Rim	Head	Bullet	Case length
.40-70	16	.531	.453	.410	2.40
.44-60	16	.600	.497	.457	1.93
.44-70	16	.600	.495	.457	2.20
.44-100	17	.497	.496	.458	2.81
.45-70	17	.601	.505	.458	2.10
.50-50	17	.655	.552	.521	1.39
.50-70	17	.661	.558	.525	1.77
.50-100	17	.655	.552	.520	2.31
.55-100	17	.716	.585	.560	1.93
.55 Shot	18	.715	.586		2.31
.64 Shot	18	.825	.682		2.45

SHARPS

	Page	Rim	Head	Bullet	Case length
.40-50 Straight	26	.553	.454	.403	1.88
.40-50 Necked	26	.580	.500	.403	1.72
.40-70 Straight	26	.534	.453	.403	2.50
.40-70 Necked	26	.595	.501	.403	2.25
.40-90 Straight	26	.545	.476	.403	3.25
.40-90 Necked	26	.602	.505	.403	2.63
.40-90-3.277	26	.608	.534	.401	3.277
.44-60	27	.630	.515	.447	1.89
.44-77	27	.625	.515	.447	2.25
.44-90-2 ¼"	27	.625	.515	.447	2.25
.44-90-2 5/8"	27	.625	.517	.446	2.63
.44-100 or 105, 2 5/8"	27	.625	.517	.446	2.63
.45-75-2.1"	27	.608	.505	.457	2.10
.45-80-2.1"	28	.608	.505	.457	2.10
.45-2 4/10"	28	.598	.505	.451	2.40
.45-2 6/10"	28	.598	.505	.451	2.60
.45-2 7/8"	28	.598	.505	.451	2.87
.45-3 ¼"	28	.598	.505	.451	3.25
.50-70 or 80	29	.660	.564	.512	1.75
.50-2"	29	.658	.569	.514	2.00
.50 2/12"	29	.663	.564	.512	2.50
.50-3 ¼"	29	.665	.565	.512	3.25
.52-70	29	.662	.565	.535	1.57

REMINGTON

	Page	Rim	Head	Bullet	Case length
.32-30	35	.437	.378	.312	1.64
.32-40	35	.535	.453	.308	2.13
.38-40	35	.536	.453	.372	1.77
.38-50	35	.535	.454	.375	2.23
.40-45	35	.501	.448	.403	2.10
.40-60	35	.501	.448	.403	2.25
.40-70 Necked	24	.501	.448	.403	2.25
.40-70 Reloadable	35	.501	.448		2.25
.44-77 or 90-2 ¼"	35	.625	.516	.446	2.25
.44-90-2 5/8"	35	.629	517	.447	2.63
.44-90-2 7/16"	35	.6162	.512	.4482	2.432
.44-90-2.6"	35	.5887	.5051	.4495	2.60

	Page	Rim	Head	Bullet	Case length
.45-90-2.4"	35	.598	.505	.451	2.40
.45-50 Black Hills rifle	33	.619	.518	.440	1.751

FRANK WESSON

	Page	Rim	Head	Bullet	Case length
.30-30	40	.438	.379	.311	1.65
.30-40	40	.438	.379	.307	1.64
.30 Extra Long	40	.371	.322	.305	1.598
.32 Extra Long	40	.369	.321	.317	1.25
.32 Extra Long Wesson	40	.379	.324	.319	1.626
.35 Wesson	40	.498	.436		2.505
.38 Long Wesson	40	.429	.380	.370	1.62
.38 Extra Long Wesson	41	.426	.378	.378	1.990
.38 Extra Long Wesson	41	.435	.378	.378	1.891
.40-40	41	.548	.451	.415	1.622
.40-60	41	.549	.451	.406	2.25
.40-70 Wesson	41	.522	.453	.392	2.48
.44 Short	41	.449	.440	.438	0.886
.44 Long	41	.449	.439	.438	1.21
.44 Extra Long	42	.503	.439	.426	1.64
.44 Extra Long Wesson, nickeled	42	.511	.448		1.648
.42 or .44-85	42	.509	.448	.412.	2.51
.42 or .44-100	42	.509	.448	.430	3.002
.44-100-2.921"	42	.530	.448		2.921
.45-60	42	.549	.474	.449	1.975
.45 Cook's Patent	42	.532	.471	.440	1.955
55 Shot-2.009	42	.653	.561		2.009

STEVENS

	Page	Rim	Head	Bullet	Case length
.22-15-60	46	.341	.262	.211	2.00
.25-21-86	46	.378	.301	.249	2.039
.25-25-86	46	.377	.300	.249	2.379
.28-30-120	46	.423	.357	.280	2.501
.32-1.472" (S & W Rifle case)	46	.411	.349	.328	1.472
.32-25-150 Ideal	46	.411	.350	.329	1.751
.32-35-165	46	.498	.399	.321	1.879
.38-35-215 (early)	47	.484	.401	.378	1.621
.38-35-215 (late)	47	.492	.401	.372	1.621
.38-35 Shot	47	.489	.400		1.622
.38-45-210 Hart's patent, 2 pc.	47	.495	.399	.381	2.199
.38-45-210 later, brass case	47	.491	.401	.375	2.181
.38-45 Shot	47	.491	.400		2.181
.44-50 early Hart's 2-pc.	47	.539	.453	.440	1.941
.44-50 Shot	47	.548	.453		1.932
.44-65 early	48	.539	.451	.415	2.451
.44-65 late	48	.539	.451	.415	2.470
.44-65 Shot	48	.533	.451		2.451

BALLARD

	Page	Rim	Head	Bullet	Case length
.32 Long	51	.367	.321	.317	0.96
.32 Extra Long	51	.369	.321	.317	1.24

	Page	Rim	Head	Bullet	Case length
.32-40-165	51	.506	.423	.321	2.13
.38-30	51	.441	.379	.363	1.63
.38-50-250	51	.503	.424		2.00
.38-55-255	51	.506	.421	.379	2.12
.40-63	51	.555	.470	.403	2.38
.40-65	52	.598	507	.403	2.38
.40-85 or 90	52	.545	.478	.403	2.95
.40-90 Everlasting	52	.566	.478	.404	3.08
.44 Long	52	.506	.441	.439	1.09
.44 Extra Long	52	.506	.441	.439	2.10
.44-75 Everlasting	52	.599	.495		2.52
.44-100	52	.598	.497	.444	2.80
.45-100 Everlasting	52	.597	.497		2.80
PEABODY					
,40-70 What Cheer	71	.661	.580	.408	1.76
.40-90 What Cheer	71	.660	.583	.408	2.00
.44-95 What Cheer	71	.665	.580	.443	2.31
.45-50 Sporting	71	.632	.517	.453	1.54
.45 Carbine	71	.665	.581	.451	1.62
.45-2.2"	71	.739	.662	.452	2.20
WINCHESTER					
1866					
.44 Henry	75	.522	.445	.423	0.88
1867					
.46 O. F. Winchester	73	.632	.536	.468	1.44
Experimentals					
.32-80-165	73	.550	.478	.318	3.00
.32-70 U. S. Navy	73	.479	.476	.318	2.52
.35-2.619 unidentified	73	.605	.504	.353	2.619
.70-150	73	.870	.805	.751	2.18
1873					
.32-20	76	.405	.353	.312	1.32
.38-40	76	.520	.465	.401	1.30
.44-40	76	.523	.469	.428	1.31
1876					
.40-60	77	.630	.506	.403	1.87
.45-60	77	.629	.507	.454	1.89
.45-75	77	.617	.559	.454	1.89
.45-90-500	77	.617	.559	.455	1.89
.50-95	77	.627	.561	.512	1.94
1885					
.22-13-45	80	.343	.294	.228	1.39
.38-90-217	80	.557	.476	.375	3.25
.40-110-260	80	.651	.543	.403	3.25
.45-125-300	80	.601	.534	.456	3.25
.50-140-473	80	.664	.564	.511	3.25
1886					
.31-62	85	.605	.549	.319	2.102
.33-200	85	.610	.508	.333	2.11
.38-56-255	85	.606	.506	.375	2.10
.38-70-255	85	.601	.506	.378	2.31
.40-65-260	85	.604	.505	.405	2.10

	Page	Rim	Head	Bullet	Case Length
.40-70-330	85	.604	.504	.405	2.40
.40-75-260 Expr.	86	.604	.502	.405	2.40
.40-82-260	86	.604	.502	.405	2.40
.45-70-405 or 500	86	.600	.501	.457	2.10
.45-82-405	86	.598	.500	.457	2.40
.45-85-300	86	.598	.500	.457	2.40
.45-90-300	86	.598	.501	.457	2.40
.46 W. C. F.	87	.620	.545	.458	2.406 & 2.30
.50-100-450	87	.606	.550	.512	2.40
.50-105-300	87	.605	.550	.509	2.39
.50-110-300	87	.607	.551	.512	2.41
1892					
.25-20-86	88	.405	.348	.257	1.33
.32-20-80 or 100	88	.405	.353	.312	1.32
.38-40-180	88	.520	.465	.400	1.32
.44-40-200	88	.525	.471	.427	1.32
1894					
.25-35-117	89	.500	.420	.257	2.04
.30-30-150 or 170	89	.500	.420	.308	2.04
.32-40-165	89	.506	.424	.320	2.13
.32-170 Special	89	.506	.422	.320	2.04
.38-55-255	89	.506	.424	.375	2.12
1895					
.38-72-275	91	.519	.460	.378	2.57
.40-72-330	91	.519	.460	.405	2.60
.35-250	91	.539	.457	.358	2.40
.405-300	91	.543	.461	.412	2.58

BULLARD

	Page	Rim	Head	Bullet	Case Length
.32-40-150	94	.510	.452	.315	1.85
.38-45-190	94	.525	.454	.373	1.80
.40-70-232	94	.606	.505	.412	2.09
.40-75-258	94	.606	.505	.412	2.10
.40-90-300	94	.623	.568	.412	2.04
.45-85-295	94	.599	.501	.456	2.15
.50-115-300	94	.619	.585	.512	2.19

MARLIN

	Page	Rim	Head	Bullet	Case Length
.40-60-260	97	.604	.503	.404	2.10

COLT

	Page	Rim	Head	Bullet	Case Length
,40-60-260	101	.604	.503	.404	2.10

SAVAGE

	Page	Rim	Head	Bullet	Case Length
303.	105				

SINGLE SPECIMENS	Page	Rim	Head	Bullet	Case length
.25-20-86 Single Shot	107	.378	.315	.257	1.63
.32-13/16"	107	.369	.321	.316	0.82
.32 S & W Rifle	107	.411	.349	.328	1.47
.38 Long	107	.440	.378	.373	1.60

	Page	Rim	Head	Bullet	Case Length
.44 Evans Old Model	107	.512	.441	.420	1.00
.44 Evans New Model	107	.514	.447	.420	1.59
.50 so-called Meigs	107	.550	.503	.50	1.28
.42 Russian Berdan No. 2	108	.635	.515	.431	2.24
.43 Spanish	108	.635	.521	.433	2.24
.43 Spanish carbine	108	.632	.519	.433	1.90
.446 Brazilian Comblain	108	.681	.587	.449	2.10
.45-70-405 or 500	108	.605	.503	.457	2.10
.45 1 5/16	108	.674	.565	.457	1.45
.40-2 7/16" Roper	109	.545	.476		2.42
.45-70 Van Choate	109	.633	.515	.479	2.24
.45-78 Wolcott	109	.606	.507	.447	2.20
.45 Brown Standard	109	.597	.509	.452	2.62
.45-85 Ward Burton	109	.757	.656	.459	1.81
.45 Holden					

MILITARY COMMERCIAL

	Page	Rim	Head	Bullet	Case Length
.50 Carbine	110	.664	.564	.510	1.42
.50-70 Springfield	110	.664	.564	.510	1.77
.58 Carbine	110	.736	.641	.610	1.08
.58 Musket	110	.742	.643	.600	1.65
.58 Roberts	110	.741	.656	.610	1.37
.58 Dangerfield & Lefever	110	.720	.638	.592	2.13
.60 ?	110	.774	.658		1.625

PRIMITIVES	Page
European and Japanese paper	114-115
Linen case	
.50 Johnston & Dow	116
.50 Perry carbine	116
.50 Jenks carbine	116
.50 Cosmopolitan carbine	116
.52 Sharps carbine	116
.54 Starr carbine	116
Johnston & Dow patent	
.36 Remington navy	116
.44 Colt army	116
.50 rifled musket	116
.574 Springfield or Enfield	116
.58 rifled musket	116
.705 (17.9mm) Austrian musket	117
Roberts Bartholow patent	
.32 Colt pocket revolver	117
.36 Colt navy	117
.44 Colt army	117
.52 Sharps	117
.54 rifled musket	117
.58 rifled musket	117
Doremus & Budd patent	
.31 revolver	117
.44 Colt army	117

	Page
.52 Hall carbine	117
.54 rifled musket	117
.52 Sharps	117
.58 rifled musket	117
.69 musket	117
Miscellaneous primitives	
U. S. Mefford sub cal. projectile	118
12 gauge Needham needle fire	118
.54 Marston	118
.54 Rochaz-Lindner	118
.56 English Sharps	118
.58 Lindner	118
.58 Lindner yarn stopper	118
13.7mm von Lenk gun cotton	118
12mm Lenoir horizontal pinfire	118
.64 Colt tinfoil	118
16.26mm Galy-Cazalat	118
.64 Pauly	119
Genhart tube ignition for pistol	119
.46 Maynard, 1856	119
Demondion tube ignition	119
16 gauge Beringer shotshell	119
.16.26 Pauly, 1816	119
14.5mm Pauly, 1812	119
16.26mm Pauly 1812	119
9mm Cent Garde pinfire, 1858	119

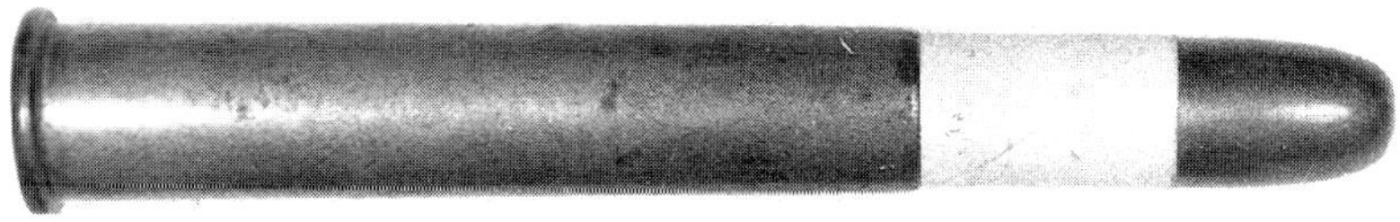